Encounters with Chaos and Fractals

Encounters with Chaos and Fractals, Third Edition provides an accessible introduction to chaotic dynamics and fractal geometry. It incorporates important mathematical concepts and backs up the definitions and results with motivation, examples, and applications.

The third edition updates this classic book for a modern audience. New applications on contemporary topics, like data science and mathematical modeling, appear throughout. Coding activities are transitioned to open-source programming languages, including Python.

The text begins with examples of mathematical behavior exhibited by chaotic systems, first in one dimension and then in two and three dimensions. Focusing on fractal geometry, the authors introduce famous, infinitely complicated fractals. How to obtain computer renditions of them is explained. The book concludes with Julia sets and the Mandelbrot set.

The Third Edition includes:

- More coding activities incorporated in each section with expanded code to include pseudo-code, with specific examples in MATLAB® (or its open-source cousin Octave) and Python.
- Additional exercises—many updated—from previous editions.
- Proof-writing exercises for a more theoretical course.
- Revised sections to include historical context.
- Short sections added to explain applied problems in developing mathematics.

This edition reveals how these ideas are continuing to be applied in the 21st century, while connecting to the long and winding history of dynamical systems. The primary focus is the beauty and diversity of these ideas. Offering more than enough material for a one-semester course, the authors show how these subjects continue to grow within mathematics and in many other disciplines.

Denny Gulick is Professor Emeritus in the Department of Mathematics at the University of Maryland. His research interests include operator theory and fractal geometry. He earned a PhD from Yale University.

Jeff Ford is a visiting assistant professor of mathematics at Gustavus Adolphus College. He earned his bachelor's from Gustavus Adolphus College, his master's in mathematics from Minnesota State University-Mankato, and his PhD in mathematics from Auburn University.

Textbooks in Mathematics
Series editors:
Al Boggess, Kenneth H. Rosen

https://www.routledge.com/Textbooks-in-Mathematics/book-series/
CANDHTEXBOOMTH

Encounters with Chaos and Fractals

Third Edition

Denny Gulick and Jeff Ford

CRC Press
Taylor & Francis Group
Boca Raton London New York

CRC Press is an imprint of the
Taylor & Francis Group, an **informa** business

A CHAPMAN & HALL BOOK

Third edition published 2024
by CRC Press
2385 NW Executive Center Drive, Suite 320, Boca Raton FL 33431

and by CRC Press
4 Park Square, Milton Park, Abingdon, Oxon, OX14 4RN

CRC Press is an imprint of Taylor & Francis Group, LLC

© 2024 Denny Gulick and Jeff Ford

First edition published by McGraw-Hill College 1992
Second edition published by Chapman and Hall/CRC Press 2012

Library of Congress Cataloging-in-Publication Data

Names: Gulick, Denny, author.
Title: Encounters with chaos and fractals / authors, Denny Gulick and Jeff Ford.
Description: Third edition. | Boca Raton : CRC Press, 2024. | Includes bibliographical references and index.
Identifiers: LCCN 2023037143 | ISBN 9781032677866 (hardback) | ISBN 9781032678818 (paperback) | ISBN 9781032678757 (ebook)
Subjects: LCSH: Chaotic behavior in systems. | Fractals.
Classification: LCC Q172.5.C45 G85 2024 | DDC 515/.39--dc23/eng/20231025
LC record available at https://lccn.loc.gov/2023037143

ISBN: 978-1-032-67786-6 (hbk)
ISBN: 978-1-032-67881-8 (pbk)
ISBN: 978-1-032-67875-7 (ebk)

DOI: 10.1201/9781032678757

Typeset in Latin Modern
by KnowledgeWorks Global Ltd.

Publisher's note: This book has been prepared from camera-ready copy provided by the authors.

Contents

Preface to the Second Edition

The far-reaching interests in chaos and fractals are outgrowths of the computer age. On the one hand, the notion of chaos is related to dynamics, or behavior, of physical systems. On the other hand, fractals are related to geometry, and appear as delightful but infinitely complex shapes on the line, in the plane, or in space. *Encounters with Chaos and Fractals* is designed to give an introduction both to chaotic dynamics and to fractal geometry.

During the past fifty years, the topics of chaotic dynamics and fractal geometry have become increasingly popular. Applications have extended to disciplines as diverse as electric circuits, weather prediction, orbits of satellites, chemical reactions, analysis of cloud formations and complicated coast lines, and the spread of disease. A fundamental reason for this popularity is the power of the computer, with its ability to produce complex calculations and create fascinating graphics. The computer has allowed scientists and mathematicians to solve problems in chaotic dynamics that hitherto seemed intractable, and to analyze scientific data that in earlier times appeared to be either random or flawed. Fractals, on the other hand, are basically geometric, but depend on many of the same mathematical properties that chaotic dynamics do.

Mathematics lies at the foundation of chaotic dynamics and fractals. The very concepts that describe chaotic behavior and fractal graphs are mathematical in nature, whether they be analytic, geometric, algebraic, or probabilistic. Some of these concepts are elementary, and others are sophisticated. There are many books that discuss chaos and fractals in an expository manner, as there are treatises on chaos theory and fractal geometry written at the graduate level. In writing *Encounters with Chaos and Fractals*, we have a goal of providing a readable introduction to chaotic dynamics and fractal geometry at a modest level of sophistication – specifically, for anyone who has a knowledge of calculus. The book includes the important mathematical concepts associated with chaotic dynamics and fractal geometry, and supports the definitions and results with motivation, examples, and applications where feasible.

Encounters with Chaos and Fractals includes seven chapters. Chapter 1 focuses on the functions of one variable and explores the fundamental notions of periodicity and bifurcation. Chapter 2 introduces the concepts of chaos and strong chaos for functions of one variable, along with the related topics of conjugacy and transitivity. These first two chapters lay the groundwork for Chapters 3 and 4 and offer examples of most of the mathematical behavior exhibited by chaotic systems in higher dimensions. They also rely on mathematics only through one-dimensional calculus.

In Chapters 3 and 4, we study chaotic dynamics for functions of more than one variable. The analysis involves vectors and matrices at an elementary level. A brief review of matrices is included in Section 3.1 for those who desire it. Chapter 3 focuses on chaotic dynamics for functions of two variables, and discusses the famous Hénon and Smale functions to illustrate the concepts. Chapter 4 analyzes solutions of systems of differential equations that relate to chaotic motion. Special attention is placed on the dynamics of the pendulum and the system of differential equations employed by Edward Lorenz to model weather prediction. Chapter 4 presumes a basic knowledge of systems of differential equations normally obtained in an elementary differential equations course.

Chapters 5–7 are devoted to fractal geometry, which came to prominence in the 1970s thanks to the computer expertise and creativity of Benoit Mandelbrot, who was an IBM fellow working at the Thomas J. Watson Research Center. The introductory Chapter 5 introduces the infinitely complicated shapes, called "fractals," such as the Cantor set, and Sierpiński gasket, as well as so-called "space-filling curves." It then turns to different notions of a dimension that need not be an integer. In Chapter 6, we discuss how to obtain computer renditions of the fractals, including those discussed in Chapter 5. Chapter 7 is the final chapter, introducing the famous Julia sets and the Mandelbrot set.

The book contains more than enough material for a one-semester course. For the one-semester course at the University of Maryland, we cover virtually all of Chapters 1–4 and then include selected topics from Chapters 5–7. At the end of each section of the book is a supply of exercises designed to reinforce topics and give added insight to the concepts presented in the section. In addition to the regular exercises, there are, where feasible, exercises utilizing computer programs (written in Python), which we have included in the Appendix. Answers to selected odd-numbered exercises are placed at the end of the book.

For the original edition of the book, *Encounters with Chaos*, I acknowledge with great appreciation the many people who have assisted me in the preparation of this book. I would like to express my deepest gratitude to former colleague Celso Grebogi and current colleague James Yorke, who helped introduce me to chaotic dynamics. To the manuscript reviewers for that edition I give my thanks: Daniel Drucker, Jim Hoste, Guan-Hsong Hsu, Edward Packel, Richard Parris, Philip Straffin, and Steven Strogatz. Special thanks go to Bau-Sen Du and Helena Nusse, who read several versions and made literally hundreds of pertinent suggestions; Helena Nusse also provided several figures. I appreciate help from colleagues Joseph Auslander, Kenneth Berg, Robert Burckel, Patrick Fitzpatrick, Alan Garfinkel, Arturo Lopes, Chris Rorres, Dan Rudolph, and Garrett Van Meter. I also appreciate the assistance of Barbara Gulick, John Montroll, Edgar Rummel, and Art Matrix of Ithaca, New York. Next, I would like to thank those who helped me technically, especially Jim

Hoste, who prepared the instructor's manual for the original edition and unearthed errata and obscurities as he proceeded. Finally, for that edition I appreciate the efforts of the staff at McGraw-Hill, including Margery Luhrs and Mel Haber, and especially the mathematics editor, Richard Wallis.

For the second edition, *Encounters with Chaos and Fractals*, I am very thankful for corrections and suggestions made by Robert Burckel, Bau-Sen Du, John Horváth, Jim Hoste, Jeffrey Nunemacher, Clark Robinson, and Donald Teets. Additionally, I give special thanks to Scott Rome, who provided a substantial number of computer graphs for the new edition, and to Manjit Bhatia, who read large portions of the manuscript and made numerous suggestions and corrections.

I also give my warmest thanks to Bob Stern, Marsha Pronin, Michele Dimont, and Samar Haddad at Taylor & Francis Group/CRC Press.

Finally, my heartfelt thanks go to my wife for her help in seeing me through the preparation of this second edition.

Denny Gulick
College Park, Maryland

Introduction

Encounters with Chaos and Fractals is an introduction to the study of the relatively new mathematical fields "chaotic dynamics" and "fractal geometry." In chaotic dynamics one analyzes objects subject to an unpredictable, but not random, behavior. We say that such behavior is chaotic. In fractal geometry one studies geometric shapes, mainly in the plane, that are created by infinite processes and are often very complex.

In order to gain a little perspective on chaotic dynamics, let us turn the clock back a little more than a century. Until then, it was thought that the motion of planets could be completely understood provided that the equations modeling their motion were accurately prescribed. Indeed, during the 17th century Johannes Kepler used the calculations of Tycho Brahe to convince himself that the planets in our solar system move in elliptical orbits around the sun. From Kepler's fundamental principles of motion, Isaac Newton proved that any planet under the influence of only the sun moves in an elliptical orbit. Thus Newton solved the "two-body problem," which refers to the analysis of the motion of one body under the influence of a single other body.

Rarely, however, is a planet subject to the force of a single other object. Indeed, the earth's motion is affected in a nontrivial way by the gravitational forces of the sun, moon, and even other planets, in our solar system, most notably Jupiter. Likewise, the orbit of our own moon is influenced by the earth as well as the sun. Therefore, Newton's results on the two-body problem gave at best only an approximate solution to the motion of a planet subject to more than one gravitational force.

During the 1890s, the great French mathematician Henri Poincaré also studied mathematical aspects of planetary motion. He focused on the "three-body problem," in which the motion of one body is influenced by precisely two other bodies. During his work on this problem, he developed a new kind of mathematics, which today is called "topology" (or rubber-sheet geometry). After much effort, Poincaré proved that there is no simple solution to the three-body problem. In other words, Poincaré proved that it is in general impossible to give a simple prescription for the orbit of one body influenced by two other bodies.

In addition, Poincaré realized that if one takes two different readings of the position of a planet at a given moment, then no matter how close the readings are, after enough time the orbits of the planet corresponding to the two different readings are expected to separate away from one another. Accurate long-range prediction of the orbit of the moon is thus impossible. This is one of the basic features of chaotic behavior. At the beginning of the 20th century, this idea was particularly offensive to scientists and mathematicians, not only because it contradicted their intuition of the regularity of nature and the universe, but also because they had no tools with which to analyze motion that is unpredictable.

With the advent of the computer, analysis of the motion of planets and other moving systems has taken on a new life. Computers can make trillions of calculations in the blink of an eye. As a result, scientists and mathematicians have been able to gain an understanding of physical and theoretical entities that until recently were beyond their reach; they have also been able to create fractal pictures and landscapes that adorn calendars and covers of magazines.

During the 1960s, the noted MIT meteorologist Edward Lorenz tried to model weather patterns by means of mathematical equations. Of course, in order to make reasonable predictions of weather one needs to use innumerable variables (such as temperature, humidity, and wind velocity, at thousands of locations both on earth and in the atmosphere), so one is inevitably faced with a great many differential equations. In order to make calculations possible, Lorenz refined a known system of three differential equations relating to convection, and turned to his Royal McBee LGP-30 electronic computer that was equipped with 6-place accuracy.

One time Lorenz substituted a 3-place approximation for the 6-place number he had earlier used as an initial condition. To his astonishment he found that the predicted weather pattern obtained from the 3-place initial condition was far different from the earlier predicted weather pattern. Thinking that it was a fluke, he repeated the process several times. Each time the McBee responded in a similar fashion. Additional analysis led Lorenz to conclude that a slight alteration in the initial conditions could result in enormous differences in the output. Thus weather patterns exhibit a particular kind of chaotic behavior that has been termed the "butterfly effect," because in theory the flapping of butterfly wings in, say, Rio de Janeiro, could precipitate a tornado in Texas several weeks later. It follows that although one might be able to predict the weather with reasonable accuracy in the short term, long-range predictions of weather are futile. Nevertheless, short-range predictions are meaningful and become more accurate as new and more variables are inserted into the system of equations employed.

Turning to fractal geometry, we note the complicated shapes that mathematicians such as Georg Cantor and Waclaw Sierpiński found amazing and confounding around 1900. Again with the advent of the computer, it became easier to analyze the figures, replicate them, and create new ones. In the 1970s, Benoit Mandelbrot, who was an IBM fellow, discovered a most fascinating and complex shape that became known as the Mandelbrot set. He also coined the term "fractal" for sets on the real line, in the plane, or in space whose dimensions are not realistically any given integer. A decade later Michael Barnsley provided algorithms for constructing a wide variety of fractals on the computer.

We hope that those who read this book will appreciate not only the beauty of chaotic dynamics and fractal geometry, but also the diversity of applications. The subject continues to grow as a mathematical area and in disciplines outside of mathematics.

Preface

It's been a privilege to work on the third edition of such a seminal text. It's always tough when you're asked to try to improve on something that has served those who came before you. It is our hope that the changes have improved the usability of the text without changing the overall feel of the book.

We've revised some of the organization to try and make things flow more smoothly. We've added more modern applications, as well as additional historical notes. The goal was to push how these ideas are continuing to be applied in the 21st century, while connecting to the long and winding history of dynamical systems.

We've added quite a few exercises, again to see how the ideas of what is important in this field have changed, while providing more opportunity for diversity and inquiry in the kinds of questions we ask our students.

Finally, the MATLAB® code has been replaced with Python and expanded to include new explorations. Our goal here is to make the book accessible to a wider audience and to show the flexibility of using open-source software for this sort of programming. It's an important and effective way to introduce students to dynamical systems by looking at the results of experiments. It is our hope that this new approach helps you bring this fabulous text to a new audience.

1

Periodic Points

Suppose that f is a function and x is in the domain of f. Then the iterates of x, which consist of $x, f(x), f(f(x)), f(f(f(x))), \ldots$, form the orbit of x. The kinds of orbits that are possible for various values of x and for different functions constitute the basis for the study of chaos.

Chapter 1 is devoted to the study of those x's for which some iterate is again x. Such a point is called a "periodic point." For example, if $f(x) = x^2 - 1$, then we have $f(0) = -1$ and $f(-1) = 0$, so that $f(f(0)) = 0$, and thus 0 is a periodic point of f. Sections 1.1–1.3 lay the foundation for a detailed analysis of periodic points. Sections 1.4–1.5 focus on families of functions and discuss how periodic points can vary from function to function within a family of functions. In those sections, we examine in detail two illustrious families of functions: the tent family and the quadratic family. In Section 1.6 we discuss various kinds of bifurcations, where the quality or number of periodic points changes. Section 1.7 is devoted to the consequences of a function having periodic orbits containing only three elements. Finally, in Section 1.8 we discuss the largest number of periodic orbits a given function can have, each of which attracts all nearby points.

Chapter 1 is a prelude to the study of chaotic functions, defined in Chapter 2.

1.1 Iterates of Functions

Suppose that we key .5 into a calculator and then repeatedly depress the x^2-button. The calculator would display the numbers

$$0.5, \quad 0.25, \quad 0.0625, \quad 0.00390625, \quad 0.0000152858, \ldots \quad (1.1)$$

one after another. What is the calculator giving us with this sequence of numbers? If we let $f(x) = x^2$ and let $x_0 = 0.5$, then the sequence in (1.1) consists of

$$x_0, \quad f(x_0), \quad f(f(x_0)), \quad f(f(f(x_0))), \ldots$$

These numbers are called iterates of x_0 for f.

DOI: 10.1201/9781032678757-1

Definition 1.1. Let f be a function and let x_0 and $f(x_0)$ be in the domain of f. Then

$$f(x_0) = \textbf{the first iterate of } x_0 \textbf{ for } f$$

$$f(f(x_0)) = \textbf{the second iterate of } x_0 \textbf{ for } f$$

More generally, if n is any positive integer, if a_n is in the domain of f, and if a_n is the nth iterate of x_0 for f, then $f(a_n)$ is the $(n+1)$st iterate of x_0 for f. The total collection of first iterate, second iterate, etc., is called the **iterates of x_0 for f**.

For convenience, we will adopt the notation

$$f^{[2]}(x_0) \text{ for } f(f(x_0)), \qquad f^{[3]}(x_0) \text{ for } f(f(f(x_0)))$$

and more generally,

$$f^{[n]}(x_0) \text{ for the } n\text{th iterate of } x_0 \text{ for } f$$

We call the sequence $\{f^{[n]}(x_0)\}_{n=0}^{\infty}$ of iterates of x_0 the **orbit** of x_0. Sometimes we will write x_n for $f^{[n]}(x_0)$. In that case, $\{x_n\}_{n=0}^{\infty}$ is the orbit of x_0. Throughout the book it will be understood that all iterates of each function under discussion will lie in the domain of the function. As a result, we normally will not explicitly mention this assumption.

Note: The first documented occurrence of this notation for an iterated function in the literature comes from J.F.W. Herschel in 1813. He used f^n or $f^{(n)}$, rather than $f^{[n]}$, to indicate the nth iterate of the function f. Though Herschel credits the earlier Hans Heinrich Bürmann with the origin of the notation, Herschel does not provide a citation for Bürmann's work in his paper, so we are uncertain of the year. Herschel's distinct contribution was the use of a -1 exponent on a function, $f^{(-1)}$, not to indicate the reciprocal, but rather the pre-image of a function. This notation persists today in elementary mathematics, in the use of sin^{-1} to represent the *arcsin* function.

Now we are ready to illustrate a few iterates for several functions:

$f(x)$	x	orbit of x
$f(x) = x^2$	1	1, 1, 1, 1, $\ldots$
$f(x) = x^2 - 1$	-1	$-1, 0, -1, 0, -1, \ldots$
$f(x) = x^2 + 1$	-2	$-2, 5, 26, 677, 458330, \ldots$
$f(x) = x^2 + 1/4$	0	$0, 0.25, 0.3125, 0.347656\cdots, 0.370864\cdots, \ldots$
$f(x) = 4x - 4x^2$	1/3	$0.\overline{3}\cdots, 0.\overline{8}\cdots, 0.395061\cdots, 0.955951\cdots, \ldots$

As you see, even with simple quadratic functions, the orbits seem to have very different behaviors. Indeed, the orbit of 1 for the function x^2 is constantly 1. Next, the orbit of -1 for the function $x^2 - 1$ oscillates between -1 and 0. By contrast, the orbit of -2 for $x^2 + 1$ is unbounded, and it is not clear how the iterates of 0 and $1/3$ in the last two examples behave. In time we will prove that the iterates of 0 for $x^2 + 1/4$ converge to $1/2$, and the iterates of $1/3$ for $4x - 4x^2$ spread themselves over the interval $(0, 1)$ in a seemingly unpredictable manner.

Among the many functions whose orbits one can describe are the sine and cosine functions. For the sine function (using radians), the orbit of, say, 2 begins

$$2, \, 0.9092 \cdots, \, 0.7890 \cdots, \, 0.7097 \cdots, \, 0.6516 \cdots, \, 0.6064 \cdots$$

Continuing, we find that the 50th iterate is $0.2350\ldots$, and the 100th iterate is $0.1692\ldots$, with successive iterates decreasing. This leads us to conjecture that $f^{[n]}(2)$ approaches 0 as n increases without bound. In Example 1.2.2 of Section 1.2, we will prove that this conjecture is true. Meanwhile you might calculate iterates of 2 for $\cos x$, and see if they seem to have a reasonable limit.

Iterates form the basis of the Newton-Raphson method for approximating a zero of a function; you may have encountered the method in calculus, where it is usually just called Newton's method. To describe the process, we start with a function f, and assume that near a zero z of f, the derivative of f is nonzero. Next, we select an initial value x_0 that ideally is reasonably close to z. We define x_1 by

$$x_1 = x_0 - \frac{f(x_0)}{f'(x_0)}$$

For any positive integer n, we use x_n to define x_{n+1} by the formula

$$x_{n+1} = x_n - \frac{f(x_n)}{f'(x_n)} \tag{1.2}$$

Formula 1.2 constitutes the **Newton-Raphson method**. For many of the functions encountered in calculus (such as polynomial functions), careful selection of the initial value x_0 can quickly give good estimates of a zero of the given function. Figure 1.1 shows one such example.

If we let

$$g(x) = x - \frac{f(x)}{f'(x)}$$

then $x_1 = g(x_0)$, $x_2 = g(x_1)$, and in general, $x_{n+1} = g(x_n)$. Thus the sequence $\{x_n\}_{n=1}^{\infty}$ generated by the Newton-Raphson method consists of the iterates of x_0 for g.

Example 1.1.1. Let $f(x) = x^2 - 7$. Use the Newton-Raphson method to approximate a positive number that is a zero of f, until successive approximations are within 10^{-4} of each other.

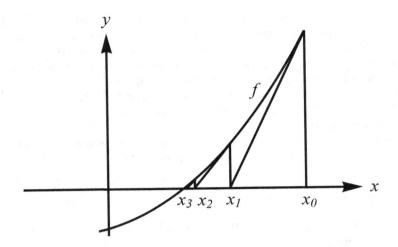

FIGURE 1.1
Iterates of the Newton-Raphson method

Solution. The positive zero z of f satisfies $z^2 - 7 = f(z) = 0$, so that $z = \sqrt{7}$. To approximate z by the Newton-Raphson method, let $x_0 = 3$, and let g be defined by

$$g(x) = x - \frac{f(x)}{f'(x)} = x - \frac{x^2 - 7}{2x} = \frac{1}{2}\left(x + \frac{7}{x}\right)$$

The initial few iterates of 3 for g are

$$x_0 = 3, \qquad x_1 = 2.6666\cdots, \qquad x_2 = 2.64583\cdots, \qquad x_3 = 2.64575\cdots$$

Since $|x_2 - x_3| < 10^{-4}$, the approximation we desire is $2.64575\cdots$, which you can check is an estimate of $\sqrt{7}$, accurate to 5 places. $\square$

In Section 1.2 we will learn why the iterates of *any* positive number approach $\sqrt{7}$ for the function g. (See Exercise 26 in Section 1.2.) Finally, we mention that an unfortunate choice of x_0 can lead to the failure of the Newton-Raphson method to approximate a zero of a given function. See Exercise 11 for an example. Thus one needs to be careful when using the method.

1.1.1   Graphical Analysis of Iterates

If we are able to render a reasonably precise graph of a given function, then we may be able to analyze graphically the orbits of various members of the domain.

First, we draw the graph of f, along with the line $y = x$. To exhibit the orbit of x_0, first locate x_0 on the x-axis. Notice that $(x_0, f(x_0))$ lies not only on the graph of f but also on the vertical line through $(x_0, 0)$ (Figure 1.2(a)).

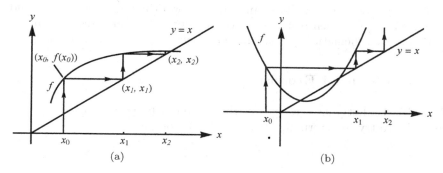

FIGURE 1.2
Graphical iteration of a function

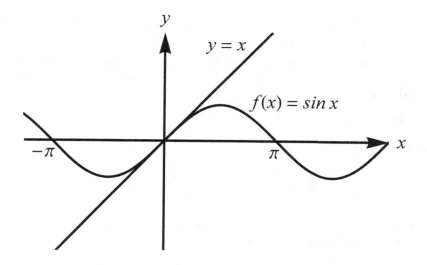

FIGURE 1.3
Looking for fixed points on the Sine function

The horizontal line through $(x_0, f(x_0))$ crosses the line $y = x$ at the point $(f(x_0), f(x_0)) = (x_1, x_1)$. By applying the same process with x_1 replacing x_0, we obtain the point (x_2, x_2). Continuing in the same manner, we can determine the location of $(x_3, x_3), (x_4, x_4)$, and thus in theory we can produce the orbit of x_0. This process is called a **graphical analysis** of the orbit of x_0 and is carried out in Figure 1.2.

Next, let $f(x) = \sin x$, the graph of which appears in Figure 1.3. With a little experimentation you should be able to convince yourself that the arrows in the graphical analysis end up in the narrow passages between the graphs of f and $y = x$, and converge to 0.

Caution: Graphical analysis can indicate how orbits behave and can guide us to a proof of a given result. However, except in special cases, we will rely on rigorous mathematical proofs and not solely on graphical analysis.

1.1.2   Section 1.1 Exercises

Exercise 1. Let $g(x) = \cos x$. Use the computer program ITERATE in the appendix to try to determine $\lim_{n \to \infty} g^{[n]}(x)$ for

(a) $x = 2$

(b) $x = -1$

(c) $x = \pi/4$

Exercise 2. Let $f(x) = 3x - 3x^2$. Calculate the first dozen iterates of $1/2$, and see if you can detect $\lim_{n \to \infty} f^{[n]}(1/2)$.

In Exercises 3–5, use graphical analysis to determine whether the iterates of the given point x seem to converge or not. If the iterates seem to converge, guess the limit.

Exercise 3. $f(x) = \cos x$; $x = 3$; $x = \pi/2$

Exercise 4. $f(x) = x - x^3/3$; $x = -1$

Exercise 5. $f(x) = \begin{cases} 2x & \text{for } 0 \le x \le 1/2 \\ 2x - 1 & \text{for } 1/2 < x \le 1 \end{cases}$; $x = 1/3$

Exercise 6. Find a function f such that $f(0)$ is the maximum value and $f^{[2]}(0)$ is the minimum value of f.

Exercise 7. Consider the function L defined by $L(x) = ax + b$, where $a \neq 1$. Determine the values of a for which $\lim_{n \to \infty} L^{[n]}(x)$ exists (as a number) for each real number of x, and find the value of that number in terms of a and b.

In Exercises 8–10, use the Newton-Raphson method, with the given value of x_0, to approximate a zero of the function f to within 10^{-4}.

Exercise 8. $f(x) = \cos x - x$; $x_0 = 1$

Exercise 9. $f(x) = x^3 - 2x - 5$; $x_0 = 2$

Exercise 10. $f(x) = x^4 + \sin x$; $x_0 = -1$

Exercise 11. Let $f(x) = 1 + x(x-1)^6$. See what happens when you try the Newton-Raphson method with each of the following initial values:

(a) 0

(b) 1

(c) 1.05

(d) 1.1

Exercise 12. Explain geometrically why the Newton-Raphson method fails to approximate the zero of the function $f(x) = x^{1/3}$.

Exercise 13. Revisit Exercises 1 and 2. Compute the inverse functions of each of these. Restrict the domain as necessary.

(a) Convergences for $g(x) = \cos(x)$ didn't seem to depend on the input. Can you compute $\lim_{n \to \infty} g^{[-n]}(x)$?

(b) What did you find for the limit in Exercise 2? Can you iterate the inverse function of that result and get close to $1/2$? How careful do you have to be with the domain?

If either of these cannot be done, explain why not.

1.2   Fixed Points

A point p whose iterates for a given function are the same point is called a "fixed point of the function." Fixed points are very important in the study of the dynamics of functions.

> **Note:** Fixed points are often called *equilibrium points* or *singular points*. The history of fixed points goes back quite awhile. The most often cited result is that of Brouwer (1912), that any mapping of a closed unit interval to itself must have a fixed point. The ideas behind this were known to Poincare, but the notion of a fixed point go back to Lipschitz and Cauchy in their study of differential equations. In fact, the modern proof of Cauchy's Existence and Uniqueness theorem for differential equations uses Banach's contraction-mapping theorem (1922), which is itself a theorem about fixed points.

Definition 1.2. Let p be in the domain of f. Then p is a **fixed point** of f if $f(p) = p$.

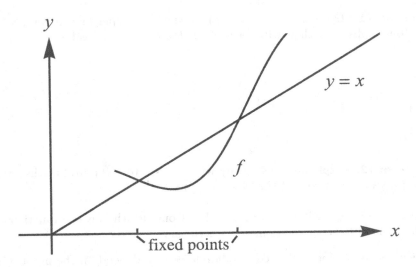

FIGURE 1.4
Fixed points cross the line $y = x$

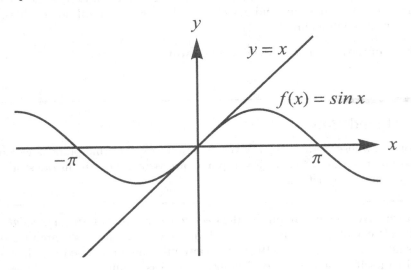

FIGURE 1.5
A single fixed point for the sine function

Graphically, a point p in the domain of f is a fixed point of f if and only if the graph of f touches (or crosses) the line $y = x$ at (p, p) (Figure 1.4).

Referring to Figure 1.5, we might conjecture that the origin is the only point at which the graph of $\sin x$ and the line $y = x$ touch each other. We will prove that this is true. In the proof we will use the Mean Value Theorem, which says that if f is continuous on $[a, b]$ and differentiable on (a, b), then there is a c in (a, b) such that

$$f'(c) = \frac{f(b) - f(a)}{b - a}, \quad \text{or equivalently,} \quad f(b) - f(a) = f'(c)(b - a)$$

Example 1.2.1. Let $f(x) = \sin x$. Show that 0 is the unique fixed point of f.

Solution. To begin, we observe that $f(x) \neq x$ if $|x| > 1$, since $|\sin x| \leq 1$ for all x. Next, if $0 < x \leq 1$, then the Mean Value Theorem implies the existence of z between 0 and x such that

$$\sin x = \sin x - \sin 0 = f'(z)(x - 0) = x \cos z$$

Since $0 < \cos z < 1$ for such z, it follows that $0 < \sin x = x \cos z < x$, true for all $x > 0$. The fact that $f(-x) = -f(x)$ implies that $x < \sin x < 0$ for all $x < 0$. Finally, $\sin 0 = 0$, so we conclude that 0 is the unique fixed point of f. $\square$

The next theorem will be very important to us. For convenience, we will write $f^{[n]}(x) \to p$ for "$f^{[n]}(x)$ approaches p" (as n increases without bound).

Theorem 1.3. Suppose that f is continuous at p, and let x and its iterates be in the domain of f. If $f^{[n]}(x) \to p$ as n increases without bound, then p is a fixed point of f.

Proof. By hypothesis, $f^{[n]}(x) \to p$, so that $f^{[n+1]}(x) \to p$. Since $f^{[n+1]}(x) \to p$, the continuity of f at p yields $f(f^{[n]})(x) \to f(p)$. However $f^{[n+1]}(x) = f(f^{[n]}(x))$, so that by substituting $f^{[n+1]}(x)$ for $f(f^{[n]}(x))$, we find that $f^{[n+1]}(x) \to f(p)$. The uniqueness of the limit of a given sequence implies that $f(p) = p$. Consequently p is a fixed point of f. $\square$

From calculus we know that a bounded sequence $\{x_n\}_{n=0}^{\infty}$ that is increasing converges to the least number z such that $x_n \leq z$ for all n. A similar statement holds for a bounded decreasing sequence, and hence for any monotone (that is, increasing or decreasing) sequence.

Corollary 1.4. Suppose that f is a continuous function defined on a closed interval. Assume that $\{f^{[n]}(x)\}_{n=0}^{\infty}$ is a bounded, monotone sequence. Then there is a fixed point p such that $f^{[n]}(x) \to p$ as n increases without bound.

Proof. By the comment above, bounded monotone sequences always converge. Thus this result is an immediate consequence of Theorem 1.3. $\square$

Now we will use Corollary 1.4 to prove that the iterates of any real x for the sine function converge to 0 — a result deduced by graphical analysis in Section 1.1.

Example 1.2.2. Let $f(x) = \sin x$. Show that the iterates of any x converge to 0.

Solution. Let x be an arbitrary number. To show that the sequence $\{f^{[n]}(x)\}_{n=0}^{\infty}$ is bounded, we observe that $-1 \le \sin x \le 1$ for each number x. Thus the sequence lies in $[-1, 1]$, and hence is bounded. Next, we observe that

$$f(-x) = \sin(-x) = -\sin x = -f(x)$$

Therefore if we can show that the sequence converges to 0 for each x in $[0, 1]$, then the same happens for each x in $[-1, 0]$, and hence for all x. Thus we only need to show that the sequence converges for each x in $[0, 1]$.

Since $f(0) = 0$, let $0 < x \le 1$. We will show next that $\{f^{[n]}(x)\}_{n=0}^{\infty}$ is a decreasing sequence. As in the solution of Example 1.2.1, the Mean Value Theorem yields a z between 0 and x such that

$$\sin x - \sin 0 = x \cos z < x$$

Therefore $0 < \sin x < x \le 1$ for $0 < x \le 1$, which means that

$$f(x) < x \quad \text{for} \quad 0 < x \le 1$$

It follows that for any $n \ge 0$,

$$f^{[n+1]}(x) = f(f^{[n]}((x)) = \sin f^{[n]}(x) < f^{[n]}(x)$$

We conclude that $\{f^{[n]}(x)\}_{n=0}^{\infty}$ is a decreasing sequence when $0 < x \le 1$. Since the sequence is also bounded, Corollary 1.4 implies that the sequence must converge to a fixed point, which by Example 1.2.1 is 0. Consequently $\{f^{[n]}(x)\}_{n=0}^{\infty}$ converges to 0 for all x. $\square$

Theorem 1.3 provides information concerning the Newton-Raphson method described in Section 1.1. Recall that the method involves calculating a sequence $\{x_n\}_{n=0}^{\infty}$ created by letting x_0 be an initial value, and defining

$$x_{n+1} = x_n - \frac{f(x_n)}{f'(x_n)}$$

Here we assume that $f'(x_n) \ne 0$ for all n. We will show that if $\{x_n\}_{n=0}^{\infty}$ converges to a number z in the domain of f, and if $f'(z) \ne 0$, then z is a zero of f. To that end, let

$$g(x) = x - \frac{f(x)}{f'(x)}$$

Then $x_1 = g(x_0)$, $x_2 = g(x_1) = g^{[2]}(x_0)$, and in general, $x_n = g^{[n]}(x_0)$. Thus $\{x_n\}_{n=0}^{\infty}$ is the sequence of iterates of x_0 for g. Theorem 1.3 tells us that if the sequence converges to z, then z is a fixed point of g. If $f'(z) \ne 0$, then

$$z = g(z) = z - \frac{f(z)}{f'(z)}$$

so that

$$\frac{f(z)}{f'(z)} = 0, \quad \text{or equivalently,} \quad f(z) = 0$$

This means that z is a zero of f, as we wished to prove.

1.2.1 Attracting and Repelling Fixed Points

By applying graphical analysis, we can see diverse behavior for the iterates of various points. Indeed, in Figure 1.6(a) the iterates of x approach the fixed point p, whereas in Figure 1.6(b) the iterates tend toward ∞ or $-\infty$. The iterates of x in Figure 1.6(c) approach the fixed point p if $x < p$ and x is close to p, whereas the iterates of x approach ∞ if $x > p$.

We are led to the following definition.

Definition 1.5. Let p be a fixed point of f.

(a) The point p is an **attracting fixed point** of f provided that there is an interval $(p - \epsilon, p + \epsilon)$ containing p such that if x is in the domain of f and in $(p - \epsilon, p + \epsilon)$, then $f^{[n]}(x) \to p$ as n increases without bound. (Such a point is also called **asymptotically stable** in the literature.)

(b) The point p is a **repelling fixed point** of f provided that there is an interval $(p - \epsilon, p + \epsilon)$ containing p such that if x is in the domain of f and in $(p - \epsilon, p + \epsilon)$ but $x \neq p$, then $|f(x) - p| > |x - p|$.

It follows from the definitions above that the fixed point in Figure 1.6(a) is attracting, and that the one in Figure 1.6(b) is repelling. That not every fixed

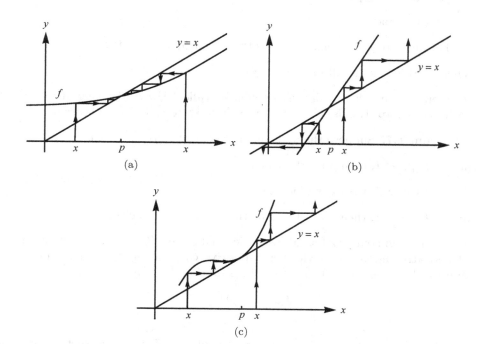

FIGURE 1.6
Attracting and repelling fixed points

point is attracting or repelling is demonstrated in Figure 1.6(c), where points to the left of p are attracted to p and points to the right of p are repelled from p. Other kinds of fixed points that are neither attracting nor repelling can occur.

Note: There is no standard definition in the literature for attracting and repelling fixed points. We have chosen definitions that seem reasonable for our purposes. For a variety of definitions that lead to equivalent results, we recommend Bhatia and Szegö, Coddington and Levinson, or Hirsch, Smale and Devaney. Details for these texts are in the references.

For most functions, it is not so easy to prove directly from Definition 1.5 that a given fixed point p is attracting (or repelling, or neither). However, if f is differentiable at p, then a useful criterion exists, which we will state and prove in Theorem 1.7. In the proof, we will need to apply the Axiom of Mathematical Induction, which is frequently called the "Law of Induction."

Theorem 1.6 (LAW OF INDUCTION). Assume that for each integer greater than or equal to an initial integer n_0, a statement, formula, or equation, $S(n)$, is given. Suppose that

i. $S(n_0)$ is true.

ii. For any integer $n > n_0$, if $S(n)$ is true, then $S(n+1)$ is true.

Then $S(n)$ is true for all integers $n \geq n_0$.

Step (ii) in the axiom is frequently called the **inductive step** or **induction hypothesis**. Now we are ready to state and prove Theorem 1.7.

Theorem 1.7. Suppose that f is differentiable at a fixed point p.

(a) If $|f'(p)| < 1$, then p is attracting.

(b) If $|f'(p)| > 1$, then p is repelling.

(c) If $|f'(p)| = 1$, then p can be attracting, repelling, or neither.

Proof. To begin our proof of (a), we notice that since $|f'(p)| < 1$, the definition of derivative implies that there is a positive constant $A < 1$ and an open interval $J = (p - \epsilon, p + \epsilon)$ such that if x is in J and $x \neq p$, then

$$\left| \frac{f(x) - f(p)}{x - p} \right| \leq A$$

Therefore $|f(x) - f(p)| \leq A|x - p|$, for all x in J. For each such x, this means that

$$|f(x) - p| = |f(x) - f(p)| \leq A|x - p|$$

so that $f(x)$ is in J because $0 < A < 1$ and x is in J. Thus $f(x)$ is at least as close to p as x is. Let x be fixed in J. If $f^{[n]}(x) = p$ for some n, then obviously $f^{[n]}(x) \to p$ as n increases without bound, so we will assume henceforth that $f^{[n]}(x) \neq p$ for all n. Next, we will use the Law of Induction to prove that

$$|f^{[n]}(x) - p| \le A^n |x - p| \quad \text{for all } n \ge 1$$

By (1.1), the inequality holds for $n = 1$. Next, we assume that (1.2) holds for a given $n > 1$. Then $f^{[n]}(x)$ is in J since $0 < A^n < A < 1$. Therefore by (1.1) with $f^{[n]}(x)$ substituted for x, and then by (1.2), we find that

$$|f^{[n+1]}(x) - p| = |f(f^{[n]}(x)) - p| \le A|f^{[n]}(x) - p| \le A(A^n |x - p|) = A^{n+1}|x - p|$$

so that $|f^{[n+1]}(x) - p| \le A^{n+1}|x - p|$. By the Law of Induction we deduce that (1.2) holds for all integers $n \ge 1$. Since $A^n \to 0$ as n increases without bound, it follows that $f^{[n]}(x) \to p$ for every x in J. Thus (a) is proved. The proof of (b) is analogous. Part (c) is addressed in Exercise 10. □

We can put Theorem 1.7 to immediate use.

Example 1.2.3. Let $\mu > 0$ be a constant, and let

$$f(x) = \mu x(1 - x) = \mu x - \mu x^2, \quad \text{for } 0 \le x \le 1$$

(a) Find the values of μ for which 0 is an attracting fixed point.

(b) Find the values of μ for which there is a nonzero fixed point.

(c) Find the values of μ for which the nonzero fixed point is attracting.

Solution. Notice that x is a fixed point of f if and only if $x = \mu x - \mu x^2$. Thus $x = 0$ is a fixed point for every $\mu > 0$. Since $f'(0) = \mu$, it follows from Theorem 1.7 that 0 is attracting if $0 < \mu < 1$ and is repelling if $1 < \mu$. By Exercise 15, 0 is also attracting if $\mu = 1$. This completes part (a) of the solution. For part (b), observe that if $x \neq 0$, then x is a fixed point only if $1 = \mu - \mu x$, or equivalently, $x = 1 - 1/\mu$. However, if $0 < \mu < 1$, then $x = 1 - 1/\mu < 0$, so x is not in the domain of f. Therefore the nonzero fixed point $1 - 1/\mu$ occurs only if $\mu > 1$, which completes part (b). For part (c), we note that

$$f'(1 - 1/\mu) = \mu - 2\mu(1 - 1/\mu) = 2 - \mu$$

Using Theorem 1.7 again, we find that $1 - 1/\mu$ is attracting if $1 < \mu < 3$ and is repelling if $\mu > 3$. Finally, it is possible to show that $1 - 1/\mu$ is attracting if $\mu = 3$ (Exercise 16). □

1.2.2 Basins of Attraction

If a fixed point p of f is attracting, then all points near to p are attracted toward p, in the sense that their iterates converge to p. The collection of all points whose iterates converge to p is called the basin of attraction of p.

Definition 1.8. Suppose that p is a fixed point of f. Then the **basin of attraction** of p consists of all x such that $f^{[n]}(x) \to p$ as n increases without bound and is denoted by B_p.

Example 1.2.4. Let $f(x) = x^2$. Find the basin of attraction B_0 of the fixed point 0.

Solution. If $|x| < 1$, then $f^{[n]}(x) = x^{(2^n)} \to 0$ as n increases without bound, so that x is in B_0. By contrast, if $|x| \geq 1$, then $|f^{[n]}(x)| \geq 1$, so that x is not in B_0. Thus B_0 consists of all x such that $|x| < 1$, that is, $B_0 = (-1, 1)$. (We could also draw the same conclusion by using graphical analysis.) $\square$

We remark that if p is a repelling fixed point, then its basin of attraction can consist of the single point p, as happens for 0 if $f(x) = 2x$. By contrast, the basin of attraction of the fixed point 0 of $\sin x$ consists of all real numbers, which in effect is what we showed in Example 1.2.2.

1.2.3 Eventually Fixed Points

Finally, we introduce the notion of eventually fixed points, which will be of use in later examples.

Definition 1.9. Let x be in the domain of f. Then x is an **eventually fixed point** of f if there is a positive integer n such that $f^{[n]}(x)$ is a fixed point of f.

A fixed point is trivially an eventually fixed point. However, if $f(x) = \sin x$, then $f(\pi) = 0$ and $f(0) = 0$, so that π is an eventually fixed point that is not a fixed point. In order not to create confusion, when we refer to x as an eventually fixed point, we will generally assume that x is not a fixed point.

Example 1.2.5. Let T be defined by

$$T(x) = \begin{cases} 2x & \text{for } 0 \leq x \leq 1/2 \\ 2 - 2x & \text{for } 1/2 < x \leq 1 \end{cases}$$

Show that $1/8$ is an eventually fixed point.

Solution. A routine check shows that

$$T(1/8) = 1/4, \quad T(1/4) = 1/2, \quad T(1/2) = 1, \quad T(1) = 0, \quad T(0) = 0$$

Therefore $1/8$ is an eventually fixed point. $\square$

The function T is called the **tent function** because of the shape of its graph (see Figure 1.7). Example 1.2.5 says that T has an eventually fixed point. One can show by an analogous argument that if $x = k/2^n$, where k and n are positive integers and $0 < k/2^n \leq 1$, then x is an eventually fixed point

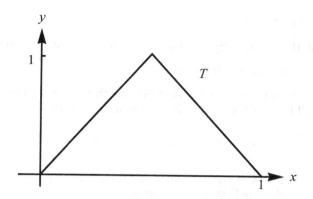

FIGURE 1.7
The tent map

of T (see Exercise 27). We will use this result when we study T in more detail in Section 1.4.

We mention that if x is an eventually fixed point of f (so that some iterate of x is a fixed point p of f), then x is automatically in the basin of attraction of p. The converse is false, however, because the iterates of points can converge to p without eventually being p. For example, if $f(x) = x/2$ for all x in R, then 1 is in the basin of attraction of f because $f^{[n]}(1) = 1/2^n \to 0$ as n increases without bound. However, 0 is not an eventually fixed point of f since $f^{[n]}(1) \neq 0$ for all n.

1.2.4 Section 1.2 Exercises

In Exercises 1–6, find the fixed points and determine whether each is attracting or repelling.

Exercise 1. $f(x) = 4x - x^2$

Exercise 2. $f(x) = x^3 - x/3$

Exercise 3. $f(x) = \sqrt{x}$

Exercise 4. $f(x) = e^{x-1}$

Exercise 5. $f(x) = \arcsin x$

Exercise 6. $f(x) = 1/x$

Exercise 7. Let $g(x) = x^2 + 1/4$. Show that if $|x| > 1/2$, then $|g^{[n]}(x)| \to \infty$ as n increases without bound.

Exercise 8. Let $f(x) = \cos x$.

(a) Show that there is exactly one fixed point p, and that it is attracting.

(b) Find the basin of attraction of p.

(c) With the value of p obtained in part (a), use the computer program NUM-BER OF ITERATES to determine the smallest n such that $|f^{[n]}(2) - p| < .001$.

Exercise 9.

(a) Let $f(x) = \sin x$. Use the computer program NUMBER OF ITERATES to determine the smallest n such that $0 < f^{[n]}(2) < .1$.

(b) The 10,000th iterate of 2 for $\sin x$ is larger than .001. Why do you suppose that the convergence to the fixed point 0 is so slow?

Exercise 10.

(a) Let $f(x) = \arctan x$. Show that $f'(0) = 1$, and that 0 is an attracting fixed point.

(b) Let $g(x) = x^2 + 1/4$. Show that $g'(1/2) = 1$, and that $1/2$ is a fixed point that is neither attracting nor repelling.

(c) Let $h(x) = x^3 + x$. Show that $h'(0) = 1$, and that 0 is a repelling fixed point.

Exercise 11. Let $f(x) = |x|/2$. Show that 0 is an attracting fixed point of f. Can you infer that 0 is attracting from Theorem 1.7? Explain why or why not.

Exercise 12. Let m be a real number, and consider the line given by $L(x) = mx + (1 - m)$.

(a) Show that 1 is a fixed point of L.

(b) Find all values of m for which 1 is attracting, and all values of m for which 1 is repelling.

(c) Find a value of m such that 1 is neither attracting nor repelling.

Exercise 13. Let a and b be constants, and let $L(x) = ax + b$.

(a) Let $|a| \neq 1$. Find the fixed point p of L, and determine the values of a for which p is attracting, and those for which p is repelling.

(b) Let $|a| = 1$. Show that any fixed point(s) that L has are neither attracting nor repelling.

Exercise 14. Let $f(x) = e^x$. Show that f has no fixed points.

Exercise 15. Let $f(x) = x - x^2$ for $0 \le x \le 1$. Show that 0 is an attracting fixed point.

Exercise 16. Let $f(x) = 3x - 3x^2$. Show that $2/3$ is an attracting fixed point.

Exercise 17. Let $f(x) = 3x(1-x)$. Using the computer program NUMBER OF ITERATES, determine the minimum number of iterates of $1/2$ that it takes to approximate the fixed point $2/3$ to within

(a) 0.1

(b) 0.05

(c) 0.01

(d) 0.005

Exercise 18. Let $f(x) = \tan x$. Show that f has an infinite number of fixed points, and classify the fixed points as attracting, repelling, or neither.

Exercise 19. Let $f(x) = (\tan x)/2$. Show that the iterates of x converge to 0 whenever x is in $(-\pi/3, \pi/3)$.

Exercise 20. Let $f(x) = x^2 + 1/4$. Use the Law of Induction to show that

(a) if $-1/2 \le x \le 1/2$, then $f^{[n]}(x) \to 1/2$ as n increases without bound.

(b) if $|x| > 1/2$, then $f^{[n]}(x) \to \infty$ as n increases without bound.

Exercise 21. Let $f(x) = 4x^2 + 1/16$. Use the Law of Induction to show that

(a) if $-1/8 \le x \le 1/8$, then $f^{[n]}(x) \to 1/8$ as n increases without bound.

(b) if $|x| > 1/8$, then $f^{[n]}(x) \to \infty$ as n increases without bound.

In Exercises 22–25, use either algebra or graphical analysis to find the largest open interval in the basin of attraction of the fixed point 0.

Exercise 22. $f(x) = \sin^2 x$

Exercise 23. $f(x) = x^4$

Exercise 24. $f(x) = x^5 + x^3$

Exercise 25. $f(x) = (.5)x(1-x)$

Exercise 26. Let $f(x) = x^2 - 7$, and let $g(x) = x - f(x)/f'(x)$. Show that if x is any positive number, then the iterates of x approach the zero of f. (Thus the Newton-Raphson method is successful for f when any initial positive value of x is picked.)

Exercise 27. Let T be the tent function defined in Example 1.2.5. Show that if x is in the interval $(0, 1)$ and has the form $x = k/2^n$, where k and n are positive integers, then x is an eventually fixed point.

Exercise 28. Let $f(x) = x^2$. Show that if $0 < |x| < 1$, then x is in the basin of attraction of 0 but is not an eventually fixed point.

Exercise 29. Let $f(x) = x^2 + 1/8$.

(a) Find the two fixed points of f, and show that one of them is attracting.

(b) Find the basin of attraction of the attracting fixed point found in (a).

Exercise 30. Suppose that $f \geq 0$, and that p is a fixed point of f. Let g be defined by $g(x) = (f(x))^{1/2}$ for all x in the domain of f. Determine under what conditions p is a fixed point of g.

Exercise 31. Suppose that f has the graph pictured in Figure 1.8, with fixed point p.

(a) Determine whether p is attracting, repelling, or neither.

(b) Find a formula for a function that has the shape pictured in Figure 1.8.

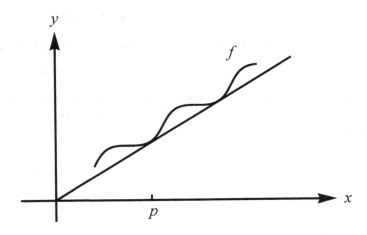

FIGURE 1.8
Graph to accompany Exercise 31 of Section 1.2

Exercise 32. Suppose that f is continuous and strictly increasing on $(-\infty, \infty)$, and 0 is the only fixed point of f. Assume also that the graph of f lies above the line $y = x$ for $x < 0$, and lies below the line $y = x$ for $x > 0$. Show that 0 is an attracting fixed point.

Exercise 33. Suppose that p is a fixed point of f, and that the graph of f is tangent to the line $y = x$ at the point $(p, f(p))$. Show that if $(p, f(p))$ is an inflection point, then p is either attracting or repelling.

Exercise 34. We will say that the fixed point p is **attracting – repelling**, or **neutral**, if p is attracting for points to one side of p and is repelling for points to the other side (and close to p).

(a) Let $f(x) = x(1 - x)$ for all real x. Find an attracting-repelling fixed point for f.

(b) Let $g(x) = x^2 + 1/4$ for all real x. Find an attracting-repelling fixed point for g.

(c) Find a continuous function h and a fixed point p such that p is neither attracting, nor repelling, nor attracting-repelling.

Exercise 35. In Theorem 1.6, we stated that, since $|f'(p)| < 1$, the definition of derivative implies that there is a positive constant $A < 1$ and an open interval $J = (p - \epsilon, p + \epsilon)$ such that if x is in J and $x \neq p$, then

$$\left| \frac{f(x) - f(p)}{x - p} \right| \leq A.$$

Prove that this is true from the definition of the derivative.

Exercise 36. In Example 1.2.5, we have an eventually fixed point of the function T. Consider what occurs when you compute $T^{[n]}(x)$ for some negative integer n, and $x \neq 1/8$. What occurs if $x = 1/8$?

1.3   Periodic Points

Periodicity is a notion common in everyday language. It refers to an event that reoccurs at a predictable interval. For example, Halley's comet has a period of approximately 76 years. Similarly, the longer a pendulum is, the longer its period is. The notion of periodicity is central to the study of dynamics.

Definition 1.10. Let x_0 be in the domain of f. Then x_0 has **period-n** (or is a **period-n point**) if $f^{[n]}(x_0) = x_0$, and if in addition, $x_0, f(x_0), f^{[2]}(x_0) \ldots, f^{[n-1]}(x_0)$ are distinct. If x_0 has period n, then the orbit of x_0, which is

$$\{x_0, f(x_0), f^{[2]}(x_0), \ldots, f^{[n-1]}(x_0)\}$$

is a **periodic orbit**, and the elements of the orbit form an **n-cycle**.

By Definition 1.10, fixed points are periodic points with period 1. If a point has period 1, then we will normally refer to it as a fixed point (rather than a periodic point).

To illustrate a 2-cycle, let $h(x) = -x^3$. Then $\{-1, 1\}$ is a 2-cycle because $h(-1) = 1$ and $h(1) = -1$. Next, we will exhibit a 3-cycle for the tent function T.

Example 1.3.1. The tent function T is given by

$$T(x) = \begin{cases} 2x & \text{for } 0 \le x \le 1/2 \\ 2 - 2x & \text{for } 1/2 < x \le 1 \end{cases}$$

Show that $\{2/7, 4/7, 6/7\}$ is a 3-cycle for T.

Solution. A routine check yields

$$T\left(\frac{2}{7}\right) = \frac{4}{7}, \qquad T\left(\frac{4}{7}\right) = \frac{6}{7}, \qquad T\left(\frac{6}{7}\right) = \frac{2}{7}$$

confirming that $\{2/7, 4/7, 6/7\}$ is a 3-cycle for T. $\square$

Not only does T have a 3-cycle; it has n-cycles for every positive integer n. This is one of the reasons why the tent function is featured in the study of dynamics. We will take a closer look at the tent function in Section 1.4.

Graphically, an n-cycle of a function is represented by a closed loop. Figure 1.9(a) shows the 2-cycle $\{-1, 1\}$ for the function $-x^3$, and Figure 1.9(b) shows the 3-cycle for the tent function in Example 1.2.1 above.

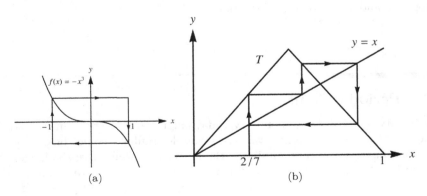

FIGURE 1.9
Examples of cycles

It is important to notice that if $f(x) = z$ and $f(z) = x$, then

$$f^{[2]}(x) = f(f(x)) = f(z) = x$$

so that x is a fixed point of $f^{[2]}$. By the same token, z is a fixed point of $f^{[2]}$. Thus, if $\{x, z\}$ is a 2-cycle for f, then x and z are both fixed points of $f^{[2]}$. Conversely, if x is a fixed point of $f^{[2]}$ that is not a fixed point of f, then there is a point z different from x such that $\{x, z\}$ is a 2-cycle of f, so that x is a period-2 point of f. Therefore

$$\{x, z\} \text{ is a 2-cycle for } f \text{ if and only if } f(x) = z \text{ and } f(z) = x,$$

where x and z are distinct fixed points of $f^{[2]}$,

For example, assume that $f(x) = x^2 - 1$, so that $f^{[2]}(x) = (x^2 - 1)^2 - 1 = x^4 - 2x^2$. Obviously 0 is a fixed point of $f^{[2]}$ that is not a fixed point of f. Thus there must be a z such that $\{0, z\}$ is a 2-cycle for f. Since $f(0) = -1$ we deduce that $\{0, -1\}$ is a 2-cycle for f. More generally, $\{x_0, x_1, x_2, \dots, x_{n-1}\}$ is an n-cycle of f if and only if the numbers $x_0, \dots, x_{n-1}$ are distinct, and x_k is a fixed point of $f^{[n]}$, for $k = 0, 1, 2, \dots, n - 1$.

1.3.1   Attracting Periodic Points

Suppose that x is a period-n point of f. Then x is a fixed point for $f^{[n]}$. Therefore we have a natural way of defining attracting and repelling periodic points.

Definition 1.11. Let x be a period-n point for a function f. Then x is an **attracting period-n point** if x is an attracting fixed point of $f^{[n]}$; also x is a **repelling period-n point** if x is a repelling fixed point of $f^{[n]}$.

Suppose that f is continuous at a period-n point x. If x is attracting (resp. repelling), then each point in $\{x, f(x), f^{[2]}(x), \dots, f^{[n-1]}(x)\}$ is an attracting (resp. repelling) period-n point, so we say that the n-cycle $\{x, f(x), f^{[2]}(x), \dots, f^{[n-1]}(x)\}$ is **attracting** (resp. **repelling**).

In particular, if $n = 2$ then the period-2 point x is attracting if and only if there is an interval $(x - \epsilon, x + \epsilon)$, such that whenever y is in $(x - \epsilon, x + \epsilon)$,

$$f^{[2n]}(y) \to x \quad \text{and} \quad f^{[2n+1]}(y) \to f(x)$$

as n increases without bound. Figure 1.10 shows a number x that is attracted to a 2-cycle of f.

Example 1.3.2. Let $f(x) = -x^{1/3}$. Show that 1 is an attracting period-2 point of f.

Solution. First, notice that $f(1) = -1$ and $f(-1) = 1$. Therefore the point 1 has period 2. Next, observe that

$$f^{[2]}(x) = f(f(x)) = -(-x^{1/3})^{1/3} = x^{1/9}, \text{ so that } (f^{[2]})'(1) = \frac{1}{9}$$

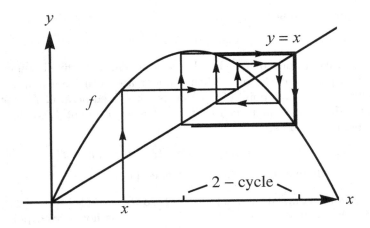

FIGURE 1.10
A point attracted to a 2-cycle

   Theorem 1.7 then implies that 1 is an attracting fixed point of $f^{[2]}$, so that
1 is an attracting period-2 point of f by Definition 1.11. □

   One could also prove that 1 is an attracting point of period 2 by showing
that there is an interval J such that whenever x is in J, then $|f^{[2]}(x)-1| <$
$|x-1|$.
   In Section 1.2 we gave a criterion for attracting and repelling fixed points
that involves the derivative. Similarly, there is a criterion for attracting and
repelling cycles that involves the derivative. Before we state it in Theorem
1.13, we have a preliminary result.

Theorem 1.12. Let $\{x, z\}$ be a 2-cycle of f. If $f^{[2]}$ is differentiable at x and
at z, then
$$(f^{[2]})'(x) = f'(x)f'(z) = (f^{[2]})'(z)$$

Proof. Using the Chain Rule and the fact that $f(x) = z$, we find that

$$(f^{[2]})'(x) = (f \circ f)'(x) = [f'(f(x))][f'(x)] = f'(x)f'(z)$$

By symmetry we have $(f^{[2]})'(z) = f'(x)f'(z)$.   □

Theorem 1.13. Let $\{x, z\}$ be a 2-cycle for f.

(a) If $|f'(x)f'(z)| < 1$, then the 2-cycle is attracting.

(b) If $|f'(x)f'(z)| > 1$, then the 2-cycle is repelling.

Proof. The result follows directly from Theorems 1.7 and 1.12, and the definition of an attracting (repelling) 2-cycle.   □

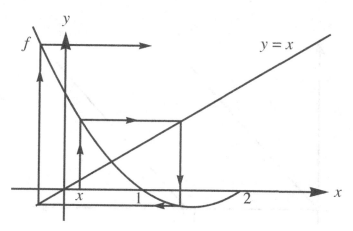

FIGURE 1.11
A 2 cycle for Example 1.3.3

If $|f'(x)f'(z)| = 1$, then we cannot conclude anything about whether the cycle $\{x, z\}$ is attracting, repelling, or neither. For example, let $f(x) = 1/x$. Then $f^{[2]}(x) = x$, so that $\{x, 1/x\}$ is a 2-cycle for each $x \neq 0, 1, -1$. Evidently the 2-cycle is neither attracting nor repelling, although $|f'(x)f'(z)| = |f^{[2]})'(x)| = 1$ for all $x \neq 0$.

When $|f'(x)f'(z)| \neq 1$, the criterion can be effective in telling if $\{x, z\}$ is attracting or repelling.

Example 1.3.3. Let $f(x) = x^2 - 3x + 2$. Show that $\{0, 2\}$ is a repelling 2-cycle.

Solution. Since $f(0) = 2$ and $f(2) = 0$, it follows that $\{0, 2\}$ is a 2-cycle. The fact that $f'(x) = 2x - 3$ implies that $f'(0) = -3$ and $f'(2) = 1$, so that

$$f'(0)f'(2) = (-3)(1) = -3$$

Therefore Theorem 1.13 implies that $\{0, 2\}$ is a repelling 2-cycle of f. $\square$

Figure 1.11 displays the graph of f, with 2-cycle $\{0, 2\}$. By analyzing the iterates of x, which is close to 0, we can see why the 2-cycle is repelling.

If $\{x, f(x), ..., f^{[n-1]}(x)\}$ is an n-cycle, then by the Chain Rule,

$$(f^{[n]})'(x) = [f'(f^{[n-1]}(x))] \, [f'(f^{[n-2]}(x))] \cdots [f'(f(x))] \, [f'(x)]$$

If the absolute value of the right-hand side of (3) is < 1 (or is > 1), then the n-cycle is attracting (or repelling). We remark that if x is a fixed point, then (3) becomes

$$(f^{[n]})'(x) = [f'(x)]^n$$

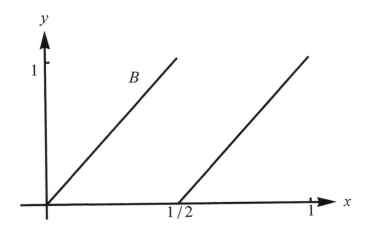

FIGURE 1.12
The baker's function

An **eventually periodic point** is a point for which some iterate is periodic. For example, let $f(x) = x^2 - 1$. It follows that 1 is an eventually periodic point, since $f(1) = 0$, $f(0) = -1$, and $f(-1) = 0$. Henceforth we will generally assume (as we do with eventually fixed points), that when we refer to a point as eventually periodic, the point is not periodic. Defining the basin of attraction of an attracting cycle is more complicated. However, informally the **basin of attraction** of an attracting cycle is the collection of points whose iterates are eventually arbitrarily close to the points in the cycle. Again using $f(x) = x^2 - 1$, we can show that the basin of attraction for the 2-cycle $\{-1, 0\}$ consists of all numbers in the interval $((1 - \sqrt{5})/2, (1 + \sqrt{5})/2)$ except those whose iterates are eventually the fixed point $(1 - \sqrt{5})/2$. (See Exercise 9.)

To give a further illustration of the notions appearing in this section, we introduce the function B, which is a close relative of the tent function:

$$B(x) = \begin{cases} 2x & \text{for } 0 \leq x \leq 1/2 \\ 2x - 1 & \text{for } 1/2 < x \leq 1 \end{cases}$$

(Figure 1.12). Then B is called the **baker's function**, in reference to the kneading of bread dough. (The function is related to the sawtooth function that is prominent in engineering. Any of you who have tried to build your own guitar effects pedals may have run across this.) Before we turn to special properties of B, we mention that a number x in $[0, 1]$ is called a **dyadic rational** if it has the form $k/2^m$ for some nonnegative integers k and m. We list three special properties of B:

1. A number x in $[0, 1]$ is eventually periodic if and only if x is rational.

2. A number x in $[0, 1]$ is eventually fixed if and only if x is a dyadic rational.

3. If p is an odd, positive integer and if k/p is reduced, then k/p is periodic, for $k = 1, 2, \ldots, p-1$.

Although the baker's function is not continuous at $1/2$, it has many interesting properties, and has been the subject of much attention in the study of dynamics. Exercises 10–18 are devoted to properties of B.

> **Note:** We can think historically about why the Baker's map has that name. Consider the domain and range of the map. In some sense $[0, 1/2]$ is stretched out to $[0,1]$ in the first iteration of the map. The analogy is that of a baker kneading dough. In kneading, dough is stretched, folded back over itself, and stretched again. So while $[0,1/2]$ is stretched to $[0,1]$, $[0,1]$ always ends up mapped to itself. The question of which points stay close to where they started or move far from where they started is an interesting one. We'll study this in more detail when we see the Smale horseshoe in Section 3.5.

1.3.2 Time Series and Periodic Points

Many of a human's or animal's biological functions can be analyzed by means of a time series, which is a graph that registers a particular variable such as voltage or pressure as a function of time. Voltage is plotted against time in an electrocardiogram (EKG) and in an electroencephalogram (EEG).

Figure 1.13 illustrates the EKG of a cat, before and after ingestion of a dose of cocaine. The EKG appearing in Figure 1.13(a) is normal, the pattern consisting of a tall and a shorter spike that repeats regularly as time passes.

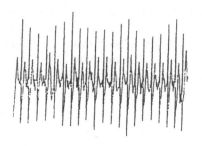
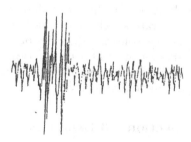

Cat's EKG before cocaine dose   Cat's EKG after cocaine dose
(a)   (b)

FIGURE 1.13
A cat, which was given cocaine

Cat's EEG before cocaine dose   Cat's EEG after cocaine dose
(a)   (b)

FIGURE 1.14
A cat, which was given more cocaine

The heartbeat is said to be periodic. By contrast, the EKG appearing in
Figure 1.13(b) is abnormal and irregular, and could represent the onslaught
of a life-threatening cardiac fibrillation.

In a similar vein, Figure 1.14(a) displays the EEG of a cat before ingestion
of a dose of cocaine, whereas Figure 1.14(b) shows the EEG after ingestion.
However, in contrast to the EKG that is normal and periodic before cocaine
and irregular afterwards, the normal EEG that is unpredictable and irregular
before cocaine became much more regular (though not really periodic) after
cocaine. The latter EEG is characteristic of a life-threatening brain dysfunc-
tion or an epileptic seizure.

A wide variety of human behavior can be represented with time series.
Perhaps the rate at which a student practices piano, or their performance in
a mathematics course. Any series of events that can be measured at discrete
time intervals can yield a time series. Any anything measured over discrete
time intervals can be considered as a dynamical system, and therefore could
demonstrate periodic behavior.

1.3.3   Section 1.3 Exercises

Exercise 1. Let $f(x) = -\dfrac{1}{2}x^2 - x + \dfrac{1}{2}$. Show that 1 is an attracting period-2
point.

Exercise 2. Show that $\{2/9, 4/9, 8/9\}$ is a repelling 3-cycle for the tent
function T.

Exercise 3. Let $f(x) = 1/x$. Show that if $x \neq -1, 0$, or 1, then x is a period-2 point.

Exercise 4. Let $f(x) = 1/(1-x)$. Show that if $x \neq 0$ or 1, then x is a period-3 point.

Exercise 5. Let $f(x) = 3.2x - 3.2x^2$. Use a calculator or the program ITERATE to find a 2-cycle, and show that it is attracting.

Exercise 6. Let $f(x) = 3.84x - 3.84x^2$. Use a calculator or the program ITERATE to find a period-3 point, and determine whether it is attracting or not.

Exercise 7. Let T be the tent function. Find a point that is not periodic but is eventually periodic with period

(a) 3

(b) 4

(c) 5

Exercise 8. Let $f(x) = \cos x$. Determine whether f has any period-n points with $n > 1$.

Exercise 9. Let $f(x) = x^2 - 1$. Show that the basin of attraction of the 2-cycle $\{-1, 0\}$ consists of all numbers in the interval $((1-\sqrt{5})/2, (1+\sqrt{5})/2)$ except those whose iterates are eventually the fixed point $(1-\sqrt{5})/2$.

(a) Let $B = ((1-\sqrt{5})/2, (1+\sqrt{5})/2)$ and show $f(B) \subset B$.

(b) Identify all fixed points in B.

(c) Show that 0 and -1 are each fixed points of the function $f^{[2]}$. Use graphical analysis to find the basin of attraction for these fixed points.

(d) These exercises should give you the intuition for why this is true. For $x \in B$, describe what can occur for $f^{[n]}(x)$ as $n \to \infty$. Perform the same computations for $x_1 \notin B$. Can you analytically prove what your graphical analysis supported?

Exercises 10–18 involve the baker's function B.

Exercise 10. Sketch the graphs of $B^{[2]}$ and $B^{[3]}$.

Exercise 11. Find the fixed points and the period-2 points of B.

Exercise 12. Determine whether the following points are fixed, eventually fixed, periodic, or eventually periodic, and indicate their periods if they are periodic.

(a) 3/7

(b) 3/16

(c) 1/10

(d) 1/11

Exercise 13. For each positive integer n, determine the number of fixed points of $B^{[n]}$.

Exercise 14. Show that x in $[0,1]$ is an eventually fixed point of B if and only if x is a dyadic rational.

Exercise 15. Show that x in $[0,1]$ is an eventually periodic point of B if and only if x is rational.

Exercise 16. Show that if p is an odd positive integer and k/p is reduced, then k/p is periodic for B, for $k = 1, 2, ..., p-1$. Is the converse true? Explain your answer.

Exercise 17. Show that if $1/2 < x \leq 1$, then $B(x) \leq x^2$.

Exercise 18. Let m be an arbitrary positive integer, and assume that x is not a dyadic rational. Show that there is an integer $n \geq m$ such that

(a) $B^{[n]}(x) < 1/2$

(b) $B^{[n]}(x) > 1/2$

Exercise 19. Let c be a constant, and let $f(x) = x^3 - 3x + c$. Determine the values of c for which $\{0, c\}$ is a 2-cycle. Is such a 2-cycle attracting for the corresponding value of c? Explain your answer.

Exercise 20. Let $f(x) = ax^3 - bx + 1$, where a and b are constants. Determine the values of a and b for which $\{0, 1\}$ is an attracting 2-cycle.

Exercise 21. Let m be a positive integer. Prove that if $(f^{[m]})'(x) = 0$, then there is an iterate x_k of x such that $f'(x_k) = 0$.

Exercise 22. Let $|f'(x)| < 1$ for all x. Show that f cannot have any period-2 points. Can it have any fixed points? Explain your answer.

Exercise 23.

(a) Let f be increasing. Show that there are no period-n points for $n > 1$.

(b) Let f be decreasing. Show that there are no period-n points for $n > 2$.

(c) Find an example of a decreasing function f that has a fixed point and a period-2 point.

Exercise 24. Let f be a linear function. Show that there are no period-n points for $n > 2$.

Exercise 25. Suppose we have time series data $\{f(1), f(2), \ldots, f(N)\}$.

(a) We will have to make assumptions about the data in order to proceed. For this case, assume

$$f(n) + a_1 f(n-1) + a_1 f(n-2) = e(t).$$

a_1 and a_2 are real coefficients, and $e(t)$ is called the *noise term*.

(b) Use this to show how you can compute $f(n+h)$ for arbitrary n.

(c) Consider the data set $\{f(1), f(2), \ldots, f(10)\} =$

$$\{1, 1, 0.3, 0.23, 0.083, 0.0543, 0.02203, 0.013063, 0.0057123, 0.00318383\}.$$

You can assume this uses the same format as 25(a) and that $e(t) = 0$. Find the values for a_1 and a_2, and predict $f(11)$.

1.4 Families of Functions

In Example 1.2.3 of Section 1.2 we discussed the function f defined by

$$f(x) = \mu x(1-x), \text{ for } 0 \leq x \leq 1$$

where μ was a fixed positive constant. We found that f has one or two fixed points, depending on whether $0 < \mu \leq 1$ or $\mu > 1$. Thus the number of fixed points depends on the value of μ. In order to emphasize the fact that the function depends on μ, we will henceforth designate f by Q_μ, so that

$$Q_\mu(x) = \mu x(1-x), \text{ for } 0 \leq x \leq 1$$

The family $\{Q_\mu\}$ for $\mu > 0$ is the **quadratic family**, or **logistic family**, so named because each of the functions in the family is a quadratic function. A collection of functions such as $\{Q_\mu\}$ is called a **parametrized family of**

functions (or a **one-parameter family**), and μ is the parameter for the family. Other parametrized families that we will encounter are

$$g_\mu(x) = x^2 + \mu, \text{ for all } x$$

$$T_\mu(x) = \begin{cases} 2\mu x & \text{for } 0 \le x \le 1/2 \\ \\ 2\mu(1-x) & \text{for } 1/2 < x \le 1 \end{cases} \qquad \text{where } 0 < \mu \le 1$$

$$E_\mu(x) = \mu e^x, \text{ for all } x$$

$$S_\mu(x) = \mu \sin x, \text{ for } 0 \le x \le \pi$$

Notice that μ is constant and x is the variable for each function in the parametrized families listed above. For the family $\{T_\mu\}$, μ is restricted to the interval $(0, 1]$ in order that the range of the functions T_μ will be contained in the domain $[0, 1]$.

There are names for the families listed above. The family $\{T_\mu\}$ is the **tent family** because the functions T_μ are from the same mold as the tent function T. Also $\{E_\mu\}$ is the **exponential family**, and $\{S_\mu\}$ is the **sine family**. We give the family $\{g_\mu\}$ no special name; later we will show that the family $\{g_\mu\}$ is a close relative of the quadratic family $\{Q_\mu\}$. When we wish to refer to a general parametrized family rather than a specific one, we will denote it by $\{f_\mu\}$.

The way in which the orbits in a parametrized family change as the parameter varies is called the **dynamics** of the family. In the present section we will study the dynamics of $\{g_\mu\}$ and $\{T_\mu\}$; we will devote the entire Section 1.5 to the dynamics of $\{Q_\mu\}$.

1.4.1   The Family $\{g_\mu\}$

The family $\{g_\mu\}$ consists of the functions defined by

$$g_\mu(x) = x^2 + \mu, \text{ for all } x$$

which are among the simplest nonlinear differentiable functions. The dynamics of the family $\{g_\mu\}$ vary according to the value of μ. If $\mu \ge 0$, we can describe in detail the orbit $\{g_\mu^{[n]}(x)\}_{n=0}^\infty$ for any real number x, whereas if $\mu < 0$, then the orbit can be very complicated. As a result, in this section we will limit the discussion to the members of $\{g_\mu\}$ for which $\mu \ge 0$.

Among nonnegative parameters for $\{g_\mu\}$, two values are especially noteworthy: $\mu = 0$ and $\mu = 1/4$. For $\mu = 0$, we have g_0, which is the simplest function in the family and is defined by $g_0(x) = x^2$ (Figure 1.15(a)). A moment's reflection reveals that the fixed points of g_0 are 0 and 1. Next, we notice that $|g_0(x)| = x^2 < |x|$ if $|x| < 1$, and $|g_0(x)| = x^2 > |x|$ if $|x| > 1$. It follows that 0 is an attracting fixed point whose basin of attraction is $(-1, 1)$, and that 1 is a repelling fixed point. Since all iterates of x approach 0 if $|x| < 1$ and are unbounded if $|x| > 1$, there can be no periodic points besides 0 and 1.

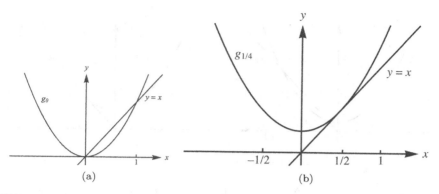

FIGURE 1.15
Graphs of the family $\{g_\mu\}$

Next, we turn to $g_{1/4}$ (Figure 1.15(b)). Notice that

$$g_{1/4}(x) - x = x^2 + \frac{1}{4} - x = \left(x - \frac{1}{2}\right)^2 \begin{cases} = 0 & \text{if } x = 1/2 \\ > 0 & \text{if } x \neq 1/2 \end{cases}$$

Therefore $g_{1/4}$ has one and only one fixed point: $1/2$. Moreover, the graph of $g_{1/4}$ lies above the line $y = x$ except at $x = 1/2$ and is tangent at the point $(1/2, 1/2)$. Using graphical analysis (or the Mean Value Theorem), one can show that the basin of attraction of the fixed point $1/2$ is $[-1/2, 1/2]$, and that $1/2$ repels points to the right, so $1/2$ is a neutral fixed point.

Because the graph of $\{g_\mu\}$ shifts upward as μ increases, and because the graph of $g_{1/4}$ is tangent to the graph of $y = x$, a glance at Figure 1.15(b) suggests that the graph of $\{g_\mu\}$ intersects the graph of $y = x$ if $0 < \mu \leq 1/4$ and does not intersect it if $\mu > 1/4$. Therefore we will divide the analysis of $\{g_\mu\}$ for the remaining positive values of μ into two groups: $0 < \mu < 1/4$ and $\mu > 1/4$.

Case 1. $0 < \mu < 1/4$

The number x is a fixed point of g_μ if and only if $x = g_\mu(x) = x^2 + \mu$, which is equivalent to $x^2 - x + \mu = 0$. Solving for x, we obtain

$$x = \frac{1}{2} - \frac{1}{2}\sqrt{1 - 4\mu} \quad \text{or} \quad \frac{1}{2} + \frac{1}{2}\sqrt{1 - 4\mu}$$

as the fixed points of g_μ. Next, we will determine which (if any) of these points is attracting. Since $g_\mu'(x) = 2x$, it follows that

$$\left| g_\mu'\left(\frac{1}{2} \pm \frac{1}{2}\sqrt{1 - 4\mu}\right) \right| = \left| 1 \pm \sqrt{1 - 4\mu} \right|$$

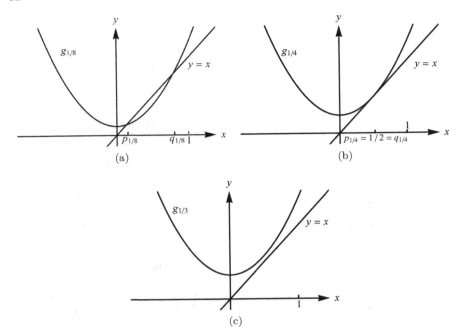

FIGURE 1.16
Comparisons of g_μ, p_μ, and q_μ

If we let

$$p_\mu = \frac{1}{2} - \frac{1}{2}\sqrt{1 - 4\mu} \quad \text{and} \quad q_\mu = \frac{1}{2} + \frac{1}{2}\sqrt{1 - 4\mu}$$

then $|g'_\mu(p_\mu)| = |1 - \sqrt{1 - 4\mu}| < 1$ for $0 < \mu < 1/4$. By Theorem 1.7, p_μ is an attracting fixed point. By contrast, $|g'_\mu(p_\mu)||1 + \sqrt{1 - 4\mu}| > 1$ for $0 < \mu < 1/4$, so that again by Theorem 1.7, q_μ is a repelling fixed point. It turns out that if $0 < \mu < 1/4$, then the basin of attraction of p_μ is the open interval $(-q_\mu, q_\mu)$, and iterates of each number x such that $|x| > q_\mu$ march off toward ∞. Thus every number other than $\pm q_\mu$ has the property that its iterates approach p_μ or are unbounded. We mention that as the parameter μ approaches $1/4$, the fixed points p_μ and q_μ are drawn toward each other, and actually coalesce when $\mu = 1/4$ (compare Figure 1.16(a) with 1.16(b)).

Case 2. $\mu > 1/4$

As μ increases beyond $1/4$, the dynamics of g_μ change dramatically because the entire graph of g_μ lies above the line $y = x$ (Figure 1.16(c)). Thus there is no fixed point. We can prove this formally by noticing that if $\mu > 1/4$, then

$$g_\mu(x) - x = x^2 - x + \mu > x^2 - x + \frac{1}{4} = \left(x - \frac{1}{2}\right)^2 \geq 0$$

so that $g_\mu(x) > x$ for all x. Moreover, since the iterates of each number x form an increasing sequence that diverges to ∞, it follows that g_μ has no periodic points.

Something very special has happened for $\mu = 1/4$, because as μ increases and passes through $1/4$, g_μ first has two fixed points and then none. We say that the family has a bifurcation at $1/4$. More generally we have the following definition.

Definition 1.14. A parametrized family $\{f_\mu\}$ has a **bifurcation** at μ_0, or **bifurcates** at μ_0, if the number or nature (attracting vs. repelling) of periodic points of f_μ changes as μ passes through μ_0. In this case μ_0 is said to be a **bifurcation point** for the family.

Note: The term "bifurcate" comes from the Latin words meaning "two branches." From Definition 1.14 we infer that our discussion in the last section shows how $\{g_\mu\}$ bifurcates at the number $1/4$. Bifurcation points signal changes in dynamics of a parametrized family. We will discuss bifurcation points for each of the parametrized families that we encounter, and will devote Section 1.6 to bifurcations.

1.4.2 The Tent Family $\{T_\mu\}$

Recall that the tent family consists of the functions T_μ defined by

$$T_\mu(x) = \begin{cases} 2\mu x & \text{for } 0 \leq x \leq 1/2 \\ 2\mu(1-x) & \text{for } 1/2 < x \leq 1 \end{cases} \qquad \text{where } 0 < \mu \leq 1$$

Figures 1.17(a)–(c) display T_μ for $\mu = 2/7, 1/2$, and $5/6$. As μ increases, the maximum height of the graph of T_μ rises, because of the factor μ in the formula for T_μ. From this observation and the three graphs in Figure 1.17 we deduce that if $0 < \mu < 1/2$, then T_μ intersects the line $y = x$ once (at 0), whereas if $1/2 < \mu < 1$, then there are two points of intersection. We are led to analyze separately the members of $\{T_\mu\}$ for which $0 < \mu < 1/2$, $\mu = 1/2$, and $1/2 < \mu < 1$. Finally, we will study T_1, which is the original tent function T and which has some very interesting features.

Case 1. $0 < \mu < 1/2$

The graph in Figure 1.17(a) shows that 0 is the only fixed point of T_μ. Since $0 < \mu < 1/2$, it follows from the definition of T_μ that if $0 \leq x < 1/2$, then

$$0 \leq T_\mu(x) = 2\mu x < x$$

and if $1/2 < x < 1$, then

$$0 \leq T_\mu(x) = 2\mu(1-x) < 1 - x < \frac{1}{2} < x$$

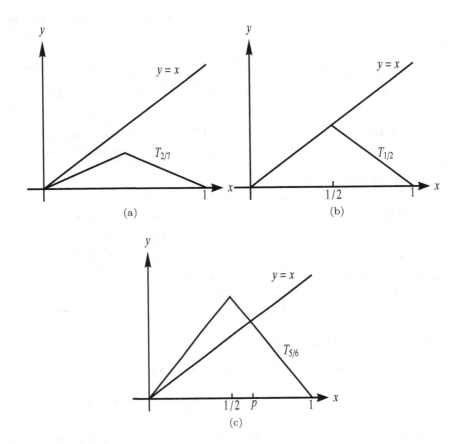

FIGURE 1.17
Exploring a bifurcation of the tent map

Consequently for any x in $[0, 1]$, the sequence $\{T_\mu^{[n]}(x)\}_{n=0}^\infty$ is bounded and decreasing. By Corollary 1.4, the sequence converges to the fixed point 0. Therefore 0 is an attracting fixed point whose basin of attraction is $[0, 1]$.

Case 2. $\mu = 1/2$

First, we notice that if $0 \le x \le 1/2$, then $T_{1/2}(x) = 2(1/2)x = x$, so that x is a fixed point of $T_{1/2}$ (Figure 1.17(b)). Next, we calculate that if $1/2 < x \le 1$, then

$$0 \le T_{1/2}(x) = 2(1/2)(1 - x) = 1 - x < 1/2$$

so that $T_{1/2}(x)$ is a fixed point of $T_{1/2}$. Consequently every point in $[0, 1]$ either is a fixed point of $T_{1/2}$ or has a fixed point for its first iterate.

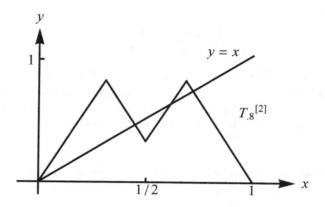

FIGURE 1.18
Squaring the tent map

Case 3. $1/2 < \mu < 1$

In addition to the fixed point 0, there is a second fixed point $p \neq 0$ that lies in $(1/2, 1]$, as you can see in Figure 1.17(c). To find p we solve the equation

$$p = T_\mu(p) = 2\mu(1 - p)$$

which yields

$$p = \frac{2\mu}{1 + 2\mu}$$

As μ increases from $1/2$ toward 1, p increases from $1/2$ toward $2/3$. Because $|T'_\mu(x)| = 2\mu > 1$ on $[0, 1]$ except at $1/2$, both 0 and p are repelling fixed points.

The period-2 points of T_μ are the fixed points of $T_\mu^{[2]}$, which is given by

$$T_\mu^{[2]}(x) = \begin{cases} 4\mu^2 x & \text{for } 0 \leq x \leq 1/4\mu \\ 2\mu(1 - 2\mu x) & \text{for } 1/4\mu < x \leq 1/2 \\ 2\mu(1 - 2\mu + 2\mu x) & \text{for } 1/2 < x \leq 1 - 1/4\mu \\ 4\mu^2(1 - x) & \text{for } 1 - 1/4\mu < x \leq 1 \end{cases}$$

The graph of $T_\mu^{[2]}$ appears in Figure 1.18, and suggests that for $1/2 < \mu < 1$, the function $T_\mu^{[2]}$ has four fixed points that can be found by solving the four equations $x = T_\mu^{[2]}(x)$ arising from the definition of $T_\mu^{[2]}$. We find that the fixed points are

$$0, \quad \frac{2\mu}{1 + 4\mu^2}, \quad \frac{2\mu}{1 + 2\mu}, \quad \text{and} \quad \frac{4\mu^2}{1 + 4\mu^2}$$

We observe that these fixed points are not equally spaced on the interval $[0, 1]$. Moreover, the first and third are the two fixed points of T_μ, so it follows that

$$\left\{ \frac{2\mu}{1 + 4\mu^2}, \ \frac{4\mu^2}{1 + 4\mu^2} \right\}$$

is a 2-cycle for T_μ. This 2-cycle is repelling, because $|(T_\mu^{[2]})'(x)| = 4\mu^2 > 1$ wherever the derivative is defined. Because the graph of $T_\mu^{[n]}$ is linear on the 2^n subintervals $[0, 1/2^n], ..., [1 - 1/2^n, 1]$, it is possible (though tedious) to describe the various n-cycles of T_μ— all of which are repelling!

Case 4. $\mu = 1$

If $\mu = 1$ then $T_\mu = T$, which is the tent function and is given by

$$T(x) = \left\{ \begin{array}{ll} 2x & \text{for } 0 \leq x \leq 1/2 \\ 2 - 2x & \text{for } 1/2 < x \leq 1 \end{array} \right.$$

The graph of T appears in Figure 1.19(a). The major difference between the graph of T and the graph of T_μ when $\mu < 1$ is the fact that the range of T

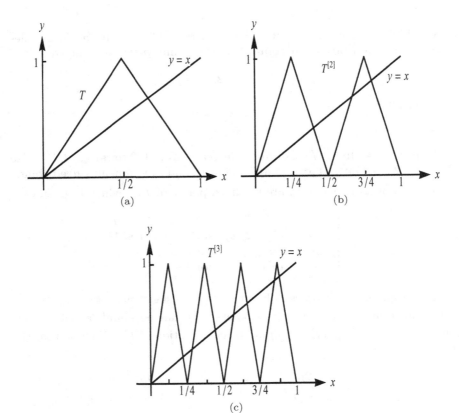

FIGURE 1.19
Multiple iterations of the tent map

fills out the whole interval $[0, 1]$. Indeed, the function T stretches the interval $[0, 1/2]$ over the entire interval $[0, 1]$, and folds the interval $[1/2, 1]$ back over the interval $[0, 1]$. It is this stretching and folding that is characteristic of many functions we will examine in this book, and which leads to the notion of chaos that we will discuss later.

As with all members of $\{T_\mu\}$, 0 is a fixed point of T. Since $x = T(x) = 2(1 - x)$ if $x = 2/3$, we know that $2/3$ is the second fixed point of T. Figures 1.19(b) and (c) indicate that $T^{[2]}$ and $T^{[3]}$ have, respectively, four and eight fixed points. Thus T has two period-2 points and six period-3 points, which we could evaluate by solving the equations $x = T^{[2]}(x)$ and $x = T^{[3]}(x)$ for x.

Rather than determine the values of the period-n points for T, we turn to the number of period-n points of T, for $n \geq 1$. The graph of $T^{[n]}$ has 2^{n-1} congruent spikes. Consequently there are 2^n fixed points for $T^{[n]}$, two in each of the subintervals $[0, 1/2^{n-1}], \ldots, [(2^{n-1} - 1)/2^{n-1}, 1]$. Some of these fixed points of $T^{[n]}$ are fixed points for $T^{[k]}$ with $k < n$; the remaining fixed points of $T^{[n]}$ join together to form n-cycles for T. The breakdown is given in the following table:

n	1	2	3	4	5
No. of fixed points for $T^{[n]}$	2	4	8	16	32
No. of period-n points for T	2	2	6	12	30
No. of n-cycles for T		2	1	2	3	6

To obtain the number of period-n points of T, we take the number, 2^n, of fixed points of $T^{[n]}$, and then subtract the total number of period-k points for all values of k for which $k < n$ and k divides n. For example, if $n = 4$, then there are $2^4 = 16$ fixed points for $T^{[4]}$. Two of these fixed points are fixed points of T, and 2 others are fixed points of $T^{[2]}$. Therefore the remaining twelve fixed points of $T^{[4]}$ are necessarily period-4 points, and form three 4-cycles for T.

1.4.3 Eventually Periodic and Periodic Points of T

In this subsection we will determine the eventually periodic points and periodic points for T. This analysis is more technical than what has preceded it, and the results are independent from what follows. In order to facilitate notation in the discussion that follows, we will allow the expression "eventually periodic point" to include periodic, eventually fixed, and fixed points.

Theorem 1.15. Let x be in the interval $(0, 1)$. Then x is eventually periodic for T if and only if x is rational.

Proof. Assume that x is eventually periodic. Either $T(x) = 2x$ or $T(x) = 2 - 2x$, so that $T(x) = $ integer $\pm 2x$. Similarly, $T^{[2]}(x) = $ integer $\pm 2^2 x$, and in general,

$$T^{[n]}(x) = \text{integer} \pm 2^n x$$

For each n, let a_n and i_n be integers such that $a_n = 2^n$ or $a_n = -2^n$, and such that $T^{[n]}(x) = i_n + a_n x$. Since x is eventually periodic, there are positive integers k and m with $k \neq m$ and such that $T^{[k]}(x) = T^{[m]}(x)$. Thus $i_k + a_k x = i_m + a_m x$. Since $k \neq m$, it follows that $a_k \neq a_m$, so that

$$x = \frac{i_m - i_k}{a_k - a_m}$$

which means that x is a rational number.

To prove the converse, we will assume for the moment that $x = k/p$ is in reduced form (so that k and p share no nontrivial factors). Assume first that p is an odd integer. Then $T(k/p) = 2k/p$ or $T(x) = 2(p - k)/p$, which means that $T(x) = $ (even integer)$/p$. There are only finitely many distinct numbers in $(0, 1)$ of the form (even integer)$/p$, so x is eventually periodic. Next, assume that $x = k/p$ with p even. In this case,

$$T(x) = \frac{2k}{p} = \frac{k}{p/2} \qquad \text{or} \qquad T(x) = 2\left(1 - \frac{k}{p}\right) = \frac{p - k}{p/2}$$

so that the denominator has become $p/2$. Continuing the process with $T(x)$ substituted for x, we find that for some positive integer i (depending on x), either $T^{[i]}(x) = $ (integer)/(odd integer), or $T^{[i]}(x) = 1$. The first possibility means that x is eventually periodic by the preceding argument, and the second possibility means that $T^{[i+1]} = 0$, so that x is eventually fixed (and hence eventually periodic). We conclude that if x is rational, then x is eventually periodic. $\qquad \square$

Theorem 1.15 implies that all rational numbers in $(0, 1)$ are eventually periodic for T. To determine which rational numbers are actually periodic, we first have two lemmas.

Lemma 1.1. Suppose that p is odd, and let $x = k/p$ be in $(0, 1)$. Then x is periodic for T if and only if k is even.

Proof. Suppose that x is periodic with period n. Since p is odd by hypothesis, and since $T(x) = 2k/p$ or $T(x) = 2(p - k)/p$, it follows that $T(x) = $(even integer)$/p$. The same is true for all iterates of x, so in particular, $x = T^{[n]}(x) = $(even integer)$/p$. Thus if x is periodic, then k must be even.

To prove the converse, assume that k is even. We will show that x is periodic. For any positive integer i, $T^{[i-1]}(x) = m/p$ for some even integer m. Thus

$$T^{[i]}(x) = \frac{4m}{p} \text{ if } T^{[i-1]}(x) \leq 1/2$$

or

$$T^{[i]}(x) = 2\left(1 - \frac{m}{p}\right) = \frac{4m+2}{p} \text{ if } T^{[i-1]}(x) > 1/2$$

Therefore as soon as we see the form of $T^{[i]}(x)$, we know whether $T^{[i-1]}(x)$ is in $[0, 1/2]$ or it is in $(1/2, 1]$.

Now recall from Theorem 1.15 that x is eventually periodic (because it is rational), so there are a least nonnegative integer i and a least positive integer n such that $n > i$ and $T^{[i]}(x) = T^{[n]}(x)$. If $i = 0$, then x is periodic with period n. Next, we will show that i cannot be positive (so that i must be 0). To obtain a contradiction, assume that $i > 0$. Then by the discussion in the preceding paragraph, both $T^{[i-1]}(x)$ and $T^{[n-1]}(x)$ lie in $[0, 1/2]$ or both lie in $(1/2, 1]$. Since T is strictly increasing on $[0, 1/2]$ and strictly decreasing on $(1/2, 1]$, we conclude that $T^{[i]}(x) = T^{[n]}(x)$ only if $T^{[i-1]}(x) = T^{[n-1]}(x)$. But that contradicts the minimality of i and n. Therefore $i = 0$, so that x is periodic. This completes the proof. $\square$

Lemma 1.2. Suppose that p is even, and let $x = k/p$ be in $(0, 1)$. Then x is not periodic for T.

Proof. Since x is assumed to be in reduced form, with p even, it follows that k must be odd, so that $x =$(odd integer)$/p$. But then $T(x) = 2k/p$ or $T(x) = 2(p-k)/p$, so that in any case, $T(x) =$(integer)$/(p/2)$. Thus the reduced form for $T(x)$, like $T^{[n]}(x)$ for any $n > 1$, cannot be (odd integer)$/p$. Thus x is not periodic. $\square$

Theorem 1.16. The rational number x in $(0, 1)$ is periodic for T if and only if x has the form (even integer)/(odd integer).

Proof. Because we assume that x is in reduced form, a moment's reflection tells us that Theorem 1.15, Lemma 1.1, and Lemma 1.2 together imply the result. $\square$

An analogous result for eventually fixed points is the following: a number x in $[0, 1]$ is eventually fixed if and only if x has the form $k/2^m$ or $k/(3 \cdot 2^m)$ for appropriate nonnegative integers k and m (see Exercise 4(a)).

Now we list a few eventually fixed, periodic, and eventually periodic points for T.

ITERATES

$$x = \frac{3}{16}: \quad \frac{3}{16} \quad \frac{3}{8} \quad \frac{3}{4} \quad \frac{1}{2} \quad 1 \quad 0 \quad 0 \quad \text{eventually fixed}$$

$$x = \frac{6}{13}: \quad \frac{6}{13} \quad \frac{12}{13} \quad \frac{2}{13} \quad \frac{4}{13} \quad \frac{8}{13} \quad \frac{10}{13} \quad \frac{6}{13} \quad \text{periodic}$$

$$x = \frac{7}{10}: \quad \frac{7}{10} \quad \frac{3}{5} \quad \frac{4}{5} \quad \frac{2}{5} \quad \frac{4}{5} \quad \frac{2}{5} \quad \frac{4}{5} \quad \text{eventually periodic}$$

There is much more to say about the tent function T. It will reappear in Section 2.2 when we discuss the concept of chaos.

1.4.4   Section 1.4 Exercises

Exercise 1. Determine which of the following are eventually fixed, which are eventually periodic, and which are periodic points for T.

(a) 3/11

(b) 10/33

(c) 5/18

(d) 6/23

(e) 3/16

Exercise 2. Find as many 5-cycles of T as you can.

Exercise 3. For T find the total number of

(a) 8-cycles

(b) 15-cycles

Exercise 4.

(a) Let x be in $[0, 1]$. Show that x is eventually fixed for T if and only if x has the form $k/2^m$ or $k/(3 \cdot 2^m)$ for nonnegative integers k and m.

(b) Show that 0.3 is eventually periodic but not periodic for T.

(c) Use the computer program ITERATE to compute the first 100 iterates of 0.3 for T. Do you notice anything strange in the behavior of the iterates the computer provides? If so, give an explanation.

Exercise 5. Show that for each x in $(0, 1)$ that is not an eventually fixed point, and for each positive integer N, there is an $n > N$ such that $T^{[n]}(x) < 1/2$.

Exercise 6. Show that $1/2$ is an eventually fixed point of $T_{\sqrt{2}/2}$.

Exercise 7. Let n be an arbitrary positive integer, and let μ be fixed. Find the maximum value of $T_\mu^{[n]}$, under the condition that μ is in

(a) $(0, 1/2)$

(b) $(1/2, 1)$

Exercise 8. Assume that $1/2 < \mu \le 1$, and let $h(\mu) = $ relative minimum value of $T_\mu^{[2]}$ on the interval $(0, 1)$.

(a) Find a formula for $h(\mu)$.

(b) Find the maximum value of $h(\mu)$.

Exercise 9. Let $1/2 < \mu < 1$. Prove that the iterates under T_μ of each x in $(0, 1)$ are eventually "trapped" in the interval $[2\mu(1 - \mu), \mu]$, in the sense that there is an n (depending on x) such that if $k \geq n$, then $2\mu(1-\mu) \leq T_\mu^{[k]}(x) \leq \mu$.

Exercise 10. Consider the function g_μ, where $0 < \mu < 1/4$. Show that the basin of attraction of the fixed point p_μ is the open interval $(-q_\mu, q_\mu)$.

Exercise 11. Find the fixed points for each of these families of functions in terms of μ.

- $Q_\mu(x) = \mu x(1 - x)$, for $0 \leq x \leq 1$

- $g_\mu(x) = x^2 + \mu$, for all x

- $T_\mu(x) = \begin{cases} 2\mu x & \text{for } 0 \leq x \leq 1/2 \\ \\ 2\mu(1 - x) & \text{for } 1/2 < x \leq 1 \end{cases}$ where $0 < \mu \leq 1$

- $E_\mu(x) = \mu e^x$, for all x

- $S_\mu(x) = \mu \sin x$, for $0 \leq x \leq \pi$

Exercise 12. For each of the families of functions above, is there a value for μ where the number of fixed points changes?

Exercises 11–15 concern the family $\{E_\mu\}$, where $E_\mu(x) = \mu e^x$ for all x and $\mu > 0$.

Exercise 13.

(a) Show that $E_{1/e}(x) \geq x$ for all x, and that $E_{1/e}$ has a single fixed point. Find the fixed point.

(b) Find $\lim_{n \to \infty} E_{1/e}^{[n]}(x)$ for all x.

Exercise 14. Let $\mu > 1/e$. Find $\lim_{n \to \infty} E_\mu^{[n]}(x)$ for all x, and thereby show that E_μ has no periodic points.

Exercise 15. Let $0 < \mu < 1/e$. Show that E_μ has two fixed points. Denote them by p_μ and q_μ, with $q_\mu < p_\mu$. Determine which fixed point is attracting and which is repelling.

Exercise 16. Let $0 < \mu < 1/e$. Show that E_μ has no periodic points that are not fixed points.

Exercise 17. Show that $\lim_{\mu \to 0^+} q_\mu = 0$ and $\lim_{\mu \to 0^+} p_\mu = \infty$.

Exercises 16–21 concern $\{E_\mu\}$, where $\mu < 0$.

Exercise 18.

(a) Show that E_μ has a unique fixed point p_μ for each $\mu < 0$.

(b) Find the fixed point of E_{-e}.

Exercise 19. Show that p_μ is repelling if $\mu < -e$ and is attracting if $\mu > -e$.

Exercise 20. Show that the maximum value of $(E_\mu^{[2]})'$ occurs for $x = -\ln(-\mu)$.

Exercise 21. Let $-e < \mu < 0$. Show that the single fixed point of E_μ has basin of attraction $(-\infty, \infty)$, and hence that there are no period-n points for $n > 1$.

Exercise 22.

(a) Show that $E_{-e}^{[2]}$ has the unique fixed point p_μ found in Exercise 18(a).

(b) Show that the graph of $E_{-e}^{[2]}$ is tangent to the line $y = x$ at (p_μ, p_μ).

Exercise 23. Let $\mu < -e$. Show that E_μ has one fixed point and one 2-cycle, and no other cycles.

1.5   The Quadratic Family

Consider a population of organisms for which there is a constant supply of food and limited space, and no predators. Many insect populations in the temperate zones fit this description at certain times in their history. In order to model the populations in successive generations, let N_n denote the population of the nth generation, and adjust the numbers so that the capacity of the environment equals 1, which means that $0 \le N_n \le 1$. One formula that has gained widespread fame is

$$N_{n+1} = \mu N_n (1 - N_n), \text{ for } 0 \le N_n \le 1$$

Sometimes the equation in (1) is called a "logistic equation," after a differential equation studied by the Belgian mathematician P. F. Verhulst 150 years ago. The parameter μ indicates the rate at which the population grows when it is very small.

> **Note:** It is worth digressing here to investigate where this came from. Earlier population models arose from differential equations. For a population N at a continuous time t, a constant rate of growth would be modeled by $\frac{dN}{dt} = rN$, for some constant r. Verhulst's model assumed that the population grown was proportional to the current population, and the difference between a population and a known carrying capacity of the environment. The original model was $\frac{dN}{dt} = rN(K - N)$, or equivalently, $\frac{dN}{dt} = rNK(1 - N/K)$. This differential equation had solutions which could yield a continuous dynamical system. Equation 1 comes from the discretization of this model, taking into account discrete rather than continuous time intervals.

Two properties of the equation in (1) are relevant to the study of population dynamics:

i. If the population is 0 at generation n, then the population remains 0.

ii. The population grows when N_n is small, and declines when N_n is large.

Property (ii) is reasonable, because when the population is small there is ample food and space, so the population can grow without hindrance. However, when the population is sufficiently large (that is, close to 1), the new generations are smaller because of food shortage and overcrowding.

The continuous version of (1) is the quadratic function given by

$$Q_\mu(x) = \mu x(1 - x) = \mu x - \mu x^2, \text{ for } 0 \le x \le 1$$

which appeared in Section 1.2. The present section is devoted to a detailed analysis of the quadratic family $\{Q_\mu\}$, not only because of its importance to the study of population dynamics, but also because its members, which are very simple polynomials, can have very complicated dynamics and can exhibit many of the characteristics that are associated with the study of chaotic dynamics.

In our study of the family $\{Q_\mu\}$, we will restrict the values of the parameter μ to be in $(0, 4]$. To see why we make this restriction, note that $Q'_\mu(x) = \mu - 2\mu x$, so that $Q'_\mu(x) = 0$ if $x = 1/2$. Since $Q''_\mu(x) = -2\mu$, we know that $Q_\mu(1/2) = \mu/4$ is an extreme value of Q_μ if $\mu \ne 0$. In order that $Q_\mu(1/2)$ lie in the domain $[0, 1]$ when $\mu \ne 0$, we must have $0 < \mu/4 \le 1$, that is, $0 < \mu \le 4$. It follows that if $0 < \mu \le 4$, then the range of Q_μ is contained in the domain of Q_μ. Figure 1.21 illustrates this function.

Although we know from Section 1.2 that Q_μ has one fixed point when $0 < \mu \le 1$ and two fixed points when $1 < \mu \le 4$, we have not discussed possible periodic points for Q_μ that are not fixed points. We will address this issue carefully below. Since it turns out that the dynamics of Q_μ change noticeably as μ passes through each of the integers 1, 2, and 3, we will split our discussion into four cases: $0 < \mu \le 1$, $1 < \mu \le 2$, $2 < \mu \le 3$, and $3 < \mu \le 4$.

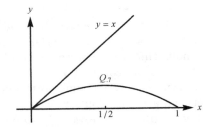

FIGURE 1.20

Case 1. $0 < \mu \leq 1$

This first case is the easiest. Since

$$0 < Q_\mu(x) = \mu x(1 - x) < \mu x \leq x \text{ for } 0 < x < 1$$

it follows that $\{Q_\mu^{[n]}(x)\}_{n=0}^\infty$ is a positive, decreasing sequence, which by Corollary 1.4 converges to the fixed point 0. We conclude that the basin of attraction of 0 is the interval $[0, 1]$, so that there are no periodic points other than the lone fixed point 0.

The same result could be achieved by noticing that the graph of Q_μ lies below the line $y = x$ because $Q_\mu(0) = 0$ and $|Q_\mu'(x)| = |\mu - 2\mu x| \leq 1$ for all x in $[0, 1]$. Figure 1.20 supports these conclusions.

Case 2 $1 < \mu \leq 2$

By Example 1.2.3 of Section 1.2, we know that Q_μ has the two fixed points, namely 0 (which is repelling) and $1 - 1/\mu$ (which is attracting). If you apply graphical analysis to the graphs in Figure 1.23, you could convince yourself that the basin of attraction of $1 - 1/\mu$ is the whole interval $(0, 1)$. To prove it rigorously, we will denote $1 - 1/\mu$ by p_μ, and let $0 < x < p_\mu = 1 - 1/\mu$. Then

$$\frac{1}{\mu} < 1 - x, \quad \text{so that} \quad 1 < \mu(1 - x), \quad \text{and thus} \quad x < \mu x(1 - x)$$
$$= Q_\mu(x)$$

Since Q_μ is increasing on $[0, p_\mu]$, this means that if $0 < x < p_\mu$, then

$$x < Q_\mu(x) < Q_\mu(p_\mu) = p_\mu$$

Consequently $\{Q_\mu^{[n]}(x)\}_{n=0}^\infty$ is a bounded increasing sequence when $0 < x < p_\mu$, so by Corollary 1.4 converges to a fixed point, which must be p_μ. Analogously one can show that if $p_\mu < x < 1/2$, then $\{Q_\mu^{[n]}(x)\}_{n=0}^\infty$ is a bounded decreasing sequence, which also converges

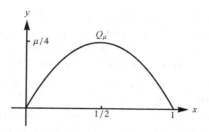

FIGURE 1.21

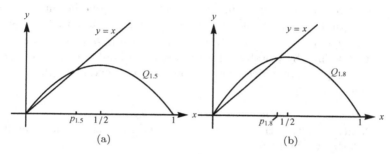

FIGURE 1.22
Changing the fixed point on Q_μ

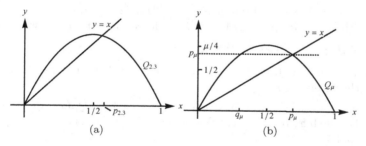

FIGURE 1.23
Describing the fixed points for Q_μ

to p_μ (see Exercise 1). Finally, if $1/2 \leq x < 1$, then $0 < Q_\mu(x) \leq 1/2$ (with $Q_\mu(x) = 1/2$ only if $\mu = 2$ and $x = 1/2$). By the above analysis, $\{Q_\mu^{[n]}(x)\}_{n=0}^\infty$ converges to p_μ. Therefore when $1 < \mu \leq 2$, the basin of attraction of p_μ is the open interval $(0, 1)$, and there are no periodic points other than the two fixed points. Two potential values for μ are illustrated in Figure 1.22.

Case 3. $2 < \mu \leq 3$

As μ increases from 1 toward 2, $p_\mu = 1 - 1/\mu$ increases from 0 toward $1/2$. Similarly, as μ increases from 2 toward 3, p_μ increases from $1/2$ to $2/3$. (Figure 1.23(a)). We will show that if $2 < \mu < 3$, then the basin of attraction of p_μ is once again the interval $(0, 1)$.

Let q_μ denote the unique number in $(0, 1/2)$ such that $Q_\mu(q_\mu) = Q_\mu(p_\mu)$ (Figure 1.23(b)). Notice that p_μ and q_μ are symmetric with respect to the line $x = 1/2$, so that $q_\mu = 1/2 - (1/2 - 1/\mu) = 1/\mu$. Our first goal is to show that any x in the interval $(0, 1)$ has an iterate in the interval (q_μ, p_μ).

Toward that end, fix x in $(0, q_\mu)$. Notice that on the interval $(0, q_\mu)$ the graph of Q_μ lies above the line $y = x$ (Figure 1.23(b)), so that $x < Q_\mu(x)$. If $\{Q_\mu^{[n]}(x)\}_{n=0}^{\infty}$ were an increasing sequence contained in $(0, q_\mu)$, then by Corollary 1.4 the sequence would need to converge to a fixed point, of which there is none in the interval $(0, q_\mu]$. We deduce that

(a) if $0 < x < q_\mu$, then x has an iterate $> q_\mu$

Next, by noting that $Q_\mu(q_\mu) = p_\mu = Q_\mu(p_\mu)$ and glancing at Figure 1.23(b), we find that

(b) if $q_\mu < x \leq p_\mu$, then $p_\mu \leq Q_\mu(x) \leq \mu/4$

In addition, by using the fact that $q_\mu < Q_\mu(\mu/4)$ (see Exercise 3), and the fact that Q_μ is decreasing on $[1/2, 1]$, we infer that

(c) if $p_\mu < x \leq \mu/4$, then $q_\mu < Q_\mu(\mu/4) \leq Q_\mu(x) < p_\mu$

Finally, since $p_\mu < \mu/4$ and Q_μ is decreasing on $[p_\mu, 1]$, we know that

(d) if $\mu/4 < x < 1$, then $0 < Q_\mu(x) < p_\mu$

We conclude from (a)–(d) that if $0 < x < 1$, then x has an iterate in the interval (q_μ, p_μ).

Next, we will show that if x is in (q_μ, p_μ), then $|Q_\mu(x) - p_\mu| < |x - p_\mu|$. For such x, by the Mean Value Theorem there is a z in (q_μ, p_μ) such that

$$|Q_\mu(x) - p_\mu| = |Q_\mu(x) - Q_\mu(p_\mu)| = |Q_\mu'(z)||x - p_\mu|$$

If we can show that $|Q_\mu'(q_\mu)| < 1$, then it is easy to see once again from Figure 1.23(b) that $|Q_\mu'(z)| < 1$, and hence that $|Q_\mu(x) - p_\mu| < |x - p_\mu|$ for all x in (q_μ, p_μ).

However, $|Q_\mu'(q_\mu)| = |\mu - 2\mu(1/\mu)| = |\mu - 2| < 1$ since $2 < \mu < 3$ by assumption. We conclude from the entire discussion above that if x is *any* number in $(0, 1)$, then x has some iterate y in (q_μ, p_μ), and for any such iterate y, we have $|Q_\mu(y) - p_\mu| < |y - p_\mu|$. Thus not only is p_μ an attracting fixed point for Q_μ if $2 < \mu < 3$, but the basin of attraction is $(0, 1)$. It is also possible to show that if $\mu = 3$, then p_μ is attracting with the same basin of attraction: $(0, 1)$.

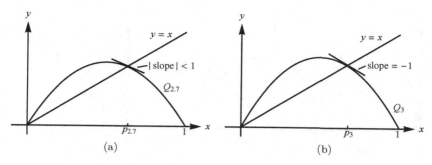

FIGURE 1.24
Slopes of the tangent line to Q_μ at a fixed point

It is noteworthy that if $2 < \mu < 3$, then as iterates of a given x in $(0, 1)$ approach p_μ, they oscillate to the left and to the right of p_μ. To see this, let q_μ be the unique number in $(0, 1/2)$ such that $Q_\mu(q_\mu) = Q_\mu(p_\mu)$ (Figure 1.23(b)). Noting that $Q_\mu(q_\mu) = p_\mu = Q_\mu(p_\mu)$ and glancing at Figure 1.23(b), we find that

$$\text{if } q_\mu < x < p_\mu, \quad \text{then } p_\mu < Q_\mu(x) \le \mu/4$$

By using the fact that $q_\mu < Q_\mu(\mu/4)$ (see Exercise 3), and the fact that Q_μ is decreasing on $[1/2, 1]$, we conclude that

$$\text{if } p_\mu < x \le \mu/4, \quad \text{then } q_\mu < Q_\mu(\mu/4) \le Q_\mu(x) < p_\mu$$

From the two sets of inequalities it follows that if x lies in the interval (q_μ, p_μ), then $Q_\mu(x) > p_\mu$, and vice versa, so that the iterates of x oscillate around p_μ.

Case 4. $3 < \mu \le 4$

We have analyzed the dynamics of Q_μ when $0 < \mu \le 3$. Now we turn to our last case: $3 < \mu \le 4$. As before, 0 and $p_\mu = 1 - 1/\mu$ are fixed points of Q_μ. Since

$$Q'_\mu(p_\mu) = Q'_\mu(1 - 1/\mu) = \mu - 2\mu(1 - 1/\mu) = 2 - \mu$$

it follows that as μ increases to 3, $Q'_\mu(p_\mu)$ decreases to -1 (Figures 1.24(a)–(b)). When μ increases further, $Q'_\mu(p_\mu) < -1$, so that p_μ is a repelling fixed point, signifying that the quadratic family bifurcates when $\mu = 3$.

From the preceding discussion, if $\mu > 3$, then both fixed points 0 and p_μ are repelling. But consider the following questions: If $3 < \mu \le 4$, do the iterates of other points in $(0, 1)$ converge, or oscillate, or

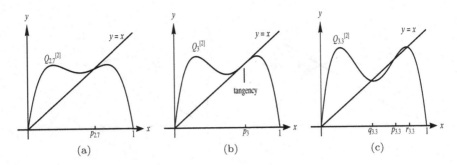

FIGURE 1.25
Squaring the quadratic family of maps

have no pattern? And are there periodic points different from 0 and p_μ? The analysis of $\{Q_\mu\}$ becomes more and more complicated as μ increases from 3 toward 4. The information we will obtain concerning the dynamics of the family when $3 < \mu \leq 4$ will come from an analysis of the dynamics of $Q_\mu^{[2]}$.

Figures 1.25(a)–(c) display the graphs of $Q_\mu^{[2]}$ for $\mu = 2.7$, 3, and 3.3. From the graphs it appears that as μ increases from 2.7 to 3.3, the middle trough descends and pierces the line $y = x$. In particular, when $\mu = 3$ the graph is tangent to the line $y = x$ at the point (p_μ, p_μ), and when $\mu = 3.3$ the graph intersects the line $y = x$ four times. We conclude that in addition to two fixed points it has when $\mu < 3$, $Q_\mu^{[2]}$ appears to be endowed with two new fixed points, q_μ and r_μ, when $\mu > 3$. Since Q_μ has only two fixed points, $\{q_\mu, r_\mu\}$ would need to be a 2-cycle for Q_μ. Could there be more than one such 2-cycle for Q_μ? To answer this question, we notice that if Q_μ had two 2-cycles, then including the two fixed points of Q_μ and two points each for the two cycles, $Q_\mu^{[2]}$ would have at least six fixed points. But that would mean that the 4th degree polynomial $Q_\mu^{[2]}(x) - x$ would have at least six distinct roots. However, since this polynomial has degree four, it is impossible for it to have more than four roots. Therefore Q_μ can have at most one 2-cycle.

Next, we will show formally that if $\mu > 3$, then Q_μ does indeed have a 2-cycle $\{q_\mu, r_\mu\}$. We will also derive formulas for numerical values of q_μ and r_μ.

Example 1.5.1. Let $\mu > 3$. Show that Q_μ has a 2-cycle $\{q_\mu, r_\mu\}$, and find formulas for the numerical values of q_μ and r_μ.

Solution. Suppose that Q_μ has a 2-cycle, which we will write as $\{q, r\}$. We will find formulas for q and r, and then show that they actually form a 2-cycle.

To that end we notice that the 2-cycle $\{q, r\}$ satisfies

$$r = Q_\mu(q) = \mu q(1 - q) \quad \text{and} \quad q = Q_\mu(r) = \mu r(1 - r)$$

It follows that

$$r - q = \mu q(1 - q) - \mu r(1 - r) = \mu(q - r) - \mu(q^2 - r^2)$$

Since $q \neq r$, we can divide through by $r - q$, which yields

$$r + q = \frac{1}{\mu} + 1, \quad \text{or equivalently,} \quad r = \frac{1}{\mu} + 1 - q$$

From (2) we obtain

$$r^2 = \mu qr(1 - q) \quad \text{and} \quad q^2 = \mu qr(1 - r)$$

Then (4) yields

$$r^2 - q^2 = \mu qr(r - q)$$

so that

$$r + q = \mu qr$$

Applications of (3), (5), and (3), respectively, yield

$$\frac{1}{\mu} + 1 = r + q = \mu qr = \mu q(\frac{1}{\mu} + 1 - q)$$

which simplifies to

$$\mu^2 q^2 - \mu^2 q - \mu q + \mu + 1 = 0$$

Solving this equation for q by means of the quadratic formula, we find that the roots are

$$\frac{1}{2} + \frac{1}{2\mu} \pm \frac{1}{2\mu}\sqrt{(\mu - 3)(\mu + 1)}$$

If we let q_μ be the smaller and r_μ the larger of the two values, then

$$q_\mu = \frac{1}{2} + \frac{1}{2\mu} - \frac{1}{2\mu}\sqrt{(\mu - 3)(\mu + 1)} \quad \text{and} \quad r_\mu = \frac{1}{2} + \frac{1}{2\mu} + \frac{1}{2\mu}\sqrt{(\mu - 3)(\mu + 1)}$$

It is easy to check that $0 < q_\mu < 1$ and $0 < r_\mu < 1$, and that $\{q_\mu, r_\mu\}$ is a 2-cycle for Q_μ. Consequently we have completed the proof. $\square$

We are ready to determine the values of μ for which $\{q_\mu, r_\mu\}$ is attracting.

Theorem 1.17. Let $3 < \mu < 4$. The 2-cycle $\{q_\mu, r_\mu\}$ is attracting for Q_μ if $3 < \mu < 1 + \sqrt{6}$.

Proof. For simplicity we will use q and r for q_μ and r_μ, respectively. Since $\{q, r\} = \{q_\mu, r_\mu\}$ is a 2-cycle and $Q'_\mu(x) = \mu - 2\mu x$, it follows from Theorem 1.12 that

$$(Q_\mu^{[2]})'(q) = Q'_\mu(q)\, Q'_\mu(r) = (\mu - 2\mu q)(\mu - 2\mu r) = \mu^2 - 2\mu^2(q + r) + 4\mu^2 qr$$

In order to be able to write the right side as a function of μ alone, we use (3) to substitute for $q + r$; after that we use (5) and then (3) to substitute for μqr. We obtain

$$(Q_\mu^{[2]})'(q) = \mu^2 - 2\mu^2(\frac{1}{\mu} + 1) + 4\mu(\frac{1}{\mu} + 1) = -\mu^2 + 2\mu + 4$$

Therefore $|(Q_\mu^{[2]})'(q)| < 1$ if and only if $|\mu^2 - 2\mu - 4| < 1$. This inequality is equivalent to $-1 < (\mu - 1)^2 - 5 < 1$, which yields $3 < \mu < 1 + \sqrt{6}$. We conclude that the 2-cycle $\{q_\mu, r_\mu\}$ is attracting if $3 < \mu < 1 + \sqrt{6}$. □

The fact that the 2-cycle $\{q_\mu, r_\mu\}$ is attracting if $\mu = 1 + \sqrt{6}$ is harder to prove, as is the fact that if $3 < \mu < 1 + \sqrt{6}$, then the basin of attraction of the 2-cycle $\{q_\mu, r_\mu\}$ consists of all x in $(0, 1)$ except the fixed point p_μ and the points whose iterates are eventually p_μ. We will not prove these results. We mention, however, that they imply that for $3 < \mu \leq 1 + \sqrt{6}$, the only periodic points of Q_μ are the repelling fixed points 0 and p_μ, and the period-2 points q_μ and r_μ that form an attracting 2-cycle.

Since

$$|(Q_\mu^{[2]})'(q_\mu)| = |\mu^2 - 2\mu - 4| < 1 \quad \text{if and only if} \quad 3 < \mu < 1 + \sqrt{6}$$

it follows that if $\mu > 1 + \sqrt{6}$, then $|(Q_\mu^{[2]})'(q_\mu)| > 1$, so that $\{q_\mu, r_\mu\}$ becomes a repelling 2-cycle. As you might suspect, a new, attracting 4-cycle is born as μ increases beyond $1 + \sqrt{6}$. As μ increases still further, the attracting 4-cycle becomes repelling, and a new 8-cycle is born. The process continues indefinitely as μ increases.

One might imagine that the values of μ at which new 2^k-cycles emerge would march unboundedly toward ∞. However, this turns out to be not true. Let μ_k be the bifurcation point defined by

$\mu_k = $ maximum value of μ for which Q_μ has an attracting 2^k-cycle, for

$\quad k = 0, 1, 2, \ldots$

From our previous results, we know that Q_μ has an attracting fixed point for $0 < \mu \leq 3$ and an attracting 2-cycle for $3 < \mu \leq 1 + \sqrt{6}$. Therefore

$$\mu_0 = 3 \quad \text{and} \quad \mu_1 = 1 + \sqrt{6}$$

However, numerical values of μ_k for $k > 1$ are not so easy to determine. What we can say, though, is that if $\mu_0 < \mu \le \mu_1$, then Q_μ has 2 fixed points and a 2-cycle

if $\mu_1 < \mu \le \mu_2$, then Q_μ has 2 fixed points, a 2-cycle and a 2^2-cycle

if $\mu_2 < \mu \le \mu_3$, then Q_μ has 2 fixed points, a 2-cycle, a 2^2-cycle, and a 2^3-cycle

In general, if $\mu_{n-1} < \mu \le \mu_n$, then Q_μ has a 2^k-cycle for $k = 0, 1, 2, \ldots, n$.

It is known, but difficult to prove, that $\mu_{k+1} \approx 1 + \sqrt{3 + \mu_k}$ for $k = 2, 3, \ldots$, and that the sequence $\{\mu_k\}_{k=0}^\infty$ has a limit μ_∞ given by

$$\mu_\infty \approx 3.61546 \cdots$$

The number μ_∞ is sometimes called the **Feigenbaum number** for the quadratic family $\{Q_\mu\}$, named after the physicist Mitchell Feigenbaum, who in the mid-1970's conjectured that the bifurcation points had a limit and found a very precise value for it. (See Feigenbaum, 1978 or 1983, for accounts of his discovery.)

The surprising part of the story is yet to come. Let

$$d_k = \frac{\mu_k - \mu_{k-1}}{\mu_{k+1} - \mu_k}, \quad \text{for } k = 2, 3, 4, \ldots$$

Since $\mu_k - \mu_{k-1}$ represents the distance between μ_k and μ_{k-1}, it follows that d_k compares distances between successive pairs of μ_k's. Feigenbaum found that the sequence $\{d_k\}_{k=1}^\infty$ converges to a number we will denote d_∞, where

$$d_\infty \approx 4.669202 \cdots$$

What is astonishing is that this constant d_∞ seems to be universal. That is, for many families of one-humped functions like the family of quadratic functions, bifurcations occur in such a regular fashion that the distances between successive pairs of bifurcation points approach the very same value d_∞! It is for this reason that d_∞ is called a **universal constant**. More particularly, it is referred to as the **Feigenbaum constant**, because Feigenbaum was the first to discover it and its universality.

We conclude by noting that the quadratic family $\{Q_\mu\}$ is one of the most illustrious parametrized families. Its functions are easy to describe, and have properties many more complicated functions have. Moreover, there is an enormous wealth of information concerning the family, spurred in part by the captivating article in the magazine *Nature* by Robert May (1975). Books by Pierre Collet and Jean-Pierre Eckmann (1980) and by Chris Preston (1983) give detailed analysis of functions like quadratic functions. We will once again study properties of the family when we investigate properties of chaos in Chapter 2.

1.5.1 Section 1.5 Exercises

Exercise 1. Let $1 < \mu \le 2$. Prove that if $p_\mu < x < 1/2$, then $\{Q_\mu^{[n]}(x)\}_{n=0}^\infty$ is a bounded decreasing sequence that converges to p_μ.

Exercise 2. Let $2 < \mu < 3$. Show that $-\mu^2 x^2 + (\mu^2 + \mu)x - \mu - 1$ has no (real) roots.

Exercise 3. Let $2 < \mu < 3$. Show that $q_\mu < 1/2 < Q_\mu(\mu/4)$. (*Hint:* Let $h(\mu) = Q_\mu(\mu/4)$, and find the range of h for μ in the interval $(2, 3)$.)

Exercise 4.

(a) Show that $(Q_\mu^{[2]})'(1/2) = 0$ for $0 < \mu < 4$.

(b) Since $1/2$ is a zero of $(Q_\mu^{[2]})'$ by part (a), we are able to write $(Q_\mu^{[2]})'(x)$ in the form $\mu^2(x - 1/2)(ax^2 + bx + c)$. Find the values of a, b, and c in terms of μ.

(c) Use part (b) to show that if $0 < \mu < 2$, then $Q_\mu^{[2]}$ has a unique relative extreme (maximum) value for x in $(0, 1)$.

Exercise 5. Let $2 < \mu \le 4$. Find the relative minimum value of $Q_\mu^{[2]}$ in the interval $(0, 1)$.

Exercise 6.

(a) Let $2 \le \mu \le 4$. Find the maximum value of $Q_\mu^{[n]}$ for any positive integer n.

(b) Let $0 < \mu < 2$. Describe the difficulties, if any, that you encounter when trying to find the maximum value of $Q_\mu^{[n]}$.

Exercise 7. Let $3 < \mu \le 4$.

(a) Show that $0 < q_\mu < 1$.

(b) Show that $Q_\mu(q_\mu) = r_\mu$ and $Q_\mu(r_\mu) = q_\mu$

Exercise 8. Let $0 < \mu_0$, and let $\epsilon > 0$. Show that if $|\mu - \mu_0| < \epsilon$, then

$$|Q_\mu(x) - Q_{\mu_0}(x)| < \epsilon$$

for all x in $[0, 1]$.

Exercise 9.

(a) Let $0 < \mu < 1$. Find c in $(0, 1)$ such that $(Q_\mu^{[n]})(x) < c^n x$ for all x in $(0, 1)$.

(b) Let $\mu = 1$. Show that no such c in $(0, 1)$ exists.

Exercise 10. Let $1 < \mu < 2$. Find c in $(0, 1)$ such that

$$|Q_\mu^{[n]})(x) - p_\mu| < c^n |x - p_\mu|$$

for all x such that $p_\mu < x < 1/2$. (*Hint:* Prove the result by using the law of induction.)

Exercise 11. Let $M_0 = 3$, and let $M_{k+1} = 1 + \sqrt{3 + M_k}$ for $k = 0, 1, 2, \ldots$. Also let

$$M_\infty = \lim_{k \to \infty} M_k$$

It is known, but hard to prove, that $M_k \approx \mu_k$ for all k, and that $M_\infty \approx \mu_\infty$.

(a) Use induction to find M_∞.

(b) Compare the numerical value of M_∞ found in (a) to that of μ_∞.

Exercise 12. Using the definition of M_k in Exercise 11, let

$$D_k = \frac{M_k - M_{k+1}}{M_{k+1} - M_k}, \text{ for } k = 1, 2, 3, \ldots$$

Show that $\lim_{k \to \infty} D_k = 1 + \sqrt{17}$.

Exercise 13. We can generalize the idea of a Feigenbaum constant. We will see period-doubling bifurcation in the next section, which is where the work originated. We have two tables below, showing the maximum value of a parameter μ for which there is a periodic orbit of a certain size. As μ increases, the value μ_n is given when the period doubles. So for example, in the first table μ_1 is the largest value of μ that does not yield a period-2 orbit.

$$f(x) = \mu - x^2$$

n	Period	μ_n
1	2	0.75
2	4	1.25
3	8	1.3680989
4	16	1.3940462
5	32	1.3996312
6	64	1.4008286
7	128	1.4010853
8	256	1.4011402

$$Q_\mu(x) = \mu x(1 - x)$$

n	Period	μ_n
1	2	3
2	4	3.4948975
3	8	3.5440903
4	16	3.5644073
5	32	3.5687594
6	64	3.5696916
7	128	3.5698913
8	256	3.5699340

For both of these functions, compute as many value of the ratio $\frac{\mu_{n-1}-\mu_{n-2}}{\mu_n - \mu_{n-1}}$. What do you notice about the numbers you approach as n increases? Is thie equivalent to the method used earlier to compute d_∞?

In the study of a family of functions such as $\{Q_\mu\}$ and $\{T_\mu\}$, values of μ at which the family bifurcates play a prominent role. After all, these values indicate where periodic points arise or disappear, as well as where periodic points become or cease to be attracting. This section is devoted to bifurcations. First, we consider a method of displaying bifurcation points on a graph. Then we discuss two basic kinds of bifurcations: period-doubling bifurcations and tangent bifurcations.

1.6   Bifurcation Diagrams

One method of displaying the points at which a parametrized family of functions $\{f_\mu\}$ bifurcates is called a bifurcation diagram and is designed to give information about the behavior of higher iterates of arbitrary members of the domain of f_μ for all values of the parameter μ.

The **bifurcation diagram** of $\{f_\mu\}$ is a graph for which the horizontal axis represents values of μ and the vertical axis represents higher iterates of the variable (normally x). For each value of μ, the diagram includes (in theory) all points of the form $(\mu, f_\mu^{[n]}(x))$, for values of n larger than, say, 50 or 100. The reason we only use the higher iterates of x is that the diagram is designed to show eventual behavior of iterates, such as convergence or periodicity or unpredictability.

Now we will study the bifurcation diagram of the quadratic family $\{Q_\mu\}$. In order to give as much detail as we can in the bifurcation diagram, we have split the diagram into two parts, $0 < \mu \leq 1 + \sqrt{6}$ in Figure 1.26 and $1 + \sqrt{6} \leq \mu \leq 4$ in Figure 1.27. The bifurcation diagram was obtained on the computer by letting x be $1/2$, taking increments of $1/1000$ for μ in the interval $[0, 4]$, and plotting all the points of the form $(\mu, Q_\mu^{[n]}(x))$, for $201 \leq n \leq 700$. You can check that if another value of x were chosen, or if the increments remained small but were altered, or if the range of n were changed, then the corresponding bifurcation diagram would have been indistinguishable from the one pictured in Figures 1.26 and 1.27.

To analyze the diagram, notice in Figure 1.26 that for $0 < \mu < 1$, the points in the diagram lie on the x-axis, because iterates of all x in the domain $[0, 1]$ are attracted to 0. Next, for $1 < \mu \leq 3$, the curve represents the points of the form $(\mu, p_\mu) = (\mu, 1 - 1/\mu)$, since p_μ attracts the iterates of all x in $(0, 1)$. When $3 < \mu \leq 1 + \sqrt{6}$, iterates of all x in $(0,1)$ that are not eventually fixed are attracted to the 2-cycle $\{q_\mu, r_\mu\}$, so for such values of μ there are two curves, one with points of the form (μ, q_μ) and the other with points of

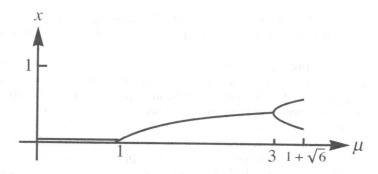

FIGURE 1.26
Partial bifurcation diagram for $\{Q_\mu\}$

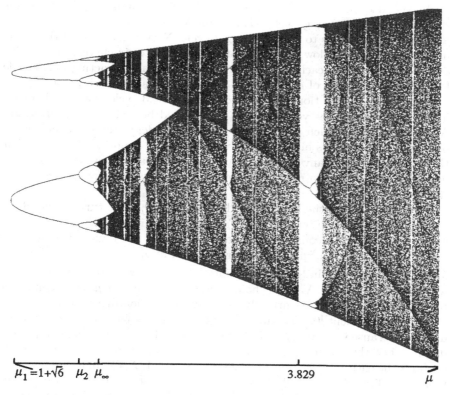

FIGURE 1.27
Remainder of the bifurcation diagram for $\{Q_\mu\}$

the form (μ, r_μ). We observe that at the bifurcation point $\mu = 3$, a single curve splits into two curves representing the attracting 2-cycle that emerges at $\mu = 3$.

Turning to Figure 1.27, which commences with $\mu = 1 + \sqrt{6}$, we see that to the right of $\mu = 1 + \sqrt{6}$, four branches appear (corresponding to an attracting 4-cycle). The four branches extend until $\mu = \mu_2$, at which point eight branches start to appear (corresponding to an attracting 8-cycle). In general these branches represent the various attracting 2^k-cycles, and split or fork at points corresponding to the various bifurcation points of $\{Q_\mu\}$. The branches represent the various attracting 2^k-cycles that appear in sequence.

The heavily shaded vertical strips in Figure 1.27 represent values of μ for which the iterates of most points x in $(0, 1)$ are not eventually periodic, but spread out over a subinterval or collection of subintervals of $[0, 1]$. These darkened patches represent unpredictable, chaotic-like patterns for iterates of such values of x. The dark curves that give interesting patterns to the heavily shaded regions represent extensions (for larger values of μ) of attracting cycles, as you can verify by following the curves to the left.

In the diagram we also observe "windows" containing isolated curves that represent attracting cycles. For example, the window that appears at $\mu \approx 3.829$ corresponds to an attracting 3-cycle. Notice that toward the right-hand side of that window, the three curves divide into six curves, then twelve curves, and so forth, each set in turn representing an attracting $3(2^k)$-cycle for appropriate values of μ. As before, the μ-coordinate of each point where a fork occurs is a bifurcation point. In addition to the 3-cycle window, there are 5-cycle, 6-cycle, and 8-cycle windows that should be visible in the diagram. At the right end of each window, the branches undergo the same succession of splitting that occurs in the 3-cycle window. What is not so obvious is that there are infinitely many windows representing attracting cycles; these windows are in general too narrow to detect without a strong zoom feature on the computer screen.

Having described the bifurcation diagram for $\{Q_\mu\}$, we turn to the bifurcation diagram for the tent family $\{T_\mu\}$, shown in Figure 1.28. This diagram was obtained from a computer in the same way as the bifurcation diagram for $\{Q_\mu\}$ was. Values of μ in $[0, 1]$ were selected in increments of $1/1000$, and for a given value of x in $(0, 1)$, the points $(\mu, T_\mu^{[n]}(x))$ were plotted for each n with $1001 \le n \le 2000$. Various choices of x, increments for μ, and number of plotted points can lead to virtually identical bifurcation diagrams for $\{T_\mu\}$.

The line segment $[0, 1/2]$ on the μ-axis appears as part of the bifurcation diagram because when $0 \le \mu < 1/2$, iterates of each x in the domain approach 0. If $\mu = 1/2$, then each x in the interval $[0, 1/2]$ is a fixed point, so the diagram contains the entire vertical line $(1/2, x)$ for $0 \le x \le 1/2$. The diagram becomes more complicated for $\mu > 1/2$, because for such values of μ, the iterates of most values of x in $(0,1)$ spread out over intervals that widen as μ approaches 1. These darkened patches suggest an increasingly chaotic type of behavior for the iterates of corresponding tent functions. The apparent eye in the diagram

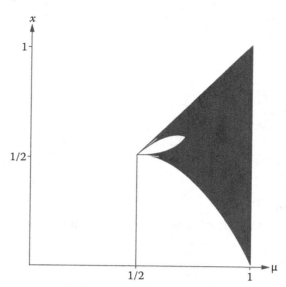

FIGURE 1.28
Bifurcation diagram for $\{T_\mu\}$

can be justified by careful analysis of the iterates of T_μ for μ in the interval $[1/2, 7/10]$.

An alternative method of displaying bifurcation points is the **orbit diagram**, which includes solid curves for attracting fixed points and attracting cycles, and dashed curves for repelling fixed points and repelling cycles. Unlike the bifurcation diagram, it does not indicate the various iterates of points in the domain. (In fact, it would perhaps make better sense to switch the names of orbit diagram and bifurcation diagram, but that would counter common usage.) Orbit diagrams for portions of $\{Q_\mu\}$ and $\{E_\mu\}$ appear in Figure 1.29.

1.6.1 Period-Doubling Bifurcations

A bifurcation at which an attracting period-n cycle becomes repelling and gives birth to an attracting $2n$-cycle is a **period-doubling bifurcation**. The bifurcations of $\{Q_\mu\}$ represented in Figure 1.29(a) are period-doubling bifurcations. Because the graph resembles a pitchfork near the point on the orbit diagram corresponding to the bifurcation, this kind of bifurcation is often called a **pitchfork bifurcation** (Figure 1.30).

Period-doubling bifurcations form one important class of bifurcations. To understand better the properties of families of functions at period-doubling bifurcation points, we will focus on Q_μ and $Q_\mu^{[2]}$ when μ is near 3.

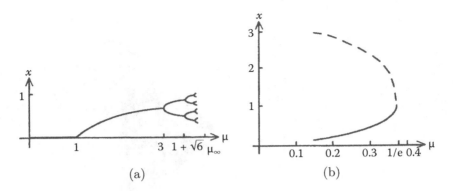

FIGURE 1.29
(a) Partial bifurcation diagram for $\{Q_\mu\}$. (b) Orbit diagram for $\{E_\mu\}$

FIGURE 1.30
"Pitchfork" bifurcation

Example 1.6.1. Show that the quadratic family $\{Q_\mu\}$ has the following properties relating to the bifurcation point $\mu = 3$:

i. $Q_3(2/3) = 2/3$, so that $2/3$ is a fixed point of Q_3.

ii. $Q'_\mu(2/3) = -1$.

iii. $Q'_\mu(2/3)$ decreases as μ increases through 3.

iv. The graph of $Q_3^{[2]}$ has an inflection point at $(2/3, 2/3)$.

Solution. First, we recall that $Q_\mu(x) = \mu x(1 - x)$, so that $Q_3(2/3) = 2/3$. Thus (i) is proved. Since $Q'_\mu(x) = \mu - 2\mu x$, it follows that $Q'_\mu(2/3) = -\mu/3$. Therefore $Q'_3(2/3) = -1$, and $Q'_\mu(2/3)$ decreases as μ increases through 3. Consequently (ii) and (iii) are both verified. To prove (iv) we calculate that

$$(Q_3^{[2]})'(x) = 9(1 - 8x + 18x^2 - 12x^3) \quad \text{and} \quad (Q_3^{[2]})''(x) = -36(3x - 2)(3x - 1)$$

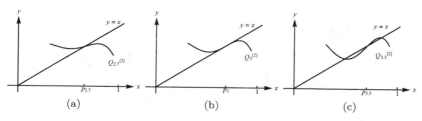

FIGURE 1.31
Investigating how bifurcations occur in the quadratic map

As a result, $(Q_3^{[2]})''$ changes sign at $2/3$, so that $(2/3, 2/3)$ is an inflection point. That finishes the proof of (iv). □

Using Example 1.6.1 as a model, we can describe general properties of families of functions that accompany period-doubling bifurcations. To that end, let $\{f_\mu\}$ be a parametrized family, and assume that the family has the following properties:

i. $f_{\mu*}$ has a fixed point p.

ii. $f'_{\mu*}(p) = -1$

iii. The graph of $f_\mu^{[2]}$ crosses the line $y = x$ when $\mu < \mu^*$, is tangent to the line $y = x$ when $\mu = \mu^*$, and snakes around the line $y = x$ when $\mu > \mu^*$. (See Figures 1.31(a)–(c), which are portions of Figures 1.25(a)–(c), respectively.)

Properties (i) and (ii) indicate that the graph of f_μ is positioned correctly with respect to the line $y = x$ when μ is near to μ^*. However, it is the rotation of the graphs of $f_\mu^{[2]}$ around the point (p, p) as μ passes through μ^*, as described in (iii), that signals the period-doubling bifurcation of f_μ at μ^*. From Example 1.2.1 and Figure 1.31, we see that (i)–(iii) hold if $f_\mu = Q_\mu, \mu^* = 3$, and $p = 2/3$.

1.6.2 Tangent Bifurcations

A bifurcation with a far different character occurs at $\mu = 1/e$ for the exponential family E_μ. To support this assertion, let us first recall that

$$E_\mu(x) = \mu e^x, \quad \text{where } \mu > 0$$

In Figures 1.32(a)–(c), when μ decreases through $1/e$, the graph of $\{E_\mu\}$ descends until it is tangent to the line $y = x$ and then breaks through the line $y = x$, giving rise to two fixed points that separate from one another as μ continues to decrease. The bifurcation at $\mu = 1/e$ occurs because the graph of E_μ breaks through the line $y = x$. Since the graph of E_μ is tangent to the line $y = x$ at the point corresponding to the bifurcation point, the bifurcation is called a **tangent bifurcation**.

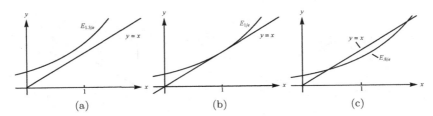

FIGURE 1.32
Investigating bifurcations in the exponential map

More generally, a family $\{f_\mu\}$ has a **tangent bifurcation** (or **saddle**, or **fold bifurcation**) at μ^* if a pair of fixed points are born as a curve in the graph of f_μ becomes tangent to and then crosses the line $y = x$ when μ passes through μ^*. It can be shown that a parametrized family $\{f_\mu\}$ has a tangent bifurcation at μ^* if

i. $f_{\mu^*}(p) = p$, so p is fixed point of f_{μ^*}.

ii. $f'_{\mu^*}(p) = 1$, so the graph of f_{μ^*} is tangent to the line $y = x$ at (p, p).

iii. $f_\mu(p)$ is a monotone function of μ near μ^*.

iv. The graph of f_μ is concave upward (or downward) near (p, p).

A major difference between period-doubling and tangent bifurcations is the derivative of f_{μ^*} at p: $f'_{\mu^*}(p) = -1$ for period-doubling bifurcations, and $f'_{\mu^*}(p) = 1$ for tangent bifurcations.

In the following example we will show that the exponential family $\{E_\mu\}$ satisfies conditions (i)–(iv) at $\mu = 1/e$.

Example 1.6.2. Show that $\{E_\mu\}$ satisfies (i)–(iv) when $\mu = 1/e$.

Solution. Since

$$(E_{1/e})'(1) = E_{1/e}(1) = \frac{1}{e}e^1 = 1$$

(i) and (ii) are satisfied with $p = 1$. The fact that $E_\mu(1) = \mu e$ verifies (iii). Finally, the graph of E_μ is concave upward because $E''_\mu(x) = E_\mu(x) = \mu e^x > 0$ for all x and all μ. Thus (iv) is proved. $\square$

The bifurcations of $\{Q_\mu\}$ that we have discussed are period-doubling. Now we will show why the family $\{Q_\mu^{[3]}\}$ has a tangent bifurcation at $\mu \approx 3.83$. Two local minima and one local maximum in Figure 1.33(a) are nearly tangent to the line $y = x$ when $\mu = 3.81$, are tangent to the line $y = x$ when $\mu \approx 3.83$ (Figure 1.33(b)), and break through the line $y = x$ as μ increases through 3.83. At the bifurcation an attracting period-3 cycle and a repelling period-3 cycle are born.

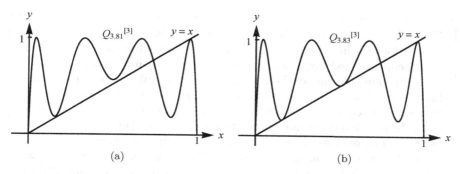

FIGURE 1.33
Tangent bifurcation in the cube of the quadratic map

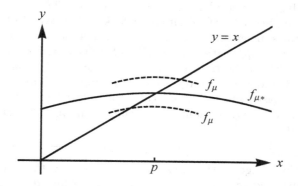

FIGURE 1.34
Demonstration of the conditions for when a bifurcation can exist at a given parameter

In general a parametrized family $\{f_\mu\}$ cannot have a bifurcation at μ^* unless f_{μ^*} has a periodic point p such that $|f'_{\mu^*}(p)| = 1$. Thus if p is a fixed point of f_{μ^*}, if $|f_{\mu^*}(p)| \neq 1$, and if both f_μ and f'_μ vary continuously as μ varies, then in effect the graph of f_μ will cross the line $y = x$ exactly once for all μ near enough to μ^*, so that μ^* is not a bifurcation point of f_μ. (We have not defined what it means for f_μ or f'_μ to vary continuously as μ varies, but all the families we have considered have this property.) Figure 1.34 demonstrates this type of condition, in which the slope of f_{μ^*} is in $(0, 1)$ and the graph of f_μ crosses the line $y = x$ exactly once for all μ close to μ^*.

The implication is that if the family $\{f_\mu\}$ bifurcates at μ^*, and if f_{μ^*} has a fixed point p, then $|f'_{\mu^*}(p)| = 1$. If the derivative is -1, then there may be a period-doubling bifurcation at μ^*; if the derivative is 1, then there may be a tangent bifurcation at μ^*. The books by Devaney (2003) and Guckenheimer and Holmes (1983) contain further information on other types of bifurcations.

Note: It is worth noting that scientists have encountered period-doubling bifurcations in many different experiments. Some fifty years ago, a Russian chemist named Boris Belousov observed interesting oscillations in the concentration of bromide when he mixed together sulfuric acid, potassium bromate, cerium sulfate, and malonic acid. At a later date, details of the reaction were confirmed by another Russian, Anatol Zhabotinskii. Nowadays, the reaction is normally called the B-Z reaction. During the reaction, bromate ions oxidize to form bromine, after which cerium oxidizes and makes the solution change from red to blue. Then autocatalysis makes the color switch back from blue to red. If the reaction is continued for a period of time by continually pumping in new reactants, the colors can flip back and forth between red and blue, displaying not only immense complexity but also period-doubling toward chaos. The entertaining article by Stephen Scott (1989) entitled, "Clocks and Chaos in Chemistry," in *New Scientist* magazine, discusses the B-Z reaction in more detail.

A more recent application involving biology is described in the article, "Estimating the Stochastic Bifurcation Structure of Cellular Networks," by Song et al. Biological bifurcations are an interesting area of study, as there are stochastic elements that can randomly affect the behavior. A good text on stochastic dynamical systems is Pikovsky, Rosenblum, and Kurths (2001).

A bifurcation diagram that looks rather similar to the diagram for the quadratic family occurs in the study of periodically forced nonlinear circuits (Figure 1.35). In the figure, the junction voltage is shown as a function of

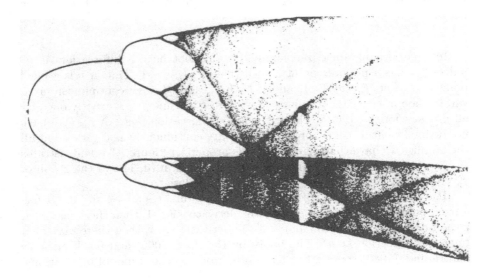

FIGURE 1.35
Bifurcation diagram for periodically forced nonlinear circuits

the drive voltage, and shows the onset of chaos through period-doubling. This diagram appeared in an article by Robert Van Buskirk and Carson Jeffries (1985). While this diagram is built from circuit data, the periodic orbits are apparent, just as they are in our diagrams generated from functions.

1.6.3   Section 1.6 Exercises

Exercise 1. Let $f_\mu(x) = x^2 + 1/4 + \mu$, for all x and all real μ. Show that there is a bifurcation at $\mu = 0$, and classify it as a period-doubling bifurcation, a tangent bifurcation, or neither.

In Exercises 2–3, determine a bifurcation for the given parametrized family, and classify it as period-doubling bifurcation, tangent bifurcation or neither.

Exercise 2. The exponential family $\{E_\mu\}$, where $\mu < 0$

Exercise 3. $\{f_\mu\}$, where $f_\mu(x) = \mu \arctan x$ for all x with $\mu > 0$

Exercise 4. Let $x_1, x_2, \ldots, x_8$ denote the eight fixed points of $Q_\mu^{[3]}$ when $3.83 < \mu < 3.84$. Assume that $x_1 < x_2 < \cdots < x_8$. Using Figure 1.33(b) as a guide, determine which of the x_k's form the attracting 3-cycle for Q_μ.

Exercise 5. Use the computer program BIFURCATION to approximate (to the nearest thousandth) the interval on the μ axis for which there is an attracting 5-cycle window for the quadratic family.

Exercise 6. Use the computer program BIFURCATION to approximate (to the nearest thousandth) the interval on the μ axis for which there is an attracting period-10 window for the quadratic family.

Exercise 7. Use the computer program BIFURCATION to make a bifurcation diagram for $\{Q_\mu\}$ in which the increments are $1/100$ and the initial point is $x = 1/2$. How does the portion between $1 + \sqrt{6}$ and 4 compare with Figure 1.27?

Exercise 8. Alter the computer program PLOT to plot the graph of $Q_\mu^{[2]}$. Observe the behavior of the graph as μ increases from 1 to 4.

Exercise 9. Alter the computer program PLOT to plot the graph of $Q_\mu^{[8]}$. Observe the behavior of the graph as μ increases from 3 to 4.

Exercise 10. Alter the computer program PLOT to plot the graphs of Q_μ and $Q_\mu^{[2]}$ simultaneously. Determine a value of μ in $[3, 4]$ such that the middle portion of the graph of $Q_\mu^{[2]}$ looks like an inverted copy of Q_μ.

Exercise 11. Consider the function $T_{0.6}$. Alter the computer program ITERATE to approximate (to within one thousandth) the largest subinterval J of $[0.5, 0.6]$ such that the higher iterates of $T_{0.6}(x)$ lie outside J for all x in J.

Exercise 12. Using the BIFURCATION program, find the windows that have a period 1 orbit and a period 3 orbit. Show that the largest windows between those that contain periodic orbit contain first a period 6 orbit, and then a period 5.

Exercise 13. You have now found the window that includes periodic orbit in order 1,6,5,3. Zoom in further, and see if you an find the periods to fill in the gaps.

$$1, ?, ?, 6, ?, ?, 5, ?, 3.$$

Do you see any pattern? Theorem 1.20 in the upcoming section will offer some insight into what you have discovered.

1.7   Period-3 Points

In Section 1.5 we discovered values of μ for which Q_μ has fixed points and 2^k-cycles for all positive integers k less than any given positive integer n, but no other cycles. In other words, there are values of μ such that Q_μ has points of certain periods but no points of other periods. In the present section, we will describe what the presence of a period-3 point, or any period-n point, implies about the existence of other periodic points. The answer will derive from two famous theorems, those of Li and Yorke (1975) and of Sharkovsky (1964). Our first goal will be to prove a wonderful theorem due to James Yorke and his student Tien-Yien Li. It tells us that if Q_μ has a period-3 point, then Q_μ has a period-n point for every $n \geq 1$! Their ingenious proof relies on two powerful theorems from calculus: the Maximum-Minimum Theorem and the Intermediate Value Theorem. For reference we state them here without proof.

Theorem 1.18 (Maximum-Minimum Theorem). Suppose that f is continuous on the interval $[a, b]$. Then f has a maximum value and a minimum value.

Concretely, the Maximum-Minimum Theorem says that if f is continuous on $[a, b]$, then there are numbers u and v in $[a, b]$ such that if x is any number in $[a, b]$, then $f(u) \leq f(x) \leq f(v)$ (Figure 1.36). Thus

$$f(u) = \text{minimum} \quad \text{value} \quad \text{of } f \text{ on } [a, b]$$
$$f(v) = \text{maximum} \quad \text{value} \quad \text{of } f \text{ on } [a, b]$$

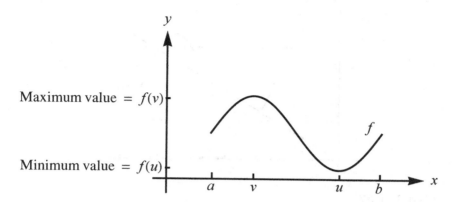

FIGURE 1.36
The maximum-minimum theorem

Theorem 1.19 (Intermediate Value Theorem). Suppose that f is continuous on the interval $[a, b]$, and let p be any number between $f(a)$ and $f(b)$. Then there is a number c in $[a, b]$ such that $f(c) = p$.

The Intermediate Value Theorem says in effect that if f is continuous on the closed interval $[a, b]$, then it cannot skip over any values between $f(a)$ and $f(b)$; in other words, the range of f contains all values between $f(a)$ and $f(b)$.

In order to make the proof of the Li-Yorke Theorem more accessible, we will prepare for it with four lemmas. The first of the lemmas is a direct consequence of the Intermediate Value Theorem.

Lemma 1.1. Let f be continuous on an interval J. Let $f(J)$ denote the collection of all values $f(x)$, for x in J. Then $f(J)$ is also an interval.

Proof. Suppose that $f(J)$ were not an interval. Then there would exist two numbers y and z in $f(J)$, with $y < z$, and a number p in (y, z) such that p is not in the range of f. By the Intermediate Value Theorem applied to $[y, z]$, the range of f must contain the entire interval $[y, z]$, and in particular must contain p. This contradiction implies that $f(J)$ is an interval. $\square$

The next lemma will be used repeatedly in the proof of the Li-Yorke Theorem.

Lemma 1.2. Let f be continuous on a closed interval J, and assume that $f(J) \supseteq [a, b]$. Then there is a closed interval K such that $J \supseteq K$ and $f(K) = [a, b]$.

Proof. Let $J = [r, s]$. Since $f(J) \supseteq [a, b]$, there are numbers x and z in J such that $f(x) = a$ and $f(z) = b$. Of all such numbers z, there is one closest to x. Call it y (Figure 1.37). That y exists is assured by the continuity of

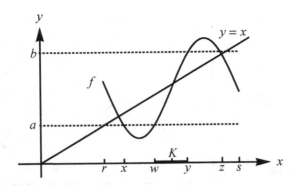

FIGURE 1.37
Visualization of the proof of Lemma 1.2

f. Similarly, there is a w between x and y that is the closest to y for which $f(w) = a$. By Lemma 1.1, the closed interval K determined by w and y (which is either $[w, y]$ or $[y, w]$) has the property that $f(K) = [a, b]$. $\square$

The proof of the following lemma uses both the Maximum–Minimum Theorem and the Intermediate Value Theorem.

Lemma 1.3. Suppose that J is a closed interval, and assume that f is continuous on J and $f(J) \supseteq J$. Then f has a fixed point in J.

Proof. By the Maximum-Minimum Theorem, f has a minimum value r and a maximum value s on J, so for suitable y and z in J,

$$r = f(y) = \text{minimum value of } f \text{ on } J, \quad s = f(z) = \text{maximum value of } f \text{ on } J$$

Since r is the minimum value of f on J, and s is the maximum value of f on J, and since $f(J) \supseteq J$, it follows that $r \leq y \leq s$ and $r \leq z \leq s$. Now let $g(x) = f(x) - x$. Then g is continuous on J since f is, and furthermore,

$$g(y) = f(y) - y = r - y \leq 0 \quad \text{and} \quad g(z) = f(z) - z = s - z \geq 0$$

By the Intermediate Value Theorem there is an x between y and z such that $g(x) = 0$, or equivalently, $f(x) = x$. Since x is in J, f has a fixed point in J. $\square$

Our final lemma tells us that if a continuous function f has a period-3 point, then it also has a fixed point and a period-2 point.

Lemma 1.4. Let f be continuous and suppose that $f(a) = b, f(b) = c$ and $f(c) = a$. Then f has a fixed point and a period-2 point.

Proof. Without loss of generality we may suppose that $a < b < c$ (Figure 1.38(a)). Since $f(b) = c$ and $f(c) = a$, we know that $f[b, c] \supseteq [a, c] \supseteq [b, c]$, so by Lemma 1.3, f has a fixed point in $[b, c]$.

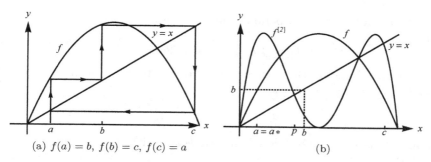

(a) $f(a) = b$, $f(b) = c$, $f(c) = a$ (b)

FIGURE 1.38
Visualization of the proof for Lemma 1.4

To show that f has a period-2 point, let

a^* = the largest number such that $a \leq a^* < b$ and $f(a^*) = b$

(Since f is continuous on $[a, b]$ and since $f(b) = c > b$, such an a^* exists.)
Then $f[a^*, b] \supseteq [b, c]$, so that

$$f^{[2]}[a^*, b] \supseteq f[b, c] \supseteq [a, c] \supseteq [a^*, b]$$

By Lemma 1.3, with $J = [a^*, b]$, there is a fixed point p of $f^{[2]}$ in $[a^*, b]$ (Figure 1.38(b)). Since $a^* < b < c$, $f(a^*) = b$ and $f(b) = c$, we know that $p \neq a^*$ and $p \neq b$. It follows from the definition of a^* that if $a^* < x < b$, then $f(x) > b$, so that $f(p) > b > p$. Therefore p is not a fixed point of f, so p is a period-2 point of f. (You can check in Figure 1.38(b) that p is indeed a period-2 point of f.) $\qquad\square$

Now we are ready to state and prove the theorem of Li and Yorke.

Theorem 1.20 (Li-Yorke Theorem). Suppose that f is continuous on the closed interval J, with $J \supseteq f(J)$. If f has a period-3 point, then f has points of all other periods.

Proof. Assume that $a < b < c$ and that $f(a) = b$, $f(b) = c$, and $f(c) = a$. By Lemma 1.4, f has points of period 1 and 2, and by assumption, f has points of period 3. Now let $n > 3$. We will show that f has points of period n. The idea of the proof is to show that there is a point p in $[b, c]$ such that

i. $f^{[k]}(p)$ lies in $[b, c]$ for $k = 1, 2, 3, \ldots, n - 2$.

ii. $f^{[n-1]}(p)$ lies in (a, b).

iii. $f^{[n]}(p) = p$, and lies in $[b, c]$.

Then automatically p will have period n.

The proof that f has a period-n point proceeds as follows. Let $J_0 = [b, c]$. Since

$$f(J_0) = f[b, c] \supseteq [a, c] \supseteq [b, c] = J_0$$

Lemma 1.2 assures the existence of a closed interval J_1 such that $J_0 \supseteq J_1$ and $f(J_1) = [b, c] = J_0$. Next,

$$f^{[2]}(J_1) = f(J_0) \supseteq J_0$$

so by Lemma 1.2 there is a closed interval J_2 such that $J_1 \supseteq J_2$ and $f^{[2]}(J_2) = J_0$. Now

$$f^{[3]}(J_2) = f(f^{[2]}(J_2)) = f(J_0) \supseteq J_0$$

so that again by Lemma 1.2 there is a closed interval J_3 such that $J_2 \supseteq J_3$ and $f^{[3]}(J_3) = J_0$. Inductively we obtain a nested sequence of closed intervals $J_0, J_1, J_2, \ldots, J_{n-2}$ with

$$[b, c] = J_0 \supseteq J_1 \supseteq \cdots \supseteq J_{n-2} \text{ and}$$
$$f^{[k]}(J_k) = J_0 = [b, c], \text{ for } k = 1, 2, \ldots, n - 2 \qquad (1.3)$$

In particular, $f^{[n-2]}(J_{n-2}) = [b, c]$. Therefore

$$f^{[n-1]}(J_{n-2}) = f(f^{[n-2]}(J_{n-2})) = f[b, c] \supseteq [a, c] \supseteq [a, b]$$

so that by Lemma 1.2 there is a closed interval J_{n-1} such that $J_{n-2} \supseteq J_{n-1}$ and

$$f^{[n-1]}(J_{n-1}) = [a, b] \qquad (1.4)$$

Consequently by (1.3) and (1.4),

$$f^{[n]}(J_{n-1}) = f(f^{[n-1]}(J_{n-1})) = f[a, b] \supseteq [b, c] \supseteq J_{n-2} \supseteq J_{n-1}$$

It follows from Lemma 1.3 that there is a point p in J_{n-1} (and hence in $[b, c]$) that is a fixed point of $f^{[n]}$. To show that p has period n, we first observe that $f^{[k]}(p)$ is in $[b, c]$ for $k = 0, 1, 2, \ldots, n - 2$ because for each such k, p is in J_k and $[b, c] = f^{[k]}(J_k)$ by (1). We will complete the proof that p has period n by showing that $f^{[n-1]}(p)$ is in $[a, b)$. To that end, we recall from (2) that $f^{[n-1]}(J_{n-1}) = [a, b]$. Since p is in J_{n-1}, we know that $f^{[n-1]}(p)$ is in $[a, b]$. If it were true that $f^{[n-1]}(p) = b$, then

$$p = f^{[n]}(p) = f(f^{[n-1]}(p)) = f(b) = c$$

so that $f(p) = f(c) = a$. However, since $J_1 \supseteq J_{n-2} \supseteq J_{n-1}$ and p is in J_{n-1}, it follows that $f(p)$ is in $f(J_1) = [b, c]$, so that $f(p) \neq a$. This contradiction implies that $f^{[n-1]}(p) \neq b$, so that $f^{[n-1]}(p)$ is in $[a, b)$. Therefore all the first $n - 2$ iterates of p lie in $[b, c]$, the $(n - 1)$st iterate lies in $[a, b)$, and the nth iterate lies again in $[b, c]$. Consequently p really does have period n. This completes the proof in case $f(a) = b$, $f(b) = c$ and $f(c) = a$. Because the case

3-cycle window for Q_μ

FIGURE 1.39
Bifurcation diagram for the quadratic family, in the period 3 window

in which $f(a) = c, f(b) = a$, and $f(c) = b$ is entirely similar, it is left as an exercise (see Exercise 9). □

If $3.829 \leq \mu \leq 3.840$, then the quadratic function Q_μ has period-3 points, so the Li-Yorke Theorem implies that such a function has points (and hence cycles) with every possible period. Finding such cycles is another story. For example, you might try to locate a point of period 11 for, say, $\mu = 3.83$. In Section 2.3 we will be able to show that the value of μ where the 3-cycles emerge is actually $1 + 2\sqrt{2}$. (See Exercise 3 in Section 2.3.)

In the same vein, in Section 1.4 we noted that $\{2/7, 4/7, 6/7\}$ is a 3-cycle for the tent function T, so that as a direct consequence of the Li-Yorke Theorem, T has cycles of all possible periods. You might look back at the discussion of T and see whether this fact is implied by any of our results in Section 1.4.

Why does the bifurcation diagram for Q_μ in Figure 1.39 basically show only three curves in the period-3 window, since for such μ the function Q_μ has points of all possible periods? The reason is that the 3-cycle is attracting (so that the orbits of almost all other points converge to it), whereas the other cycles are repelling. Nevertheless, Li and Yorke have shown that there are uncountably many numbers in $[0, 1]$ that are not in the basin of attraction of the attracting 3-cycle.

The Li-Yorke Theorem says that if f has a period-3 point, then it has points of all other periods. But suppose that we can only show that f has, say, a period-5 point. Then must f have points of all periods? A remarkable theorem by the Russian mathematician A. N. Sharkovsky provides a complete answer. In order to present Sharkovsky's result, we need to define the **Sharkovsky ordering** of the positive integers:

$3 \dashv 5 \dashv 7 \dashv \cdots 2 \cdot 3 \dashv 2 \cdot 5 \dashv 2 \cdot 7 \dashv \cdots 2^2 \cdot 3 \dashv 2^2 \cdot 5 \dashv 2^2 \cdot 7 \dashv \cdots \dashv 2^3 \dashv 2^2 \dashv 2 \dashv 1$

odd integers   2 · (odd integers) 2^2 · (odd integers)   powers of 2

Here $m \dashv n$ signifies that m appears before n in the Sharkovsky ordering. Thus $17 \dashv 14$ (because $14 = 2 \cdot 7$) and $40 \dashv 64$ (because $40 = 2^3 \cdot 5$ and $64 = 2^6$). Since every positive integer can be written as $2^k \cdot$ (odd integer) for a suitable nonnegative integer k and a suitable odd integer, the Sharkovsky ordering is an ordering of the collection of all positive integers. Now we are ready for the theorem.

Theorem 1.21 (Sharkovsky Theorem). Let f be a continuous function defined on the interval J, and suppose that $J \supseteq f(J)$. If f has a point with period m, then f has a point with period n for all n such that $m \dashv n$.

The original proof of this theorem was long and technical. Even though accessible proofs involving concepts from graph theory have been given (see the papers by Straffin, 1978, and Ho and Morris, 1981), we must omit the proof.

By letting $m = 3$ in Sharkovsky's Theorem, we see that the Li-Yorke Theorem (as stated above) is an immediate corollary of Sharkovsky's Theorem. Moreover, from the Sharkovsky ordering we can imagine why the period-5 window visible in the bifurcation diagram for Q_μ lies to the left of the large period-3 window.

By Sharkovsky's Theorem, if a continuous function on a closed interval has a period-5 point, then it has points of all periods except possibly 3, since 5 precedes all positive integers except 3 in the Sharkovsky ordering. That such a function need not have a period-3 point is illustrated in the following example.

Example 1.7.1. Show that the function f defined in Figure 1.40 has a period-5 point but no period-3 point.

Solution. The function f in Figure 1.40 is linear on each of the intervals $[1, 2]$, $[2, 3]$, $[3, 4]$, and $[4, 5]$. It is easy to check that $\{1, 3, 4, 2, 5\}$ is a 5-cycle of f, so that 1 is a period-5 point. To verify that f has no period-3 point, let $f[m, n]$ denote the image of the closed interval $[m, n]$. We obtain

$$f[1, 2] = [3, 5], f[3, 5] = [1, 4], f[1, 4] = [2, 5]$$

so that $f^{[3]}[1, 2] = [2, 5]$. Similarly,

$$f^{[3]}[4, 5] = [1, 4]$$

Therefore $f^{[3]}$ has no fixed points on $[1, 2]$, $[2, 3]$ or $[4, 5]$, and hence f has no period-3 points in these intervals. By contrast, $f[3, 4] = [2, 4] \supseteq [3, 4]$, so

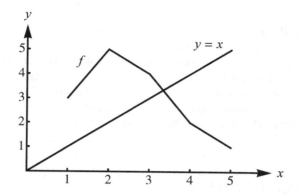

FIGURE 1.40
Piecewise linear function for Example 1.7.1

Lemma 1.3 implies that f has a fixed point p in $[3, 4]$. Next, we notice that

f is decreasing on $[3, 4]$, with image $[2, 4]$,

f is decreasing on $[2, 4]$, with image $[2, 5]$, so $f^{[2]}$ is increasing on $[3, 4]$,

f is decreasing on $[2, 5]$, with image $[1, 5]$, so $f^{[3]}$ is decreasing on $[3, 4]$.

It follows that $f^{[3]}$ can cross the line $y = x$ at most once in $[3, 4]$, and that crossing must occur at (p, p) corresponding to the fixed point p. Therefore the only fixed point of $f^{[3]}$ in the interval $[3, 4]$ is the fixed point of f. Consequently in $[3, 4]$ there is no period-3 point of f, so f has no period-3 point at all. $\square$

More generally, it can be shown that for any positive integer n, there is a continuous function with a period-n point but no period-m points for any integer m such that $m \dashv n$.

Sharkovsky's Theorem does not tell us how many period-n points a function must have for those positive integers to the right of a given integer m in the Sharkovsky ordering. The minimum number of such period-n points guaranteed to exist is known. (See, for example, the article by Bau-Sen Du, 1985.)

Note: We mention that Sharkovsky's Theorem was published in a Russian journal in 1964, and was relatively unknown to mathematicians in the western world, despite his follow-up publications in 1965 and 1968. It was not until after Li and Yorke's celebrated paper, "Period Three Implies Chaos" appeared in 1975 that his result received widespread recognition. Their major theorem included what we have stated as the Li-Yorke Theorem, as well as information about the orbits of various points. It is the information about the orbits that gave rise to the notion of "chaos," a topic we will discuss in Chapter 2.

Sharkovsky's work has been extended in more recent years. It has been proposed for random maps (Andres, 2008), non-continuous functions (Szuca 2003), and multivalued functions (Andres, Fiser, and Jutner, 2002). Sharkovsky himself reflected on his results and their implications in a publication 30 years later (Sharkovsky, 1995).

1.7.1   Section 1.7 Exercises

Exercise 1.

(a) Find an approximate value of μ such that

$$\{.149407, .488004, .959447\}$$

is (approximately) a 3-cycle for Q_μ.

(b) Show that the 3-cycle in part (a) is attracting.

Exercise 2.   Draw the graph of $f^{[3]}$ for the function f in Example 1.2.1.

Exercise 3.   Suppose that f is defined on the interval $[1, 7]$, passes through the points $(1, 4)$, $(2, 7)$, $(3, 6)$, $(4, 5)$, $(5, 3)$, $(6, 2)$, and $(7, 1)$ and is linear in between. Show that f has a period-7 point but no period-5 point.

Exercise 4.

(a) Define a function f on the interval $[1, 3]$ that is linear on $[1, 2]$ and on $[2, 3]$, and such that the point 1 has period 3.

(b) Find the numerical value of a period-2 point for your function f.

Exercise 5.

(a) Define a function that has a 1-cycle, 2-cycle and 3-cycle, but does not have any n-cycles for $n \geq 4$.

(b) Can a function such as that described in (a) be continuous? Explain why or why not.

Exercise 6.   Suppose that $|f'(x)| < 1$ for all x. Show that f cannot have any periodic points other than a unique fixed point.

Exercise 7.   Use the Intermediate Value Theorem to prove that every real number has a cube root.

Exercise 8. Use the Intermediate Value Theorem to prove that every real number is the tangent of a number in the interval $(-\pi/2, \pi/2)$.

Exercise 9. Prove the case in which $f(a) = c, f(b) = a$, and $f(c) = b$ in the Li-Yorke Theorem.

Exercise 10. Assume that f is continuously differentiable on the closed interval $[a, b]$ and let $f'(a) < A < f'(b)$. Prove that there exists a number c in the open interval (a, b) such that $f'(c) = A$. (*Hint*: Let $g(x) = f(x) - Ax$ for $a \leq x \leq b$, and use the Intermediate value Theorem.) This result is called **Darboux's Theorem** and is a kind of Intermediate Value Theorem for the derivative of a differentiable function.

Exercise 11. Use the computer program ITERATE to show that if $\mu = 1 + 2\sqrt{2}$, then Q_μ has a period-3 point.

1.8 The Schwarzian Derivative

The question we address in this section is the following: How many attracting cycles can a differentiable function have? To understand why the question might be relevant, we need only look at the window in the bifurcation diagram for $\{Q_\mu\}$ that appears in Figure 1.41. For a given μ in the interval, the six horizontal curves that dominate the window could, in theory, represent an attracting 6-cycle, two attracting 3-cycles, three attracting 2-cycles, or even six attracting fixed points. We will be able to resolve this issue by Theorem 1.23.

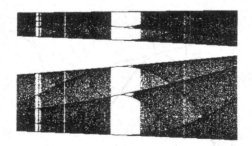

FIGURE 1.41
Window in the bifurcation diagram of Q_μ

Recall that if f is a differentiable function defined on an interval J, then x is a critical point in the interior of J if $f'(x) = 0$. The goal of this section is to prove a wonderful theorem by the American mathematician David Singer (1978) to the effect that under certain conditions, if f has n critical points, then it has at most $n + 2$ attracting cycles. The proof of this result is long (and entails seven lemmas). The hypothesis that is crucial involves what is called the "Schwarzian derivative."

Definition 1.22. Let f be defined on the interval J, and assume that the third derivative f''' is continuous on J. Define Sf by

$$(Sf)(x) = \frac{f'''(x)}{f'(x)} - \frac{3}{2} \left(\frac{f''(x)}{f'(x)} \right)^2$$

Then $(Sf)(x)$ is the **Schwarzian derivative** of f at x whenever it exists as a number or as $-\infty$ or ∞.

The Schwarzian derivative is named for the German mathematician Hermann Schwarz, who in 1869 defined it and used it in the study of complex-valued functions. For our purposes, the Schwarzian derivative has two important features:

i. Composites of functions with negative Schwarzian derivatives also have negative Schwarzian derivatives (Lemma 1.4).

ii. If a function f has a negative Schwarzian derivative, together with enough fixed points, then f has a critical point (Lemma 1.3).

To illustrate (ii), consider a continuously differentiable function f having four isolated fixed points $a, b, c,$ and d with $a < b < c < d$ (Figure 1.42). Notice that $f'(x) > 1$ and $f'(y) < 1$ for appropriate values of x and y in $[a, d]$.

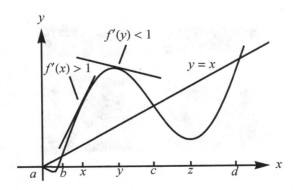

FIGURE 1.42
Example with 4 fixed points

If f has a negative Schwarzian derivative on $[a, d]$, then it turns out that there must be a number z in $[a, d]$ such that $f'(z) = 0$, that is, f has a critical point.

One can show that any polynomial the zeros of whose derivative are real and distinct has a negative Schwarzian derivative. (For the general result, see Devaney, 1989; for third degree polynomials, see Exercise 21 of this section.) We will show, more modestly, that the quadratic function Q_μ has a negative Schwarzian derivative for all values of μ.

Note: The work of Schwarz beyond what we need here is worth investigating. If you intend to study continuous dynamical systems, the notion of existence and uniqueness is vital, and Picard's work was based on the work of Schwarz. Generalizations of the Cauchy–Schwarz inequality end up being quite useful in the study of non-linear dynamical systems in control theory. So while we will not use many of Schwarz's result beyond Section 1.8, his work laid the foundation for some seminal results in dynamics, particularly for complex dynamics.

Example 1.8.1. Let $Q_\mu(x) = \mu x(1-x)$. Show that $SQ_\mu(x) < 0$ for $0 < x < 1$.

Solution. Notice that $Q'_\mu(x) = \mu(1 - 2x)$, $Q''_\mu(x) = -2\mu$, and $Q'''_\mu(x) = 0$. Therefore if $x \neq 1/2$, then by the definition of SQ_μ,

$$(SQ_\mu)(x) = -\frac{3}{2}\left(\frac{-2\mu}{\mu(1 - 2x)}\right)^2 = \frac{-6}{(1 - 2x)^2} < 0$$

Finally, $\lim_{x \to 1/2}(Sf)(x) = -\infty$, so we write $(Sf)(1/2) = -\infty$. $\square$

Having defined the Schwarzian derivative, we are ready to direct our attention toward Singer's Theorem. Throughout we will assume that the third derivative of each function under discussion exists and is continuous. The goal of the first three lemmas is to show that any function with negative Schwarzian derivative and four isolated fixed points has a critical point. We will write "$Sg < 0$" for the more complete "$Sg(x) < 0$ for all x in the domain of g."

Lemma 1.1. Let $Sg < 0$. If g' has a relative minimum value at x^*, then $g'(x^*) < 0$.

Proof. Suppose that g' has a relative minimum value at x^*. Then $g''(x^*) = 0$, so that

$$\frac{g'''(x^*)}{g'(x^*)} = (Sg)(x^*) < 0$$

However, since $g'(x^*)$ is a relative minimum value, the Second Derivative Test from calculus tells us that $g'''(x^*) \geq 0$. Consequently $g'(x^*) < 0$. $\square$

The same reasoning as we used in Lemma 1.1 implies that if $Sg < 0$, then any relative maximum value of g' must be positive. Lemma 1.1 and the

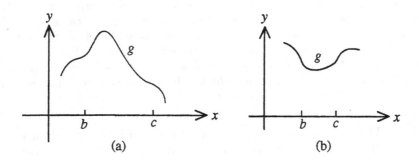

FIGURE 1.43
Graphs which illustrate the Schwarzian derivative

preceding comment imply that if $Sg < 0$, then the graph of g cannot appear as in Figure 1.43(a) because g' has a relative minimum value at b that is positive (rather than negative), and also has a relative maximum value at c that is negative (rather than positive). By contrast, the graph appearing in Figure 1.43(b) is allowed because g' has a negative relative minimum value at b and a positive relative maximum value at c.

Before we turn to Lemmas 1.2 and 1.3, we remark that if $Sg < 0$ on an open interval, then there must be an x in the interval such that $g(x) \neq x$.

Lemma 1.2. Let a, b, and c be fixed points of g, with $a < b < c$. Assume also that $Sg < 0$ on (a, c). If $g'(b) \leq 1$, then g has a critical point in (a, c).

Proof. Since $g(a) = a, g(b) = b$, and $g(c) = c$, it follows from the Mean Value Theorem that there exist an r in (a, b) and s in (b, c) such that $g'(r) = 1 = g'(s)$ (Figure 1.44). Since $g'(r) = 1 = g'(s)$ and $g'(b) \leq 1$, and since g' is

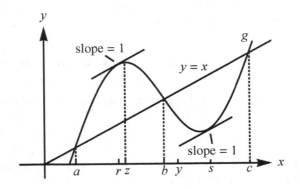

FIGURE 1.44
Illustration of Lemma 1.2

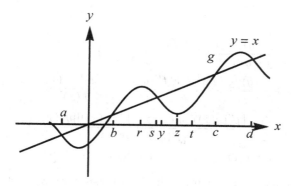

FIGURE 1.45
Illustration of Lemma 1.3

continuous on $[r, s]$, the Maximum-Minimum Theorem implies that g' has a minimum value $g'(y)$ on $[r, s]$. Since $Sg < 0$ by hypothesis, g' cannot be constant on $[r, s]$, so $g'(y) < 1$. It then follows from Lemma 1.1 that $g'(y) < 0$. Since $g'(s) = 1$ and $g'(y) < 0$, the Intermediate Value Theorem guarantees a z in (r, s) such that $g'(z) = 0$. This z is the critical point of g that we seek. $\square$

Lemma 1.3. Suppose that g has fixed points a, b, c, and d, with $a < b < c < d$. Assume also that $Sg < 0$ on $[a, d]$. Then g has a critical point in (a, d).

Proof. If $g'(b) \leq 1$, then Lemma 1.2 implies that g has a critical point in (a, c). Similarly, if $g'(c) \leq 1$, then g has a critical point in (b, d). So let us assume that $g'(b) > 1$ and $g'(c) > 1$ (Figure 1.45). Then there are r and t such that $b < r < t < c$ and such that $g(r) > r$ and $g(t) < t$ (Figure 1.45). Hence by the Mean Value Theorem there is an s in (r, t) such that $g'(s) < 1$. Since g' is continuous on $[b, c]$, g' must have a relative minimum value $g'(y)$ on (b, c). Lemma 1.1 implies that $g'(y) < 0$. Then the Intermediate Value Theorem yields a z in (y, c) such that $g'(z) = 0$. Thus z is a critical point of g. $\square$

In the proof we assumed that $g'(b) > 1$ and $g'(c) > 1$. Since b and c were fixed points, it followed that b and c were not adjacent fixed points of g.

With Lemma 1.3 we have shown that a negative Schwarzian derivative and the existence of four fixed points together imply the existence of a critical point. The next lemma analyzes the Schwarzian derivative of a composite of functions.

Lemma 1.4. Suppose that $Sf < 0$ and $Sg < 0$. Then $S(f \circ g) < 0$.

Proof. First, we use the Chain Rule to calculate that

$(f \circ g)'(x) = [f'(g(x))][g'(x)]$

$(f \circ g)''(x) = [f''(g(x))][g'(x)]^2 + [f'(g(x))][g''(x)]$

$(f \circ g)'''(x) = [f'''(g(x))][g'(x)]^3 + 3[f''(g(x))][g'(x)][g''(x)] + [f'(g(x))][g'''(x)]$

Then

$$[S(f \circ g)](x) = \frac{(f \circ g)'''(x)}{(f \circ g)'(x)} - \frac{3}{2} \left(\frac{(f \circ g)''(x)}{(f \circ g)'(x)} \right)^2$$

$$= \frac{[f'''(g)][g'(x)]^3 + 3[f''(g(x))][g'(x)][g''(x)] + [f'(g(x))][g'''(x)]}{[f'(g(x))][g'(x)]}$$

$$- \frac{3}{2} \left(\frac{[f''(g(x))][g'(x)]^2 + [f'(g(x))][g''(x)]}{[f'(g(x))][g'(x)]} \right)^2$$

$$= [(Sf)(g(x))][g'(x)]^2 + (Sg)(x) < 0$$

the last inequality following from the hypothesis that $Sf < 0$ and $Sg < 0$. This completes the proof.   □

Lemma 1.5. Suppose that $Sf < 0$. Then $Sf^{[n]} < 0$ for any positive integer n.

Proof. We will use induction. We know that $Sf^{[2]} = S(f \circ f) < 0$ by letting $g = f$ in Lemma 1.4. Therefore the result is valid for $n = 2$. Next, assume that $Sf^{[n]} < 0$ for a given positive integer n. Then $Sf^{[n+1]} = S(f \circ f^{[n]}) < 0$ by letting $g = f^{[n]}$ in Lemma 1.4. By the Law of Induction, $Sf^{[n]} < 0$ for each positive integer n.   □

Since period-n points of f are fixed points of $f^{[n]}$, Lemma 1.5 implies that arguments involving periodic points of f can reduce to arguments about fixed points of $f^{[n]}$.

Lemmas 1.6 and 1.7 assure us that if f has a finite number of critical points, then f has a finite number of period-m points, for each positive integer m.

Lemma 1.6. Let f be differentiable, and suppose that f has a finite number of critical points. Then $f^{[m]}$ has a finite number of critical points for each $m > 1$.

Proof. To start the proof, suppose that $x < y$ and $f(x) = f(y)$. Then the Mean Value Theorem implies that there is a z in the interval (x, y) such that $f'(z) = 0$. This means that z is a critical point of f. Since f has only a finite number of critical points by hypothesis, it follows that the collection of points x such that x or $f(x)$ is a critical point of f is a finite set of points. We will use this fact later in the proof.

Our proof proceeds by induction. Let x be a critical point of $f^{[2]}$. Then

$$0 = (f^{[2]})'(x) = [f'(f(x))][f'(x)]$$

so that x or $f(x)$ is a critical point of f. Since there are only a finite number of such points by our preceding comment, we deduce that $f^{[2]}$ has but a finite set of critical points. Thus the proof is complete for $m = 2$. Now we let $m > 2$, and assume that x is critical point of $f^{[m]}$. In this case,

$$0 = (f^{[m]})'(x) = [f'(f^{[m-1]}(x))][f'(f^{[m-2]}(x))] \cdots [f'(f(x))][f'(x)]$$

so that $x, f(x), \ldots,$ or $f^{[m-1]}(x)$ is a critical point of f. Using the same reasoning as for $m = 2$, we conclude that $f^{[m]}$ can have only a finite number of critical points. □

Lemma 1.7. Let f have a finite number of critical points, and assume that $Sf < 0$. Then for any positive integer m there is a finite number of period-m points of f.

Proof. Let $g = f^{[m]}$. Then $Sg < 0$ by Lemma 1.5, and g has finitely many critical points by Lemma 1.6. If g were to have an increasing (or decreasing) infinite sequence of fixed points $\{p_n\}$, then Lemma 1.3 tells us that there would be a critical point between p_1 and p_4, between p_5 and p_8, etc. Thus g would have infinitely many critical points. This contradiction proves that g has but a finite number of fixed points. Since $g = f^{[m]}$, this is tantamount to f having only finitely many points of period m. □

We are prepared to state and prove Singer's Theorem. For convenience, when we refer to cycles, we will include fixed points.

Theorem 1.23 (Singer's Theorem). Let f be defined on a closed interval J, and suppose that $J \supseteq f(J)$. Assume that $Sf < 0$, and that f has n critical points. Then f has at most $n + 2$ attracting cycles.

Proof. Let $J = [A, B]$, where $-\infty \le A$ and $B \le \infty$. Suppose that p is an attracting period-m point of f. This means that if $g = f^{[m]}$, then p is an attracting fixed point of g. We will focus on g. Let (L, R) be the largest open interval about p all of whose points are attracted to p. (We allow the possibility that $L = -\infty$ or $R = \infty$ or both.) If L and R are both interior to J, so that L and R are finite, then since g is continuous and the interval (L, R) is maximal, it follows that $g(L) = L$ and $g(R) = R$, or $g(L) = R$ and $g(R) = L$, or $g(L) = g(R)$. We will show that in each case there is a critical point of g that is attracted to p, and hence a critical point of f that is attracted to the orbit of p.

Case 1. $g(L) = L$ and $g(R) = R$

Since p is an attracting fixed point of g, we know by Theorem 1.7 that $|g'(p)| \le 1$, and therefore $g'(p) \le 1$. Next, $Sg < 0$ by Lemma 1.5. If we let $L = a, p = b$, and $R = c$ in Lemma 1.2, we find that g has a critical point x^* in the interval (L, R). Now

$$0 = g'(x^*) = (f^{[m]})'(x^*) = [f'(f^{[m-1]}(x^*))][f'(f^{[m-2]}(x^*))] \cdots [f'(x^*)]$$

Therefore one of $x^*, f(x^*), f^{[2]}(x^*), \ldots, f^{[m-1]}(x^*)$ is a critical point of f. Since x^* is in (L, R) and hence is attracted (by the iterates of g) to p, all of the iterates of x^* are similarly attracted to p. Consequently f has a critical point that is attracted (by the iterates of f) to the orbit of p.

Case 2. $g(L) = R$ and $g(R) = L$

In this case we first consider $g^{[2]}$. Notice that $g^{[2]}(L) = L$ and $g^{[2]}(R) = R$, and that p is an attracting fixed point of $g^{[2]}$. By Case 1 this means that $g^{[2]}$ has a critical point x^* in the interval (L, R) that is attracted (by the iterates of g) to p. However,

$$0 = (g^{[2]})'(x^*) = [g'(g(x^*))][g'(x^*)]$$

so that x^* or $g(x^*)$ is a critical point of g. Therefore by the argument in Case 1, there is an iterate (with respect to f) of x^* that is a critical point of f. Since x^* is in (L, R) and hence is attracted (by the iterates of g) to p, all of the iterates of x^* are similarly attracted to p. Consequently f has a critical point that is attracted (by the iterates of f) to the orbit of p.

Case 3. $g(L) = g(R)$

If $g(L) = g(R)$, then by the Mean Value Theorem there is an x^* in the interval (L, R) such that $g'(x^*) = 0$, meaning that x^* is a critical point of g. Once again some iterate of x^* not only must be a critical point of f, but also must be attracted to the orbit of p.

In each of Cases 1–3, there is a critical point of f that is attracted to the orbit of p. Since a critical point can be attracted to at most one orbit, and since by hypothesis f has only n critical points, we conclude that there are at most n attracting periodic orbits that are associated with intervals of the form (L, R) that are interior to J. Orbits associated with intervals of the form (A, R) or (L, B) add a maximum of 2 more possible attracting cycles, making the maximum possible number of attracting cycles be $n + 2$.

□

Singer's Theorem yields the following corollary, which was known to the French mathematician Gaston Julia significantly earlier than Singer's result.

Corollary 1.24. Let $0 < \mu \le 4$. Each function in the quadratic family $\{Q_\mu\}$ has at most one attracting cycle.

Proof. If $0 < \mu \le 1$, then the basin of attraction of 0 is $[0, 1]$, so 0 is the only attracting periodic point. For the remainder of the proof, assume that $1 < \mu \le 4$. Since Q_μ has the unique critical point $1/2$, Singer's Theorem implies that there can be at most 3 attracting cycles, one each associated with intervals of the form $[0, L), (L, R)$, and $(R, 1]$, where $0 < L < R < 1$. Since 0 is a repelling fixed point and since $Q_\mu(1) = 0$, neither $[0, L)$ nor $(R, 1]$ appears as a basin of attraction for cycles of Q_μ. Consequently Q_μ at most one attracting cycle.   □

Corollary 1.24 implies that Q_μ can have only one attracting cycle. Therefore, the six horizontal curves in the window featured in Figure 1.41 represent

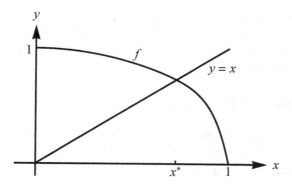

FIGURE 1.46
Attracting point illustration for Example 1.8.2

an attracting 6-cycle (rather than multiple attracting 3-cycles or 2-cycles or fixed points). Thus we have resolved the issue posed at the outset of the section with respect to functions in $\{Q_\mu\}$.

Corollary 1.24 also yields a method for detecting potential attracting cycles of a function Q_μ. The method involves calculating the iterates of the lone critical point $1/2$ of Q_μ. If the iterates appear to converge to a periodic orbit, then that orbit would be the candidate for the unique attracting periodic orbit of Q_μ. You might try this technique to search for the attracting 2-cycle of, say, the function $Q_{3.2}$. By contrast, if the iterates of $1/2$ do not seem to converge to a periodic orbit, then it may well happen that no attracting orbit exists (although it can be hard to tell because orbits can have a very large number of points). This is the case with Q_4.

Singer's Theorem indicates that a function with but one critical point and a negative Schwarzian derivative can have no more than three attracting cycles. Thus in theory there could be zero, one, two, or three attracting cycles (see Exercise 14).

By contrast, the result in Singer's Theorem need not hold if the hypothesis that $Sf < 0$ is ignored. Example 1.8.2 is devoted to a function with an attracting fixed point *and* an attracting 2-cycle.

Example 1.8.2. Let $f(x) = 1 - x^2/2 - x^{14}/2$ for $0 \leq x \leq 1$ (see Figure 1.46). Verify that f has not only an attracting fixed point but also an attracting 2-cycle in the interval $[0, 1]$. Show also that $Sf(x) > 0$ for some x in $(0, 1)$.

Solution. If $x^* \approx 0.72861$, then a routine calculation shows that x^* is an attracting fixed point for f, and that $Sf(x^*) > 0$. In addition, $\{0, 1\}$ is a 2-cycle that is attracting because $(f^{[2]})'(0) = [f'(1)][f'(0)] = 0$. $\square$

We close the discussion of attracting cycles with an observation noted by Singer. Suppose a population is in stable equilibrium that could be constant (represented by a fixed point) or could involve oscillations (hence cycles). Singer's Theorem would imply that if the stable equilibrium is, say, constant, then by changing the population size (that is, the value of x), the type of stable equilibrium cannot change.

1.8.1   Section 1.8 Exercises

Exercise 1. Let $E_\mu(x) = \mu e^x$ for all x. Show that whatever nonzero number μ is, SE_μ is a constant function. Determine that constant.

In Exercises 2–4, show that $Sf < 0$.

Exercise 2. $f(x) = \sin x$ for $0 \le x < \pi/2$

Exercise 3. $f(x) = \cos x$ for $0 < x \le \pi/2$

Exercise 4. $f(x) = x(1 + \mu(1 - x))$

In Exercises 5–6, determine whether the Schwarzian derivative of f is positive, or is negative.

Exercise 5. $f(x) = x^2$

Exercise 6. $f(x) = x^2(x - 1)$

Exercise 7. Let $f(x) = x^a$, where $a > 0$. Determine the values of a for which $Sf < 0$.

Exercise 8. Let $f_\mu(x) = x^3 + \mu x$. Show that if $\mu > 0$, then $Sf(x) > 0$ for x near to 0.

Exercise 9. Let $f(x) = (x^3/3) + x$. Determine the interval on which $Sf > 0$.

Exercise 10. Let $f_\mu(x) = x - \mu x^3$. Show that $Sf_\mu(0)$ changes sign as μ passes through 0.

Exercise 11. Let $f(x) = 7.86x - 23.31x^2 + 28.75x^3 - 13.30x^4$. (This function appeared in Singer's article.)

(a) Show that if $x^* \approx 0.7263986$, then x^* is an attracting fixed point.

(b) Show that if $y^* \approx 0.3217591$, then $\{y^*, f(y^*)\}$ is an attracting 2-cycle.

(c) How many critical points does f have? Explain your answer.

(d) Do the results of (a) and (b) contradict Singer's Theorem? Explain why or why not.

(e) Show that if $c \approx 0.3239799$, then c is a critical point.

(f) Determine whether c is attracted to the fixed point, the 2-cycle, or neither.

Exercise 12. Let $f_\mu(x) = \mu(7.86x - 23.31x^2 + 28.75x^3 - 13.30x^4)$.

(a) Show that the attracting fixed point of f_μ becomes repelling as μ increases through 1.

(b) Using a calculator or computer, draw the graphs of f_μ for $\mu = .95$, $\mu = 1$ and $\mu = 1.05$, and show that as μ increases through the interval $[.95, 1]$, an attracting 2-cycle and a repelling 2-cycle are born.

(c) Using the graphs in part (b), show that as μ increases through the interval $[1, 1.05]$, the repelling 2-cycle coalesces with the fixed point, which is repelling.

(d) Show that the family $\{f_\mu\}$ has a bifurcation point at 1.

Exercise 13. Let $f_\mu(x) = \mu \sin x$ for $0 \leq x \leq \pi$, where $0 < \mu < \pi$. Determine the maximum possible number of attracting periodic orbits of f_μ.

Exercise 14.

(a) Find a value of μ with $0 < \mu \leq 4$ such that Q_μ has no attracting cycles.

(b) Find a value of μ with $0 < \mu \leq 4$ such that Q_μ has one attracting fixed point and no other cycles.

(c) Let $g(x) = 1.2 \sin x$ on $[-p, \pi]$, where p is a fixed point of g that lies in $[0, \pi]$. Show that g has one critical point and two attracting fixed points, but no other cycles.

Exercise 15. Suppose that the differentiable function f is defined on $[0, 1]$, and that f satisfies the following conditions:

(a) $f(0) = 0 = f(1)$.

(b) $1/2$ is the only critical point of f and $f(1/2) = 1$.

(c) $Sf < 0$ on $[0, 1]$.

(d) The graph of f is concave downward on $(0, 1)$.

Determine the maximum possible number of attracting cycles of f.

Exercise 16. Suppose that Sf exists.

(a) Show that $S(af) = Sf$ for any nonzero constant a.

(b) Show that $S(f + b) = Sf$ for any constant b.

(c) Show that $S(1/f) = Sf$.

(d) Let a, b, c, and d be nonzero constants, and define the function g by

$$g = \frac{af + b}{cf + d}$$

Use parts (a)–(c) to show that $Sg = Sf$.

Exercise 17. Show that if $Sf > 0$ and $Sg > 0$, then $S(f \circ g) > 0$.

Exercise 18. Show that if $Sg < 0$, then g' cannot have a negative relative maximum value.

Exercise 19. Suppose that $Sg(x) < 0$ for all x in the interval J. Show that for each x in J, either $g''(x) \neq 0$ or $g'''(x) \neq 0$.

Exercise 20. Suppose that f has a second derivative.

(a) Find a formula for $(f^{[2]})'(x)$ in terms of x, f, and f'.

(b) Find a formula for $(f^{[2]})''(x)$ in terms of x, f, f', and f''.

(c) Suppose that $f(p) = p$ and $f'(p) = -1$. Show that $(f^{[2]})''(p) = 0$.

Exercise 21. Consider the general third-degree polynomial function defined by $f(x) = ax^3 + bx^2 + cx + d$, with $a > 0$.

(a) Let $x = z^3 - b/(3a)$. Show that $ax^3 + bx^2 + cx + d = az^3 + (c - b^2/(3a))z + r$ for an appropriate constant r. (The substitution $x = z - b/(3a)$ has been used in the process of identifying the solutions of the general cubic equation, and was first published in 1545 by the Italian mathematician Girolamo Cardano.)

(b) Let $g(z) = az^3 + (c - b^2/(3a))z + r$. Show that $g(z) = 0$ has three real roots if and only if $c < b^2/(3a)$.

(c) Let $c < b^2/(3a)$. Show that $Sg(z) < 0$ for all z.

(d) Show that $Sf(x) < 0$ for all x if and only if $Sg(z) < 0$ for all z.

Exercise 22. Let f be a degree n polynomial, with n distinct roots. Prove that the Schwarzian derivative is negative.

Exercise 23. Using the programs SCHWARZ and GRAPHICAL ANALYSIS in the appendix, investigate the orbit for $f(x) = ax^n e^{-bx}$, where $x \neq 0$, a and b are positive reals, and n is a positive integer.

(a) Show that f has a negative Schwarzian derivative

(b) Investigate different changes in the parameters a and b. Use graphical analysis to show how many critical points the function has.

2

One-Dimensional Chaos

In Chapter 1 the focus was mainly on periodic points, and more particularly, attracting periodic points. Attracting periodic points indicate a regularity, predictability, and stability in the dynamics of a function of a parametrized family of functions.

Chapter 2 is devoted to a contrasting dynamical action: points whose iterates separate from one another. This kind of behavior is symptomatic of what we call chaotic dynamics, or just plain chaos. It was only after the advent of the high-speed computer that such dynamics could be investigated and analyzed effectively.

In Section 2.1 we define the most illustrious concept in the study of chaotic dynamics: sensitive dependence on initial conditions. Sensitive dependence on initial conditions, along with the closely related notion of the Lyapunov exponent, serve as the ingredients in the definition of chaos. In Section 2.2 we turn to points whose orbits virtually fill up the whole domain space. If a function is chaotic and has this added property and enough periodic points, then it is strongly chaotic. Section 2.3 is devoted to the notion of conjugacy. If two functions are conjugate to one another, then they share many properties pertaining to chaos and strong chaos. We use conjugacy to prove the main result of the section: Q_4 is strongly chaotic. The final section concerns Q_μ for $\mu > 4$. We establish that such a Q_μ is strongly chaotic on a subset of $[0, 1]$ that is a so-called Cantor set. It is interesting that Cantor sets, which play a central role in analysis, play a like role in chaotic dynamics. Chapter 2 completes the study of chaos for functions of one variables.

2.1   Chaos

In this section we study two methods of describing the way in which iterates of neighboring points separate from one another: sensitive dependence on initial conditions and the Lyapunov exponent. These notions are fundamental to the concept of chaos, which also will appear in the present section.

DOI: 10.1201/9781032678757-2

2.1.1   Sensitive Dependence on Initial Conditions

Before defining sensitive dependence on initial conditions, we adopt a notation that henceforth will facilitate our discussion. We will write $f : A \to B$ to indicate that the domain of the function f is A and the range of f is contained in B. Thus $f : J \to J$ signifies that the domain of f is J and the range is contained in J.

Definition 2.1. Let J be an interval, and suppose that $f : J \to J$. Then f has **sensitive dependence on initial conditions at** x, or just **sensitive dependence at** x if there is an $\epsilon > 0$ such that for each $\delta > 0$, there is a y in J and a positive integer n such that

$$|x - y| < \delta \quad \text{and} \quad |f^{[n]}(x) - f^{[n]}(y)| > \epsilon$$

If f has sensitive dependence on initial conditions at each x in J, we say that f has **sensitive dependence on initial conditions on** J, or that f **has sensitive dependence on initial conditions on** J, or that f **has sensitive dependence on** J, or that f **has sensitive dependence**.

The "initial conditions" in the definition refer to the given, or initial, points x and y. The definition says in effect that a function f has sensitive dependence if arbitrarily close to any given point x in the domain of f there is a point y in the domain of f such that (at least some) iterates of x and y separate from one another. This has practical significance, because in such instances higher iterates of an approximate value of x may not resemble the true iterates of x. Thus calculator or computer calculations may be misleading.

To illustrate sensitive dependence on initial conditions, we turn to the baker's function.

Example 2.1.1. Consider the baker's function B, given by

$$B(x) = \begin{cases} 2x & \text{for } 0 \le x \le 1/2 \\ 2x - 1 & \text{for } 1/2 < x \le 1 \end{cases}$$

Show that after 10 iterates, the iterates of $1/3$ and $.333$ are farther than $1/2$ apart.

Solution. Notice that $B(1/3) = 2/3$ and $B^{[2]}(1/3) = 1/3$, so that the iterates of $1/3$ alternate between $1/3$ and $2/3$. To compare the iterates of $1/3$ and $.333$ we make the following table (where we use 3-place approximations for the iterates of $.333$):

iterates	1	2	3	4	5	6	7	8	9	10
$\frac{1}{3}$	$\frac{2}{3}$	$\frac{1}{3}$	$\frac{2}{3}$	$\frac{1}{3}$	$\frac{2}{3}$	$\frac{1}{3}$	$\frac{2}{3}$	$\frac{1}{3}$	$\frac{2}{3}$	$\frac{1}{3}$
$.333$	$.666$	$.332$	$.664$	$.328$	$.656$	$.312$	$.624$	$.248$	$.496$	$.992$

Therefore the tenth iterates of $1/3$ and $.333$ are, respectively, $1/3$ and $.992$, which are farther apart than a distance $1/2$. $\square$

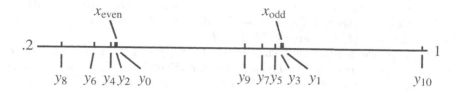

FIGURE 2.1
Iterates of points in the baker's function

Letting

$$x_n = n\text{th} \quad \text{iterate of} \quad 1/3 \quad \text{and} \quad y_n = n\text{th} \quad \text{iterate of} \quad .333$$

we display the separation of the iterates of $1/3$ and $.333$ in Figure 2.1.

The implication of Example 2.1.1 is that higher iterates of $.333$ can have little relationship to corresponding iterates of $1/3$. In fact, although we can tell precisely what the 20th or 35th iterate of $1/3$ is, there is absolutely no way of determining the 20th or 35th iterate of $.333$ without actually calculating those iterates.

Example 2.1.2. Show that the tent function T has sensitive dependence on initial conditions on $[0, 1]$.

Solution. Let x be any number in $[0, 1]$. First, we will show that if v is any dyadic rational number (that is, a number of the form $j/2^m$, with j and m integers and $j/2^m$ in lowest terms) in $[0, 1]$ and w is any irrational number in $[0, 1]$, then there is a positive integer n such that

$$|T^{[n]}(v) - T^{[n]}(w)| > \frac{1}{2}$$

Toward that goal, we recall from Section 2.4 that if $v = j/2^m$ with $1 \le j < 2^m$, then $T^{[m]}(v) = 1$ and $T^{[m+k]}(v) = 0$ for all $k > 0$. By contrast, if w is any irrational number in $[0, 1]$, then since T doubles each number in $(0, 1/2)$, there exists an $n > m$ such that $T^{[n]}(w) > 1/2$. Since $n > m$, it follows that $T^{[n]}(v) = 0$, so that (2.1) is valid. Next, let $\delta > 0$. Then there exist a dyadic rational v and an irrational number w in $[0, 1]$ such that $|x - v| < \delta$ and $|x - w| < \delta$. Therefore (2.1) implies that

$$\text{either } |T^{[n]}(x) - T^{[n]}(v)| > \frac{1}{4} \quad \text{or} \quad |T^{[n]}(x) - T^{[n]}(w)| > \frac{1}{4}$$

Thus if we let $\epsilon = 1/4$, then we have proved that T has sensitive dependence on initial conditions at the arbitrary number x, and hence on $[0, 1]$. $\square$

Basically the reason that T has sensitive dependence on initial conditions is that if $x \neq 1/2$, then $|T'(x)| = 2$, so that distances between pairs of numbers in $(0, 1/2)$ or in $(1/2, 1)$ are doubled by T.

If p is period-n point of f such that $|(f^{[n]})'(p)| < 1$, then f cannot have sensitive dependence on initial conditions at p (Exercise 5). Because this condition holds for any attracting periodic point of Q_μ whenever $0 < \mu < \mu_\infty$, it follows that Q_μ does not have sensitive dependence on initial conditions for such values of μ. However, for $\mu > \mu_\infty$, the story can be quite different. In fact, in Section 2.3 we will show that Q_4 has sensitive dependence on initial conditions.

2.1.2   Lyapunov Exponents

Although the concept of sensitive dependence on initial conditions is easy to visualize, actually determining that a given function has sensitive dependence is usually not so simple as it is for the tent function.

Before we indicate a second method of describing the separation of iterates of neighboring points, let us return to the tent function. Suppose that x is in $(0, 1)$ and is not a dyadic rational. If $\epsilon > 0$ and is small enough, then

$$|T(x + \epsilon) - T(x)| = 2\epsilon \qquad \text{and} \qquad |T^{[2]}(x + \epsilon) - T^{[2]}(x)| = 2^2 \epsilon$$

Now let n be an arbitrary positive integer. Again, if ϵ is small enough, then

$$|T^{[n]}(x + \epsilon) - T^{[n]}(x)| = 2^n \epsilon \tag{2.1}$$

Next, we divide both sides by ϵ and let ϵ approach 0. We obtain

$$|(T^{[n]})'(x)| = \lim_{\epsilon \to 0} \left| \frac{T^{[n]}(x + \epsilon) - T^{[n]}(x)}{\epsilon} \right| = 2^n \tag{2.2}$$

Thus for the tent function the derivative is related to separation of iterates of nearby points. We will now explore this idea for other functions.

Let J be a bounded interval, and consider a function $f : J \to J$ having a continuous derivative. By analogy with (2.2), we assume that for each x in the interior of J and each small enough $\epsilon > 0$, there is a number $\lambda(x)$ such that for each positive integer n,

$$|f^{[n]}(x + \epsilon) - f^{[n]}(x)| \approx [e^{\lambda(x)}]^n \, \epsilon$$

(For the tent function T, the number corresponding to $e^{\lambda(x)}$ would be 2.) This implies that

$$e^{n\lambda(x)} \approx \left| \frac{f^{[n]}(x + \epsilon) - f^{[n]}(x)}{\epsilon} \right|$$

so that

$$e^{n\lambda(x)} = \lim_{\epsilon \to 0} \left| \frac{f^{[n]}(x + \epsilon) - f^{[n]}(x)}{\epsilon} \right| = |(f^{[n]})'(x)| \tag{2.3}$$

If $(f^{[n]})'(x) \neq 0$, then by taking logarithms and dividing by n in (2.4), we obtain

$$\lambda(x) = \frac{1}{n} \ln |(f^{[n]})'(x)| \tag{2.4}$$

This leads us to make the following definition.

Definition 2.2. Let J be a bounded interval, and $f : J \to J$ continuously differentiable on J. Fix x in J, and let $\lambda(x)$ be defined by

$$\lambda(x) = \lim_{n \to \infty} \frac{1}{n} \ln |(f^{[n]})'(x)| \tag{2.5}$$

provided that the limit exists. In that case, $\lambda(x)$ is the **Lyapunov exponent of f at x**. If $\lambda(x)$ is independent of x wherever $\lambda(x)$ is defined, then the common value of $\lambda(x)$ is denoted by λ and is the **Lyapunov exponent of f**.

The definition honors the 20th century Russian mathematician A. M. Lyapunov. Although the tent function T is not continuously differentiable, the definition applies for x that are not dyadic rational. Using (2.3), we find that

$$\lim_{n \to \infty} \frac{1}{n} \ln |(T^{[n]})'(x)| = \lim_{n \to \infty} \frac{1}{n} \ln 2^n = \ln 2$$

Therefore $\lambda(x) = \ln 2$ for the tent function, whenever x is *not* a dyadic rational.

The number $\lambda(x)$ can be considered to measure "the average loss of information" of successive iterates of points near x. If y is near x, and if the iterates of x and y remain close together, then $\lambda(x)$ is apt to be negative because of the presence of the logarithm. By contrast, if the iterates separate from one another, then $\lambda(x)$ is apt to be positive. Thus the larger $\lambda(x)$ is, the greater the loss of information. For the baker's function B discussed in Example 1, we found that the iterates of $1/3$ and $.333$ separate so that the 10th iterate of $1/3$ sheds no information on the 10th iterate of $.333$. Thus for the baker's function, successive iterates of a number such as $1/3$ provide less and less information about the corresponding iterates of nearby points. For B, $\lambda = \ln 2 > 0$ (Exercise 3).

To make general calculations of $\lambda(x)$ simpler, let $x_0 = x$ and $x_k = f^{[k]}(x)$ for $k = 1, 2, \ldots$. By (2.3) in Section 2.3 and the Law of Logarithms,

$$
\begin{aligned}
\ln |(f^{[n]})'(x)| &= \ln |f'(x_{n-1})f'(x_{n-2}) \cdots f'(x_1)f'(x_0)| \\
&= \ln[|f'(x_{n-1})||f'(x_{n-2})| \cdots |f'(x_1)||f'(x_0)|] \\
&= \sum_{k=0}^{n-1} \ln |f'(x_k)|
\end{aligned}
$$

Therefore (2.6) can be rewritten as

$$\lambda(x) = \lim_{n\to\infty} \frac{1}{n} \sum_{k=0}^{n-1} \ln |f'(x_k)| \qquad (2.6)$$

We will use (2.17) in order to determine the Lyapunov exponent for quadratic functions.

Example 2.1.3. Let $Q_\mu(x) = \mu x(1-x)$, for $0 \le x \le 1$, where $1 < \mu < 3$ and $\mu \ne 2$. Show that $\lambda = \ln |2 - \mu|$.

Solution. Let x be arbitrary in $(0, 1)$. Recall from Section 1.5 that Q_μ has the fixed point $p_\mu = 1 - 1/\mu$ that attracts all points in $(0, 1)$. Therefore

$$x_k = Q_\mu^{[k]}(x) \to p_\mu$$

as k increases without bound. Since $Q'_\mu(x) = \mu - 2\mu x$, it follows that

$$Q'_\mu(x_k) \to Q'_\mu(p_\mu) = \mu - 2\mu \left(1 - \frac{1}{\mu}\right) = 2 - \mu$$

By hypothesis, $\mu \ne 2$, so that

$$\ln |Q'_\mu(x_k)| \to \ln |2 - \mu|$$

as k increases without bound. Now let $\epsilon > 0$. Because the natural logarithm is continuous and x_k approaches p_μ as k increases, there is a positive integer N such that if $k > N$, then

$$\ln |2 - \mu| - \epsilon < \ln |Q'_\mu(x_k)| < \ln |2 - \mu| + \epsilon$$

Consequently for $n > N$,

$$\frac{n-N}{n}(\ln |2-\mu| - \epsilon) = \frac{1}{n}\sum_{k=N+1}^{n}(\ln |2-\mu| - \epsilon) < \frac{1}{n}\sum_{k=N+1}^{n} \ln |Q'_\mu(x_k)|$$

$$< \frac{1}{n}\sum_{k=N+1}^{n}(\ln |2-\mu| + \epsilon) = \frac{n-N}{n}(\ln |2-\mu| + \epsilon)$$

If n is sufficiently large and much larger than N, then $(n - N)/n \approx 1$, so that the preceding inequalities reduce to

$$\ln |2-\mu| - \epsilon < \frac{1}{n}\sum_{k=N+1}^{n} \ln |Q'_\mu(x_k)| < \ln |2-\mu| + \epsilon$$

Moreover, for large enough n, we have

$$\left| \frac{1}{n}\sum_{k=0}^{N} \ln |Q'_\mu(x_k)| \right| < \epsilon$$

Therefore

$$\lim_{n \to \infty} \frac{1}{n} \sum_{k=0}^{n} \ln |Q'_\mu(x_k)| = \lim_{n \to \infty} \left(\frac{1}{n} \sum_{k=0}^{N} \ln |Q'_\mu(x_k)| + \frac{1}{n} \sum_{k=N+1}^{n} \ln |Q'_\mu(x_k)| \right)$$

$$= 0 + \ln |2 - \mu| = \ln |2 - \mu|$$

Since the final expression is independent of the number x in $(0, 1)$, we conclude that $\lambda = \ln |2 - \mu|$. This completes the solution. $\square$

For Q_μ, because $\ln |2 - \mu| < 0$ whenever $1 < \mu < 2$ or $2 < \mu < 3$, it follows that $\lambda < 0$ for such values of μ. As μ approaches 2, $\ln |2 - \mu|$ approaches $-\infty$. Consequently if we consider $-\infty$ as a legitimate value of λ, then we can conclude that $\lambda < 0$ for all μ in the interval $(1, 3)$.

Using the same kind of analysis as in the solution of Example 2.1.3, one can show that if $3 < \mu < 1 + \sqrt{6}$ and if x is not eventually periodic, then $\lambda(x) < 0$ (Exercise 7). It turns out that if $3 < \mu < \mu_\infty$, then $\lambda(x) < 0$ for all x that are not eventually periodic. However, as μ increases toward 4, λ oscillates more and more wildly between positive and negative values, as Figure 2.2 shows. Thus λ seems to have more positive values as μ increases, and Q_μ is increasingly sensitive to initial conditions. Finally, for $\mu = 4$ it is known that the Lyapunov exponent $\lambda(x) = \ln 2$ whenever $0 < x < 1$ and x is not eventually periodic.

The formula in (2.17) condenses considerably when x is eventually periodic. Indeed, if an iterate of x for a function f is eventually the fixed point p, then

$$\lambda(x) = \ln |f'(p)|$$

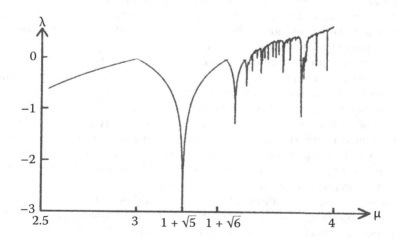

FIGURE 2.2
Lyapunov exponents for the quadratic family

(see Exercise 6). Similarly, if the iterates of x eventually join the 2-cycle $\{q, r\}$, then

$$\lambda(x) = \frac{1}{2} \ln |f'(q)f'(r)|$$

2.1.3 Chaos

The word "chaos" is familiar in everyday speech. It normally means a lack of order or predictability. Thus one says that the weather is chaotic, or that rising particles of smoke are chaotic, or that the stock market is chaotic. It is the lack of predictability that lies behind the mathematical notion of chaos. Both sensitive dependence on initial conditions and the Lyapunov exponent qualify as measures of unpredictability. Thus we have the following definition of chaos.

Definition 2.3. Let f be defined on a bounded interval. Then f is **chaotic** if it satisfies at least one of the following conditions:

1. f has a positive Lyapunov exponent at each point in its domain that is not eventually periodic

2. f has sensitive dependence on initial conditions on its domain.

> **Note:** The term "chaos" in reference to functions was first used in Li and Yorke's paper "Period Three Implies Chaos" (1975). An essential portion of their theorem that the existence of period-3 points implies the existence of points with periods of all orders was that in such a case the function had a kind of sensitive dependence on initial conditions.
>
> From Example 2.1.2 we see that the tent function T is chaotic. Similarly the baker's function B in Example 2.1.1 is chaotic. In Section 2.3 we will prove that because T is chaotic, so is Q_4.
>
> Notions of unpredictability in mathematics have been acknowledged for a long time. A century ago one of the outstanding, unsolved problems was the following: Is the solar system stable? The problem was very well known, and even excited King Oscar II of Sweden, who in 1887 offered a prize of 2500 crowns to anyone who might correctly resolve it. At the turn of the century, the great French mathematician Henri Poincaré tried to answer the simpler 3-body problem: Can one characterize the motion of a sun and two planets revolving about it? He concluded not only that there was no general solution to the 3-body problem, but also that minute differences in initial conditions for the three bodies could result in wildly divergent positions after a period of time. In other words, he concluded that the three bodies were sensitive to initial conditions. He wrote:
>
> "If we knew exactly the laws of nature and the situation of the universe at the initial moment, we could predict exactly the situation of that

same universe at a succeeding moment. But even if it were the case that the natural laws had no longer any secret for us, we could still know the situation approximately. If that enabled us to predict the succeeding situation with the same approximation, that is all we require, and we should say that the phenomenon had been predicted, that it is governed by the laws. But it is not always so; it may happen that small differences in the initial conditions produce very great ones in the final phenomena. A small error in the former will produce an enormous error in the latter. Prediction becomes impossible ..." (p. 321 of Gleick, 1987).

More recently, sensitive dependence on initial conditions, as well as the related positive Lyapunov exponent, have been observed in a wide variety of experiments. Below we discuss two such areas: weather prediction and the asteroid belt.

2.1.4 The Butterfly Effect

About sixty years ago the American meteorologist Edward Lorenz published a paper "Deterministic Turbulent Flow," in which he concluded from computer simulations that weather patterns are sensitive to initial conditions (see Lorenz, 1963). At that time the idea that ever so slight a variation in initial conditions could have profound repercussions on future weather was a startling discovery. It gave rise to the phrase "butterfly effect," which Lorenz said in reference to the apparent fact that the weather is so sensitive to initial conditions that the mere flapping of butterfly wings in Rio de Janeiro at one instant could (in theory) bring on a tornado in Texas several weeks later. An example to illustrate the sensitivity to initial conditions in weather prediction appeared in an expository article by Tim Palmer (1989).

Examples of forecasts with nearly identical initial conditions were given in the article. Two of them are reproduced as the upper pictures in Figure 2.3. Notice that those pictures are nearly identical, reflecting virtually identical initial conditions for weather prediction. However, one week later the computer model shows weather patterns that are dramatically different, as shown in the lower pictures in Figure 2.3. This is a vivid example of sensitive dependence on initial conditions! From it we can understand why forecasters would have a difficult task in long-range weather prediction. In fact, it appears that we may never be able to predict weather far into the future. We will return to Lorenz's example relating to weather prediction in Section 4.4.

2.1.5 The Asteroid Belt

A quite different application of chaotic dynamics concerns the asteroid belt, which consists of several thousand objects that orbit around the sun and lie

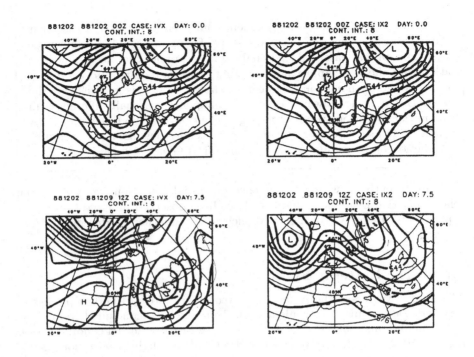

FIGURE 2.3
The butterfly effect in weather systems

between Mars and Jupiter. The largest of the asteroids is Ceres, approximately 389 kilometers in radius. When the Belters eventually get a human settlement on Ceres, a year will be 467 days. Most of the asteroids are much smaller, on the order of a kilometer in radius. By now several thousand asteroids have been cataloged.

Usually distances of such objects from the sun are given in terms of an **astronomical unit**, which by definition is the distance between the earth and the sun. Figure 2.4 is taken from an expository article of Carl Murray (1989), and shows the relative number of asteroids at various distances from the sun.

As you can see from the figure, the asteroid belt is not uniformly distributed. In 1867 the American astronomer, Daniel Kirkwood, observed gaps in the belt, now called **Kirkwood gaps**. In addition, he noted that those gaps correspond to asteroid orbits that are linked to the orbit of Jupiter. We say that an asteroid is in **resonance** with Jupiter if the ratio of the period of the asteroid to that of Jupiter is a simple fraction (such as 2:1 or 5:2 or 3:1). For example, if the ratio is 2:1, then the asteroid makes two revolutions around the sun every time Jupiter makes one (Figures 2.5(a)–(c)). When the sun,

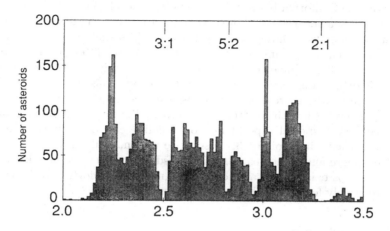

FIGURE 2.4
Average distance from the sun in astronomical units

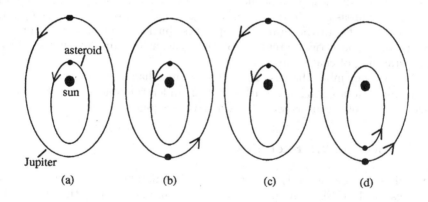

FIGURE 2.5
Resonance between asteroids and Jupiter

asteroid and Jupiter are aligned, then they are said to be in **conjunction**. Figures 2.5(a)–(d) show four possible conjunctions. It turns out that there is a maximum effect on the orbit of an asteroid if conjunction occurs when the asteroid is closest to Jupiter and simultaneously farthest from the sun, as is illustrated in Figure 2.5(d). The reason is that such an alignment gives a maximum exchange of angular momentum between the asteroid and Jupiter, resulting in potential changes in the orbit, as well as the spin, of the asteroid.

For over a hundred years astronomers tried to understand why there were gaps in the asteroid belt. Finally in 1981 Jack Wisdom, then a graduate student at the California Institute of Technology, developed and demonstrated on a computer a theory that showed that asteroids moving in resonance with Jupiter could undergo large changes in their orbits that made them unpredictable and sensitive to initial conditions (see Wisdom (1987)). In particular, asteroids in a 3:1 resonance with Jupiter could receive such perturbations that their orbits could even cross the earth's orbit. This not only addressed the question of why there is a virtual absence of asteroids at resonances such as 3:1, 5:2, and 2:1 with Jupiter, as shown in Figure 2.4. This also finally explained how a special type of meteorite called chondrite, known to have come from the vicinity of the 3:1 resonance region, could have plunged into our atmosphere and struck earth. With Wisdom's theory, we now know why Kirkwood gaps exist in the asteroid belt. And it is comforting to know that only a relatively few asteroids remain in those gaps.

2.1.6   Conclusion

Chaotic motion is so common in natural and scientific phenomena that many scientists say that it should be considered to be the rule, rather than the exception, in the study of natural phenomena. Chaotic motion has been observed in such diverse areas as fluid dynamics, ecology, meteorology, optics, the dynamics of the heart and the brain, astrophysics, buckling beams, oceanography, and nonlinear electrical circuits. For this reason it is important to gain an understanding of chaotic motion.

Always remember the wise Dr. Ian Malcolm, that the island would behave in an unpredictable manor. With a better understanding of chaotic motion, perhaps the velociraptors would never have escaped.

2.1.7   Section 2.1 Exercises

Exercise 1. Let $x = 1/3$, and let $y =$ the number that your calculator displays for $1/3$. Find the minimum positive integer n such that $|B^{[n]}(x) - B^{[n]}(y)| > 1/2$.

Exercise 2. Show that the baker's function B has sensitive dependence.

Exercise 3.

(a) Find the Lyapunov exponent associated with the baker's function B, and show that it is constant (where defined).

(b) Find the Lyapunov exponent of $B^{[2]}$.

Exercise 4. Let $B_\mu(x) = \begin{cases} 2\mu x & \text{for } 0 \le x \le 1/2 \\ \mu(2x - 1) & \text{for } 1/2 < x \le 1 \end{cases}$, where $0 < \mu < 1$.

(a) Determine the values of μ for which B_μ has sensitive dependence on initial conditions.

(b) Determine the Lyapunov exponent of B_μ.

Exercise 5. Suppose that p is a period-n point of f such that $|(f^{[n]})'(p)| < 1$. Show that f cannot have sensitive dependence on initial conditions at p.

Exercise 6. Suppose that f is differentiable on the interval J.

(a) Suppose that the iterates of x are eventually the fixed point p. Show that $\lambda(x) = \ln|f'(p)|$.

(b) Suppose that the iterates of x are eventually the cycle $\{p, q, r\}$. Find a formula for $\lambda(x)$.

Exercise 7. Consider the quadratic function Q_μ, with $3 < \mu < 1 + \sqrt{6}$.

(a) Let x be in $(0, 1)$ but not eventually fixed. Show that if $\lambda(x)$ is finite, then

$$\lambda(x) = \frac{1}{2} \ln |\mu^2 - 2\mu - 4| < 0$$

(b) Suppose that x is eventually fixed. Find $\lambda(x)$.

Exercise 8.

(a) Find the value of $Q_4^{[n]}(1/2)$ for $n \ge 2$.

(b) Use the computer program ITERATE, with any necessary alterations, to find a value of y and a positive integer n such that

$$\left| \frac{1}{2} - y \right| \le \frac{1}{100} \quad \text{and} \quad |Q_4^{[n]}(1/2) - Q_4^{[n]}(y)| > \frac{1}{2}$$

(This should convince you that Q_4 has sensitive dependence at $1/2$.)

2.2 Transitivity and Strong Chaos

The iterates of a fixed point do not wander at all; they remain the same point. At the other end of the spectrum are points whose iterates wander all over the domain of the function. Functions with such points are called "transitive."

Definition 2.4. Suppose that J is an interval and $f : J \to J$. Then f is **transitive** if for any pair of nonempty open intervals U and V that lie inside J, there is a positive integer n such that $f^{[n]}(U)$ and V have a common element.

It seems unlikely that we would be able to prove transitivity of a given function directly from the definition of transitivity. We need a manageable criterion for transitivity. The criterion we will present utilizes the notion of density of a subset of J.

A subset A of the interval J is **dense** in J if A intersects every nonempty open subinterval of J. Since every nonempty open interval of reals contains rational numbers, it follows that the collection of rationals in the interval $[0, 1]$ is dense in $[0, 1]$ (Exercise 3). By contrast, the interval $[0, 1/2]$ is not dense in the interval $[0, 1]$ because $[0, 1/2]$ does not intersect open intervals like $(1/2, 1)$.

Example 2.2.1. Show that the set P of periodic points of the tent function T is dense in $[0, 1]$.

Solution. Let U be an open subinterval in $[0, 1]$, with $U = (a, b)$. Let $d = b - a$, and let n be an odd integer such that $n > 2/d$. Since

$$\frac{k}{n} - \frac{k-1}{n} = \frac{1}{n} < \frac{d}{2}$$

and since U has length d, it follows that two successive numbers in the group $1/n, 2/n, \ldots, (n-1)/n$ lie in U. Of the two successive numbers, one of them must have the form (even integer)/(odd integer). By Theorem 1.15 such a number is periodic for T, so is in P. Therefore P is dense in $[0, 1]$. $\square$

The definition of density leads to a criterion for transitivity.

Theorem 2.5. Suppose that J is a closed interval and $f : J \to J$. Then f is transitive if and only if there is an x in J whose orbit is dense in J.

Proof. We will prove that if there is a dense orbit, then f is transitive; the other half of the proof utilizes advanced mathematics and is omitted.

Suppose that x has a dense orbit in J, and let U and V be arbitrary nonempty open subsets of J. Because of the density of the orbit of x, there is a positive integer k such that $y = f^{[k]}(x)$ is in U. Using the density of the orbit of x a second time, we find a positive integer n such that $f^{[n]}(y) = f^{[k+n]}(x)$ is in V. Then y is in U and $f^{[n]}(y)$ is in V, which by the definition of transitivity implies that f is transitive. $\square$

Our next goal is to show that the tent function T is transitive. Even with the help of the criterion in Theorem 2.5, we have much work to accomplish before we can prove the transitivity of T in Theorem 2.9.

First, let

$$S = \text{the collection of all sequences of } 0\text{'s and } 1\text{'s}$$

Next, let us call a sequence of the form $x_1 x_2 x_3 \cdots \overline{0} \cdots$, in which all terms to the right of a given term are 0, a **finite sequence**. Let

$A \quad = \quad$ the collection of all sequences of 0's and 1's that are not finite sequences

and define the function $h : [0, 1] \rightarrow A$ by

$$h(x) = \text{the sequence } x_0 x_1 x_2 \cdots, \text{ where } x_n = \left\{ \begin{array}{ll} 0 & \text{if } 0 \leq T^{[n]}(x) \leq 1/2 \\ \\ 1 & \text{if } 1/2 < T^{[n]}(x) \leq 1 \end{array} \right.$$

We will show that h defines a correspondence between the set of numbers in $[0, 1]$ that are not dyadic rationals and the set A. (Note that $h(x)$ is a finite sequence if and only if x is a dyadic rational.)

The terms $x_0, x_1, x_2, \ldots$ of $h(x)$ indicate where in $[0, 1]$ the point x lies. In particular,

x_0 tells whether x lies in the left or right half of $[0, 1]$. Call the correct half J_0.

x_1 tells whether x lies in the left or right half of J_0. Call the correct half J_1.

x_2 tells whether x lies in the left or right half of J_1. Call the correct half J_2.

The sets $J_3, J_4, \ldots$ are defined analogously. It follows that $x_0 x_1$ indicates in which fourth of the interval $[0, 1]$ the point x lies, $x_0 x_1 x_2$ indicates in which eighth of the interval $[0, 1]$ the point lies, and so forth. In general, $x_0 x_1 x_2 \cdots x_n$ indicates in which subinterval of $[0, 1]$ of length $1/2^{n+1}$ the number x is located. Figure 2.6 gives the initial terms of the sequences with appropriate subintervals of $[0, 1]$.

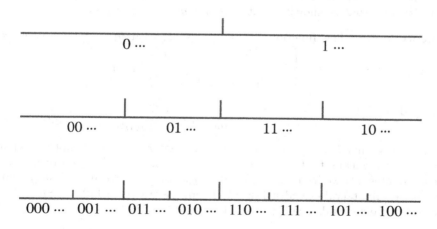

FIGURE 2.6
Initial terms of the sequence for Theorem 2.5

Example 2.2.2. Determine $h(3/11)$.

Solution. Let $x = 3/11$. We make the following table of values of $T^{[n]}(3/11)$ and corresponding x_n:

n	0	1	2	3	4	5	6
value of $T^{[n]}(\dfrac{3}{11})$	$\dfrac{3}{11}$	$\dfrac{6}{11}$	$\dfrac{10}{11}$	$\dfrac{2}{11}$	$\dfrac{4}{11}$	$\dfrac{8}{11}$	$\dfrac{6}{11}$
x_n	0	1	1	0	0	1	1

Since $T^{[6]}(3/11) = T(3/11)$, it follows that $3/11$ is eventually repeating, so that

$$h(\frac{3}{11}) = 011001\overline{11001}\cdots$$

where the digits with an "overbar" indicate the group of digits that is repeated in the sequence $h(3/11)$. $\square$

The sequence $h(3/11)$ is called a **repeating sequence** because of the group of digits 11001 that is repeated. By Theorem 1.15 every rational number x in the interval $[0, 1]$ is eventually periodic for T, which means that $h(x)$ is a repeating sequence the way $h(3/11)$ is.

By the definition of h,

$$\text{if} \quad h(x) = x_0 x_1 x_2 \cdots, \quad \text{then} \quad h(T(x)) = x_1 x_2 x_3 \cdots \qquad (2.7)$$

By induction we deduce that for any positive integer,

$$\text{if} \quad h(x) = x_0 x_1 x_2 \cdots, \quad \text{then} \quad h(T^{[n]}(x)) = x_n x_{n+1} x_{n+2} \cdots \qquad (2.8)$$

The next step in proving that T is transitive involves showing that h is **one-to-one**, that is, showing that if $h(x) = h(y)$, then $x = y$.

Theorem 2.6. The function $h : [0,1] \to A$, with $h(x) = $ the sequence $x_0 x_1 x_2 \cdots$, where $x_n = \begin{cases} 0 & \text{if } 0 \le T^{[n]}(x) \le 1/2 \\ 1 & \text{if } 1/2 < T^{[n]}(x) \le 1 \end{cases}$ is one-to-one.

Proof. Let

$$h(x) = x_0 x_1 x_2 \cdots \quad \text{and} \quad h(y) = y_0 y_1 y_2 \cdots$$

and suppose that $h(x) = h(y)$. We will prove that $x = y$. The hypothesis that $h(x) = h(y)$ means that $x_k = y_k$ for $k = 0, 1, 2, \ldots$. Consequently for any positive integer n we have $x_0 x_1 x_2 \cdots x_n = y_0 y_1 y_2 \cdots y_n$, so that x and y lie in the same subinterval of width $1/2^{n+1}$. Therefore $|x - y| \le 1/2^{n+1}$. Since n is arbitrary, we conclude that $|x - y| = 0$, that is, $x = y$. $\square$

Although Theorem 2.6 associates every number in $[0, 1]$ with a sequence of 0's and 1's, it is not true that, conversely, every sequence of 0's and 1's corresponds to a number in $[0, 1]$ by means of h. (See Exercise 5.) The reason is that h is not defined symmetrically on $[0, 1/2]$ and on $[1/2, 1]$: the term x_n in the sequence $x_0 x_1 x_2 \cdots x_n$ is 0 if $T^{[n]}(x)$ is in the closed interval $[0, 1/2]$, whereas x_n is 1 if $T^{[n]}(x)$ is in the half-open interval $(1/2, 1]$.

Despite the fact that not every sequence of 0's and 1's corresponds to a number in $[0, 1]$, we will prove that every sequence that is not a finite sequence corresponds to a number in $[0, 1]$. But to prove this, we need a simplified version of the famous theorem due to two 19th century European mathematicians: Eduard Heine and Emile Borel.

Theorem 2.7. (Heine-Borel Theorem). Suppose that $B_0, B_1, \ldots$ is a sequence of closed, bounded intervals of reals such that $B_n \supseteq B_{n+1}$ for all n. Then there is a point common to all the B_n's.

The proof utilizes notions usually studied in advanced analysis, so we omit it. (A proof of a version called the nested interval theorem can be found in the book by Fitzpatrick.) The sequence $B_0, B_1, B_2, \ldots$ appearing in the Heine-Borel Theorem is called a **nested sequence**, since B_{n+1} is "nested" inside B_n for each n.

The theorem says that a nested sequence of closed, bounded intervals of reals has at least one common point. Thus if

$$J_n = \left[\frac{1}{2} - \frac{1}{2n}, \ \frac{1}{2} + \frac{1}{2n} \right], \ \text{for } n = 1, 2, 3, \ldots$$

then each interval is closed and bounded, and the sequence of intervals is nested. Consequently the Heine-Borel Theorem indicates that there is a point common to all of the J_n's. Of course that point is $1/2$! We emphasize that a nested sequence of sets can contain infinitely many common points. However, if the sequence does not consist of closed intervals or does not consist of bounded intervals, then the intersection may well be empty. (See Exercises 1 and 2.)

We say that a map $f : D \to E$ is **onto** E if for any element z in E, there is an element x of D such that $f(x) = z$. In other words, the range of f contains E (and possibly contains points in addition to the points in E).

In proving that h maps $[0, 1]$ onto A, we will use the notation $\{x \text{ in } [0, 1] : x_0 = z_0\}$, which is to be read "the set of all x in $[0, 1]$ such that $x_0 = z_0$."

Theorem 2.8. $h[0, 1]$ contains A.

Proof. Let a be in A, and let $a = \{z_n\}_{n=0}^{\infty}$. We will show that $h(x) = a$ for some x. Next, let J_0 be the smallest closed interval containing $\{x \text{ in } [0, 1] : x_0 = z_0\}$, so that J_0 is either the interval $[0, 1/2]$ or the interval $[1/2, 1]$. In addition, let J_1 be the smallest closed interval containing $\{x \text{ in } [0, 1] : x_0 x_1 = z_0 z_1\}$, so that J_1 is $[0, 1/4]$, or $[1/4, 1/2]$, or $[1/2, 3/4]$, or $[3/4, 1]$. Then $J_0 \supseteq J_1$. In general, let

$$J_n \text{ be the smallest closed interval containing}$$
$$\{x \text{ in } [0,1] : x_0 x_1 x_2 \cdots x_n = z_0 z_1 z_2 \cdots z_n\}$$

Then J_n is a closed, bounded subinterval of $[0, 1]$ with length $1/2^{n+1}$, and for all n we have $J_n \supseteq J_{n+1}$. By the Heine-Borel Theorem there is a number x common to all J_n. In addition, x is unique because the length of each J_n is $1/2^{n+1}$. Since x is in J_n for each n, it follows that $x_0 x_1 x_2 \cdots x_n = z_0 z_1 z_2 \cdots z_n$ for each n, and thus $h(x) = a$.   □

Finally, we are ready for the promised theorem about the transitivity of T.

Theorem 2.9. T is transitive.

Proof. Consider the sequence

$$s = \quad 0\,1 \qquad 00\,01\,10\,11 \qquad 000\,001\,010\,011\,100\,101\,110\,111 \cdots$$
$$\quad 1blocks \quad 2blocks \qquad 3blocks$$

which is composed of blocks consisting of all singles, all pairs, all triples, etc., of 0's and 1's. Because s is not a finite sequence, by Theorem 2.8 there is an x_t in $[0, 1]$ such that $h(x_t) = s$. Thus

$$h(x_t) = 0\,1 \quad 00\,01 \quad 10\,11 \quad 000\,001\,010\,011\,100\,101\,110\,111 \cdots$$

Using the blocks of 0's and 1's in the formula for $h(x_t)$, along with (2.7) and the definitions of h and T, we find that

x_t is in $[0, 1/2]$ $\qquad$ $T(x_t)$ is in $[1/2, 1]$ $\qquad$ $T^{[2]}(x)$ is in $[0, 1/4]$

$T^{[4]}(x)$ is in $[1/4, 1/2]$ $\qquad$ $T^{[6]}(x)$ is in $[3/4, 1]$ $\qquad$ $T^{[8]}(x)$ is in $[1/2, 3/4]$

and so forth. In general, for any interval of the form

$$L = [\frac{k}{2^n}, \frac{k+1}{2^n}]$$

there is a positive integer m such that $T^{[m]}(x_t)$ is in L. Thus the iterates of x_t are dense in $[0, 1]$, so that T is transitive by Theorem 2.5.   □

Proving that T is transitive by means of dense orbits was made possible because of the use of sequences of the "symbols" 0 and 1. Employing any other pair of symbols, such as ▲ or ◆, to define the sequences would have worked in the same manner. The use of such sequences in analyzing the dynamics of functions is referred to as **symbolic dynamics**. Thus we have proved that the tent function T is transitive by employing symbolic dynamics.

From Theorem 1.15 we know that the rationals in $[0, 1]$ are the numbers that are eventually periodic under the function T. However, eventually periodic points cannot have dense orbits, because they entail but a finite number of points. Consequently the number x_t that arose in the preceding proof is irrational. Given our previous results, one might expect that each irrational number in $[0, 1]$ would have a dense orbit for T. However, this is not true. (See Exercise 7.)

It would appear that one might be able to prove the transitivity of T directly, without employing symbolic dynamics. In fact, were there infinitely many numbers of the form $2/n$, where n and $(n-1)/2$ are prime and the orbit of $2/n$ is spread evenly throughout the interval $[0, 1]$, then we could prove directly that T is transitive. (See Exercise 13.) However we are not able to prove that there are infinitely many such numbers, so we must leave it as a conjecture. The conjecture is apparently intimately related to several famous conjectures discussed in the delightful book *Solved and Unsolved Problems in Number Theory*, by Daniel Shanks (1978).

2.2.1 Strong Chaos

The function T exhibits three important features:

(i) T is chaotic (by Example 2 of Section 2.1).

(ii) T has a dense set of periodic points (by Example 1).

(iii) T is transitive (by Theorem 2.9).

In addition to the unpredictability of orbits that follows from (2.7), T has a certain regularity because of its dense set of periodic points. Finally, the transitivity tells us that the iterates of certain points spread themselves throughout the domain $[0, 1]$ of T. Thus T is more than chaotic, and we say that it is strongly chaotic.

Definition 2.10. A function f on a finite interval J is **strongly chaotic** if

(i) f is chaotic.

(ii) f has a dense set of periodic points.

(iii) f is transitive.

By our comments above, T is strongly chaotic. As we indicated when we defined sensitive dependence and Lyapunov exponents in Section 2.1, it is not so easy to prove that Q_4 is chaotic. Similarly, it is not so easy to prove that Q_4 has a dense set of periodic points, or that it is transitive. By the results of Section 2.3, the fact that T is strongly chaotic will imply that Q_4 is also strongly chaotic.

2.2.2   Section 2.2 Exercises

Exercise 1. Find a nested sequence $\{B_n\}_{n=1}^{\infty}$ of closed intervals whose intersection is empty.

Exercise 2. Find a nested sequence $\{B_n\}_{n=1}^{\infty}$ of bounded intervals whose intersection is empty.

Exercise 3.

(a) Show that the set of rationals in $[0, 1]$ is dense in $[0, 1]$.

(b) Show that the set of dyadic rationals in $[0, 1]$ is dense in $[0, 1]$.

Exercises 4-8 relate to the tent function T.

Exercise 4.

(a) Find the number x in $[0, 1]$ such that $h(x) = \overline{1} \cdots$.

(b) Find the number x in $[0, 1]$ such that $h(x) = \overline{10} \cdots$.

Exercise 5.

(a) Show that $11\overline{0} \cdots$ is not the image under h of any x in $[0, 1]$.

(b) Show that any sequence of the form $x_0 x_1 x_2 \cdots x_n 11\overline{0} \cdots$ is not the image under h of any x in $[0, 1]$.

Exercise 6. Let D consist of all irrationals in $[0, 1]$ with dense orbit for T. Show that D is dense in $[0, 1]$.

Exercise 7.

(a) Find an irrational number x in $[0, 1]$ that does not have a dense orbit.

(b) Let D^* consist of all irrationals in $[0, 1]$ whose orbits are not dense in $[0, 1]$. Show that D^* is dense in $[0, 1]$.

Exercise 8. Let $x_0 x_1 x_2 \cdots$ be a sequence of 0's and 1's. For each positive integer n, let

$$g(x_0, x_1, x_2, \ldots, x_n) = (x_0 + x_1 + x_2 + \cdots + x_n) \bmod 2$$

(a) Show that if $h(x) = x_0 x_1 x_2 \cdots$, then

$$x = \frac{g(x_0)}{2} + \frac{g(x_0, x_1)}{2^2} + \frac{g(x_0, x_1, x_2)}{2^3} + \cdots$$

(b) Use part (a) to find an approximation to within .01 of the irrational number x_t with dense orbit that appears in the proof of Theorem 2.9.

Exercise 9. Let $x = .28$ and $z = .63$. Use the computer program NUMBER OF ITERATES in order to determine an integer n such that $|Q_4^{[n]}(x) - z| < .001$.

Exercise 10. Let $x = .45$ and $z = .99$. Use the computer program NUMBER OF ITERATES in order to determine an integer n such that $|T^{[n]}(x) - z| < .001$.

Exercise 11. Show that the baker's function B is strongly chaotic. We can do this using a few steps.

- Show B is chaotic.

- Show B has a dense set of periodic points.

- Show B is transitive.

Exercise 12. Let $f(x) = \begin{cases} 3x & \text{for } 0 \leq x \leq 1/3 \\ 3x - 1 & \text{for } 1/3 < x \leq 2/3 \\ 3x - 2 & \text{for } 2/3 < x \leq 1 \end{cases}$

(a) Let A_0 denote the collection of all sequences of 0's, 1's, and 2's. Define a function $h_0 : [0, 1] \to A_0$ in the same spirit as the function h defined in this section.

(b) Show that there is a dense set of periodic points in $[0, 1]$.

(c) Show that f is strongly chaotic.

Exercise 13. Suppose that both n and $(n - 1)/2$ are prime numbers, with $n > 2$.

(a) Find the elements in the orbit of $2/n$ for T.

(b) Suppose that there are infinitely many prime numbers n with the property that $(n-1)/2$ is a prime number. Then show directly that the tent function T is transitive.

2.3   Conjugacy

Among the many features of the tent function T already discussed are transitivity and the existence of a dense set of periodic points. We might ask whether these features apply to the quadratic function Q_4. In the present section we will define the concept of conjugacy. We will show that if two functions are conjugate to one another, then one function inherits such properties as transitivity and the existence of a dense set of periodic points from the other function. In particular, this information will apply to T and Q_4, and hence will yield valuable information about Q_4.

Before we define the notion of conjugacy, we make an important definition.

Definition 2.11. Let J and K be intervals. The function $f : J \to K$ is a **homeomorphism from J onto K** provided that f is one-to-one and onto, and provided that both f and f^{-1} are continuous.

A nice way to illustrate this is with a **commutative diagram**.

$$
\begin{array}{ccc}
J & \xrightarrow{\ f\ } & J \\
\downarrow{\scriptstyle h} & & \downarrow{\scriptstyle h} \\
K & \xrightarrow{\ g\ } & K
\end{array}
$$

What this illustrates is, if you begin with an element in $j \in J$, and you apply the map f then h, you end up at the same element as if you had applied h and then g. In other words, $h(f(j)) = g(h(j))$, the same as our definition above.

Using the fact that h^{-1} is a homeomorphism whenever h is, one can prove that if $f \approx_h g$, then $g \approx_{h^{-1}} f$ (Exercise 11). Thus g is conjugate to f whenever f is conjugate to g (although the homeomorphisms h and h^{-1} are normally quite different from one another!).

Recall from calculus that if $f : J \to K$ is a continuous, one-to-one function that is onto K, then $f^{-1} : K \to J$ is a continuous function, so that f is automatically a homeomorphism. As a result, if $f(x) = x^2$ for $0 \le x \le 2$, then f is a homeomorphism of $[0, 2]$ onto $[0, 4]$ (Figure 2.7(a)). Similarly, if $g(x) = \arctan x$, then g is a homeomorphism of $(-\infty, \infty)$ onto $(-\pi/2, \pi/2)$ (Figure 2.7(b)). Finally, if $h(x) = \sin^2(\pi x/2)$ for $0 \le x \le 1$, then h is a homeomorphism of $[0, 1]$ onto itself (Figure 2.7(c)). We will use h when we discuss the relationship between T and Q_4 later in this section.

We remark that if f is a homeomorphism from J onto K, then f^{-1} is automatically a homeomorphism from K onto J. For example, since the function h defined above is a homeomorphism from $[0, 1]$ onto $[0, 1]$, it follows that h^{-1}, defined by

$$
h^{-1}(x) = \frac{2}{\pi} \arcsin \sqrt{x}
$$

is also a homeomorphism from $[0, 1]$ onto $[0, 1]$.

Suppose that $f : J \to K$ is a homeomorphism. It is an important consequence of the concept of homeomorphism that f maps open (resp. closed) subintervals interior to J onto open (resp. closed) subintervals interior to K.

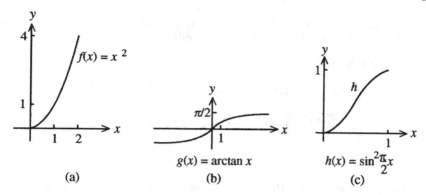

FIGURE 2.7
Examples of homeomorphisms

Moreover, if A is a dense subset of J, then $f(A)$ is a dense subset of K (Exercise 12).

We are ready to define the notion of conjugacy.

Definition 2.12. Let J and K be intervals, and suppose that $f : J \to J$ and $g : K \to K$. Then f and g are **conjugate** (to one another) if there is a homeomorphism $h : J \to K$ such that $h \circ f = g \circ h$. In this case, we write $f \approx_h g$.

Using the fact that h^{-1} is a homeomorphism whenever h is, one can prove that if $f \approx_h g$, then $g \approx_{h^{-1}} f$ (Exercise 11). Thus g is conjugate to f whenever f is conjugate to g (although the homeomorphisms h and h^{-1} are normally quite different from one another!).

Example 2.3.1. Let

$$g(x) = x^2 - \frac{3}{4} \quad \text{for} \quad -\frac{3}{2} \le x \le \frac{3}{2} \quad \text{and} \quad Q(x) = 3x(1 - x)$$
$$\text{for} \quad 0 \le x \le 1$$

Show that $g \approx_h Q$, where $h(x) = -\frac{1}{3}x + \frac{1}{2}$.

Solution. We will verify that $h \circ g = Q \circ h$. A routine check verifies that h is a homeomorphism from $[-3/2, 3/2]$ onto $[0, 1]$, and that

$$(h \circ g)(x) = h(g(x)) = h\left(x^2 - \frac{3}{4}\right) = -\frac{1}{3}\left(x^2 - \frac{3}{4}\right) + \frac{1}{2} = -\frac{1}{3}x^2 + \frac{3}{4}$$

and

$$(Q \circ h)(x) = Q(h(x)) = Q\left(-\frac{1}{3}x + \frac{1}{2}\right) = 3\left(-\frac{1}{3}x + \frac{1}{2}\right)\left(1 - \left(-\frac{1}{3}x + \frac{1}{2}\right)\right)$$
$$= -\frac{1}{3}x^2 + \frac{3}{4}.$$

Therefore $h \circ g = Q \circ h$. $\square$

You might recognize the functions g and Q in Example 2.3.1 as $g_{-3/4}$ and Q_3, respectively, from the parametrized families $\{g_\mu\}$ and $\{Q_\mu\}$.

Example 2.3.2. Let

$$Q_4(x) = 4x(1-x) \quad \text{for} \quad 0 \le x \le 1 \qquad \text{and}$$

$$T(x) = \left\{ \begin{array}{ll} 2x & \text{for } 0 \le x \le 1/2 \\ 2(1-x) & \text{for } 1/2 < x \le 1 \end{array} \right.$$

Show that $T \approx_h Q_4$, where $h(x) = \sin^2 \dfrac{\pi}{2} x$.

Solution. As mentioned earlier, h is a homeomorphism from $[0, 1]$ onto $[0, 1]$. On the one hand, we observe that

$$(Q_4 \circ h)(x) = Q_4(h(x)) = Q_4\left(\sin^2 \frac{\pi}{2}x\right) = 4\left(\sin^2 \frac{\pi}{2}x\right)\left(1 - \sin^2 \frac{\pi}{2}x\right)$$

$$= 4\left(\sin^2 \frac{\pi}{2}x\right)\left(\cos^2 \frac{\pi}{2}x\right) = \left(2\sin \frac{\pi}{2}x \cos \frac{\pi}{2}x\right)^2 = \sin^2 \pi x$$

where the last equality results from a trigonometric double-angle formula. On the other hand, we find that

$$\text{if } 0 \le x \le \frac{1}{2}, \text{ then } (h \circ T)(x) = h(2x) = \sin^2 \pi x$$

$$\text{if } \frac{1}{2} < x \le 1, \text{ then } (h \circ T)(x) = h(2 - 2x) = \left(\sin \frac{\pi}{2}(2 - 2x)\right)^2$$

$$= [\sin(\pi - \pi x)]^2 = \sin^2 \pi x$$

Consequently $(h \circ T)(x) = (Q_4 \circ h)(x)$ for $0 \le x \le 1$, so that $h \circ T = Q_4 \circ h$. $\square$

Example 2.3.2 shows that the tent function T and the quadratic function Q_4 are conjugates. This conjugacy will play a decisive role in our proof that Q_4 is strongly chaotic.

Periodic points are inherited through conjugacy, as Theorem 2.13 tells us.

Theorem 2.13. Suppose that $f \approx_h g$. Then

i. $h \circ f^{[n]} = g^{[n]} \circ h$ for $n = 1, 2, \ldots$.

ii. If x^* is a period-n point of f, then $h(x^*)$ is a period-n point of g.

iii. If f has a dense set of periodic points, then so does g.

Proof. Since $f \approx_h g$, we know that $h \circ f = g \circ h$. For the purposes of an induction proof, let us assume that $h \circ f^{[n-1]} = g^{[n-1]} \circ h$. Then

$$h \circ f^{[n]} = (h \circ f) \circ f^{[n-1]} = (g \circ h) \circ f^{[n-1]} = g \circ (h \circ f^{[n-1]}) = g \circ (g^{[n-1]} \circ h) = g^{[n]} \circ h$$

so that

$$h \circ f^{[n]} = g^{[n]} \circ h \text{ for } n = 1, 2, \ldots$$

By the Law of Induction, (i) is proved. To prove (ii), assume that x^* is a period-n point of f. Together with the fact that $f^{[n]}(x^*) = x^*$, (i) implies that

$$g^{[n]}(h(x^*)) = (g^{[n]} \circ h)(x^*) = (h \circ f^{[n]})(x^*) = h(f^{[n]}(x^*)) = h(x^*)$$

Consequently $h(x^*)$ is a period-n point of g, which proves (ii). Finally, the image $h(A)$ of the dense set A of periodic points of f contains only periodic points of g by (ii) and is dense in the range of g by a remark following the definition of homeomorphism. □

Theorem 2.13 has immediate application to the periodic points of Q_4.

Example 2.3.3. Find a 3-cycle for Q_4.

Solution. A simple calculation shows that $\{2/7, 4/7, 6/7\}$ is a 3-cycle for T. By Example 2, $T \approx_h Q_4$, where $h(x) = \sin^2(\pi x/2)$. Therefore by Theorem 2.13,

$$\{\sin^2 \frac{\pi}{7}, \ \sin^2 \frac{2\pi}{7}, \ \sin^2 \frac{3\pi}{7}\}$$

is a 3-cycle for Q_4. By calculator we find that the 3-cycle is approximately

$$\{.1882550991, \quad .611260467, \quad .950484434\}$$

□

As you can see, it would be hard to guess this 3-cycle of Q_4 by trial and error. Our next result shows that transitivity is also inherited through conjugacy.

Theorem 2.14. Let J and K be closed intervals, and let $f : J \to J$ and $g : K \to K$. If f is transitive and $f \approx_h g$, then g is also transitive.

Proof. Recall from Theorem 2.5 that a function defined on a closed interval is transitive if and only if it has a dense orbit. To prove our result, we suppose that the orbit of x for f is dense in the domain J. We will show that the orbit of $h(x)$ for g is dense in K. To that end, let U be a nonempty open subinterval of K. Because h is a homeomorphism, it follows that $h^{-1}(U)$ is an open subinterval of J. Since the orbit of x is dense in J, there is a positive integer n such that $f^{[n]}(x)$ is in $h^{-1}(U)$, so $h(f^{[n]}(x))$ is in $h(h^{-1}(U)) = U$. By (2.7) of Theorem 2.13,

$$h(f^{[n]}(x)) = (h \circ f^{[n]})(x) = (g^{[n]} \circ h)(x) = g^{[n]}(h(x))$$

Therefore $g^{[n]}(h(x))$ is in U. Since U is an arbitrary nonempty open interval in K, we have succeeded in proving that the orbit of $h(x)$ for g is dense in K, so that by Theorem 2.5, g is transitive. □

We are now equipped to prove that Q_4 is strongly chaotic.

Theorem 2.15. The function Q_4 is strongly chaotic.

Proof. First, we note that Q_4 is transitive, because T is transitive and transitivity is preserved for conjugates by Theorem 2.14. Next, one can prove that Q_4 has sensitive dependence on initial conditions by the following argument: Let z be any point in the domain $[0, 1]$ of Q_4. Then there is an x in the domain $[0, 1]$ of T such that $h(x) = z$, where h is the homeomorphism from T to Q_4 in Example 2 of this section. Because T has sensitive dependence on initial conditions, we deduce from Example 2 of Section 2.1 that for the x at hand there are a periodic point v and a point w with dense orbit, with both v and w as close to to x as we like, and with the property that there is an n such that either $|T^{[n]}(v) - T^{[n]}(x)| > 1/4$ or $|T^{[n]}(w) - T^{[n]}(x)| > 1/4$. Since homeomorphisms preserve small distances as well as large distances, it follows that $|Q_4^{[n]}(h(v)) - Q_4^{[n]}(z)|$ or $|Q_4^{[n]}(h(w)) - Q_4^{[n]}(z)|$ is not near 0. That is, Q_4 has sensitive dependence on initial conditions at the arbitrary point z in the domain of Q_4. (Also, after Example 2.3.3 of Section 2.1 we mentioned that Q_4 has a positive Lyapunov exponent.) In any case, Q_4 is chaotic. Finally, since T and Q_4 are conjugates, and T has a dense set of periodic points, the same properties hold for Q_4 by Theorem 2.13. These observations together imply that Q_4 is strongly chaotic.   □

In our discussion of $f \approx_h g$, we have previously assumed that h is a homeomorphism. There are important cases in which the homeomorphism is also a linear function. We say that f and g are **linearly conjugate** if there is a linear homeomorphism h such that $f \approx_h g$.

Theorem 2.16. Let

$$f(x) = ax^2 + bx + c \qquad \text{and} \qquad g(x) = rx^2 + sx + t$$

where $a \neq 0$ and $r \neq 0$, and where

$$c = \frac{b^2 - s^2 + 2s - 2b + 4rt}{4a} \tag{2.9}$$

Then f and g are linearly conjugate to one another, with associated homeomorphism given by

$$h(x) = \frac{a}{r}x + \frac{b-s}{2r} \tag{2.10}$$

Proof. Let $h(x) = dx + e$. We will prove that there are constants $d \neq 0$ and e such that $h \circ f = g \circ h$. Now $(h \circ f)(x) = (g \circ h)(x)$ if $h(f(x)) = g(h(x))$, that is, if

$$d(ax^2 + bx + c) + e = r(dx + e)^2 + s(dx + e) + t$$

or equivalently, if

$$dax^2 + dbx + (dc + e) = rd^2x^2 + (2rde + sd)x + (re^2 + se + t)$$

Collecting coefficients of like powers of x, and using the hypothesis that $r \neq 0$, we find that

x^2 terms: $\quad da = rd^2$, so that if $d \neq 0$, then $d = \dfrac{a}{r}$

x terms: $\quad db = 2rde + sd$, so that if $d \neq 0$, then $e = \dfrac{b-s}{2r}$

constant terms: $\quad dc + e = re^2 + se + t$

Substituting for d and e in the last equation, we obtain=

$$\frac{a}{r}c + \frac{b-s}{2r} = r\left(\frac{b-s}{2r}\right)^2 + s\,\frac{b-s}{2r} + t$$

which yields

$$4ac + 2b - 2s = b^2 - 2bs + s^2 + 2bs - 2s^2 + 4rt$$

Using the fact that $a \neq 0$, we can solve for c to obtain

$$c = \frac{b^2 - s^2 + 2s - 2b + 4rt}{4a}$$

which is (2.7). We conclude that for this value of c, f and g are linearly conjugate, with h as given in (2.8). $\qquad\square$

In effect, Theorem 2.16 says that each quadratic function g is linearly conjugate to a suitable vertical shift of any other quadratic function f. The shift of f must be such that the constant term c of f satisfies (2.7).

Example 2.3.4. Let $g_c(x) = x^2 + c$ and $Q_\mu(x) = \mu x(1-x)$, with $0 < \mu \leq 4$. Find c and h so that $g_c \approx_h Q_\mu$ and h is linear.

Solution. We use Theorem 2.16 with the following substitutions:

$$a = 1, \quad b = 0, \quad c = c, \quad r = -\mu, \quad s = \mu, \quad t = 0$$

Using (2.7), we find that

$$c = \frac{b^2 - s^2 + 2s - 2b + 4rt}{4a} = -\frac{\mu^2}{4} + \frac{\mu}{2}$$

Then $g_c \approx_h Q_\mu$, where by (2.8), h satisfies

$$h(x) = \frac{a}{r}x + \frac{b-s}{2r} = -\frac{1}{\mu}x + \frac{1}{2}$$

This completes the solution. $\qquad\square$

In a similar fashion, one can show that the following are pairwise conjugate (with appropriate domains and values of μ, r, c, and C):

$$Q_\mu(x) = \mu x(1-x), \quad f_\mu(x) = (2-\mu)x - \mu x^2, \quad K_c(x) = c - x^2, \quad F_C(x) = 1 - Cx^2$$

(See Exercise 4.)

2.3.1   Section 2.3 Exercises

Exercise 1. Let $f_\mu(x) = (2 - \mu)x - \mu x^2$ for x in some interval L. Determine L so that $f_\mu \approx_h Q_\mu$. (Notice that μ is the same for both functions.)

Exercise 2. Let $0 < \mu \leq 4$, and let $K_c(x) = c - x^2$ for x in an interval L. Find L and c so that Q_μ and K_c are linearly conjugate.

Exercise 3. Let $F_C(x) = 1 - Cx^2$ for all real x.

(a) Assume that Q_μ is defined for all real x. Show that for a given μ, there is a constant C such that F_C and Q_μ are linearly conjugate.

(b) By altering the computer program BIFURCATION, show that F_C has a 3-cycle if $C > 7/4$, and does not appear to have a 3-cycle if $C < 7/4$. (Thus $7/4$ is a special bifurcation point of F_C.)

(c) Use (b) to show that Q_μ has a 3-cycle if $\mu : 1 + 2\sqrt{2} \approx 3.828427\ldots$.

Exercise 4. Show that for any given c, there is an appropriate C such that the functions K_c and F_C appearing in Exercises 2 and 3 are linearly conjugate.

Exercise 5. Using the fact that Q_4 and T are conjugates, approximate a period-n point of Q_4, where

(a) $n = 4$

(b) $n = 5$

Exercise 6. Let $f_\mu(x) = (2 - \mu)x - \mu x^2$, for $0 \leq x \leq 1$, where $3 < \mu < 3.25$. Use Exercise 1 to show that a 2-cycle exists for f_μ, and find it.

Exercise 7. Let $f(x) = x^3$, and let $g_\mu(x) = x - \mu x^3$, where μ is a positive constant. Show that there is no nontrivial homeomorphism h such that $f \approx_h g_\mu$.

Exercise 8. Is there a linear function h on $[0, 1]$ such that Q_4 is linearly conjugate to the tent function T? Explain why or why not.

Exercise 9. Suppose that $0 < \mu < 4$. Does there exist a λ in $(0, 1)$ such that Q_μ and T_λ are linearly conjugate? Explain why or why not.

Exercise 10. Suppose that $f \approx_h g$ and $g \approx_H k$, where h and H are linear. Show that f and k are linearly conjugate.

Exercise 11. Suppose that $f \approx_h g$. Show that $g \approx_{h^{-1}} f$.

Exercise 12. Let $f : J \to K$ be a homeomorphism.

(a) Let U be an open interval interior to J. Prove that $f(U)$ is an open interval interior to K.

(b) Prove that if A is dense in J, then $f(A)$ is dense in K.

Exercise 13. Let $\{f_\mu\}$ be a parametrized family of functions, and assume that x_μ is a fixed point of f_μ for all μ. Define a parametrized family $\{g_\mu\}$ such that for each μ, $f_\mu \approx_h g_\mu$ and 0 is a fixed point of g_μ.

Exercise 14. Let $g : [0,1] \to [0,1]$ be continuous, and let $f(x) = \dfrac{1}{3} g(3x)$ for $0 \le x \le 1/3$.

(a) Find a linear function h such that $f \approx_h g$.

(b) Let F be defined by

$$
F(x) = \begin{cases} 2/3 + f(x) & \text{for } 0 \le x \le 1/3 \\ f(1/3)(2 - 3x) & \text{for } 1/3 < x \le 2/3 \\ x - 2/3 & \text{for } 2/3 < x \le 1 \end{cases}
$$

(Figure 2.8). Show that F has one fixed point, and that all other periodic points of F have double the periods of the corresponding periodic points of g. (Hint: Use part (a) to show that $f \approx_h g$, and then show that $F^{[2n]}(x) = f^{[n]}(x)$ for all x in $[0, 1/3]$.)

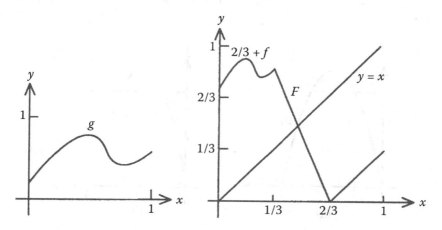

FIGURE 2.8
Illustration of the functions in Exercise 14

2.4   Cantor Sets

In Sections 1.5 and 2.3 we studied the functions Q_μ for $0 < \mu \leq 4$. Now in this final section of Chapter 2 we will study the behavior of Q_μ for $\mu > 4$. The first substantial difference we notice is that if $\mu > 4$, then the range of μ is not totally contained in $[0, 1]$ (Figure 2.9(a)). In our study we will introduce several kinds of sets, such as closed, perfect, uncountable, and Cantor sets. As a result, the section is more technical and theoretical than those that have preceded it.

Throughout the section we will assume that $\mu > 4$. The graph of Q_μ extends above the line $y = 1$, so that in order to be able to describe the iterates of all x in the domain of Q_μ, we need to enlarge the domain. This we do by taking the domain of Q_μ to be $(-\infty, \infty)$ whenever $\mu > 4$ (Figure 2.9(b)). Then surely the range of Q_μ is contained in the domain! If $x < 0$, then

$$Q_\mu(x) = \mu x(1 - x) < x < 0 \qquad (2.11)$$

so that $\{Q_\mu^{[n]}(x)\}_{n=1}^{\infty}$ is a negative, decreasing sequence. If it converged, its limit would need to be a fixed point of Q_μ by Corollary 1.4. However, by (1) there are no negative fixed points. Thus

$$\lim_{n \to \infty} Q_\mu^{[n]}(x) = -\infty \text{ for all } x < 0 \qquad (2.12)$$

If $Q_\mu(x) > 1$, then $Q_\mu^{[2]}(x) = Q_\mu(Q_\mu(x)) < 0$, so that by (2), $\lim_{n \to \infty} Q_\mu^{[n]}(x) = -\infty$. Thus if any iterate of x is greater than 1, then the successive iterates approach $-\infty$. Figures 2.10(a)–(b) show the behavior of $Q_\mu^{[2]}$ and $Q_\mu^{[3]}$ when $\mu = 5$, and give an indication of the unboundedness of iterates.

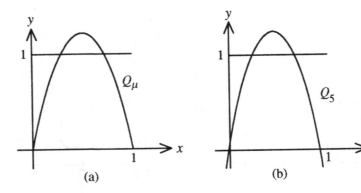

(a)   (b)

FIGURE 2.9
Quadratic maps with $\mu > 4$

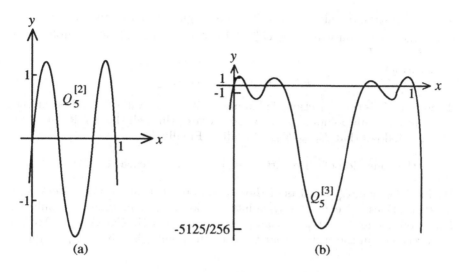

FIGURE 2.10
Multiple iterations of Q_5

Although many iterates converge to $-\infty$, it is those x in $[0,\, 1]$ whose iterates remain in $[0,\, 1]$ that we will study. The set of numbers in $[0,\, 1]$ for which $0 \leq Q_\mu(x) \leq 1$ consists of two closed subintervals which we will call J_{11} and J_{12} (Figure 2.11(a)). The set on which $0 \leq Q_\mu^{[2]} \leq 1$ consists of the

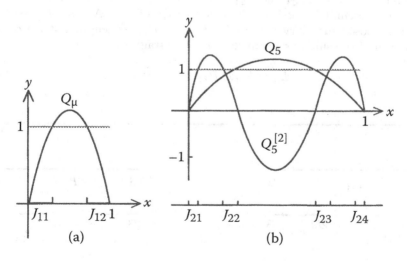

FIGURE 2.11
Multiple iterations of Q_5

four closed subintervals J_{21}, J_{22}, J_{23}, J_{24} (Figure 2.11(b)). In general, at the nth stage the set on which $0 \leq Q_\mu^{[n]} \leq 1$ consists of the 2^n closed subintervals $J_{n1}, J_{n2}, \ldots, J_{n2^n}$.

Next, let

$$K_n = J_{n1} \cup J_{n2} \cup \cdots \cup J_{n2^n}$$

Therefore K_n is the collection of all numbers in J_{nk} for various k. Alternatively, K_n consists of all points in $[0, 1]$ whose iterates through the nth iterate lie in $[0, 1]$. It follows that $K_n \supseteq K_{n+1}$ for all n. Finally, let

C_μ = the collection of all points in $[0, 1]$ that are in each K_n, for $n = 1, 2, \ldots$

By its definition, C_μ consists of those points in $[0, 1]$ all of whose iterates lie in $[0, 1]$. Does C_μ contain any points? It does, for it contains not only the fixed points 0 and $1 - 1/\mu$, but all periodic points of Q_μ (Exercise 9). We will discover later in the section that C_μ contains points that are not periodic.

2.4.1   The Cantor Ternary Set

You may have noticed that the set C_μ is described by the same process, for every value of $\mu > 4$. In order to study the special features of the set C_μ, we will focus on one single set C in the interval $[0, 1]$ that not only is described in a way analogous to the description of the sets C_μ, but whose defining subintervals all have the same size at each stage.

Specifically, we define C by first selecting the two closed subintervals P_{11} and P_{12} of length $1/3$ shown in Figure 2.12, then selecting the four closed subintervals P_{21}, P_{22}, P_{23} and P_{24} of length $1/3^2$ (Figure 2.12). The process continues indefinitely (as with the construction of C_μ). At the nth stage we select 2^n closed subintervals $P_{n1}, P_{n2}, P_{n3}, \ldots, P_{n2^n}$ of length $1/3^n$, two each from each of the subintervals in the $(n-1)$st stage.

If

$$L_n = P_{n1} \cup P_{n2} \cup \cdots \cup P_{n2^n} \tag{2.13}$$

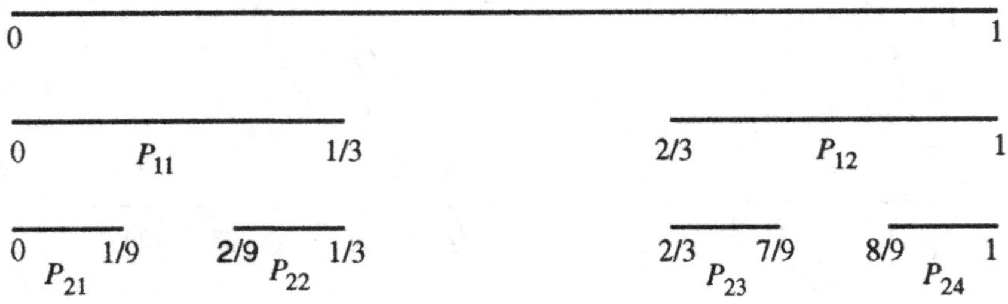

FIGURE 2.12
The beginning of the Cantor ternary set

then we define C as follows:

$C =$ the collection of all points in $[0, 1]$ that are in each L_n, for $n = 1, 2, \ldots$

The set C is called the **Cantor ternary set**. The word "ternary" means third, and indicates that at each stage the middle third of each remaining open interval is deleted. We will see later on that representing the elements of this set in ternary, rather than decimal numbers, can help with understanding the uncountability of the set.

Note: The name Cantor honors the Russian-born German mathematician Georg Cantor, though it was first discovered by Henry Smith (1874). Cantor formalized the definitions in 1883. Cantor's work in set theory, and the definition of infinity, was groundbreaking at the time. The work on infinite cardinals is an interesting representation of how higher mathematics is discovered, formalized, and accepted. The notion that two sets could be infinite, but with different cardinality, was derided by many of Cantor's peers. It eventually came to be accepted, and now, over 140 years later, it's commonly taught to undergraduates.

In the discussion below we will identify several basic features of C. Because of the similar ways that C and C_μ are constructed, with a minimum of effort we will then be able to associate features of C with corresponding features of C_μ for $\mu > 4$.

The first property concerns closed sets. A set is said to be **closed** if it contains all its limit points. In other words, a set D is closed if whenever $\{x_i\}_{i=1}^\infty$ is a sequence of points in D such that $\lim_{i \to \infty} x_i = x$, then x is also in D. It is evident that any closed interval is a closed set. In addition, any finite point set is closed, as is the union of a finite number of closed intervals (Exercise 1). Also the set consisting of the sequence $\{1/n\}_{n=1}^\infty$ and 0 is closed. By contrast, the set of rationals in the interval $[0, 1]$ is not closed because, for example, the irrational number $.50550555055550555550\cdots$ is the limit of the rational numbers $.5, .505, .5055, .505505, \ldots$.

Now we will show that C is closed.

Lemma 2.1. C is closed.

Proof. By our remarks above, for each n the set L_n is closed since it is the union of a finite collection of closed intervals. Next, let $\{x_i\}_{i=1}^\infty$ be a sequence in C with $\lim_{i \to \infty} x_i = x$. By the definition of C, the sequence is in L_n for each n. Since L_n is closed, it follows that x is also in L_n. Consequently x is in C. Therefore every limit point of a sequence in C is in C, so that C is closed. $\square$

The next property is total disconnectedness. A set D is **totally disconnected** if it contains *no* nonempty open intervals. Thus finite point sets, the rationals in the interval $(0, 1)$, and the sequence $\{1/n\}_{n=1}^{\infty}$ are totally disconnected. In the proof of Lemma 2.2 we will use the fact that every nonempty open interval has positive length, and hence its length is larger than $1/3^n$ for an appropriate positive integer n.

Lemma 2.2. C is totally disconnected.

Proof. for each n, we conclude from the remark preceding the lemma that C cannot contain any nonempty open intervals. Thus C is totally disconnected. $\square$

The third property involves perfect sets. A set D of real numbers is **perfect** provided that it is closed and each of its points is a limit of other points in D. Any closed interval with more than one point is perfect. By contrast, any closed set with points that are isolated from one another is not perfect.

Lemma 2.3. C is perfect.

Proof. We know that C is closed, by Lemma 2.1. Let x be in C. Then for each positive integer n, there is a positive integer $k(n)$ in $[1, 2^n]$ depending on n and such that x is in $P_{nk(n)}$. By the definition of $P_{nk(n)}$,

$$P_{nk(n)} \supseteq P_{(n+1)k(n+1)} \cup P_{(n+1)m(n+1)}$$

where $m(n + 1)$ denotes one of the $k(n + 1)$, and where the two sets on the right-hand side are closed and disjoint, and where x is in $P_{(n+1)k(n+1)}$. For each n let y_n be in $P_{(n+1)m(n+1)}$. Since $P_{(n+1)k(n+1)}$ and $P_{(n+1)m(n+1)}$ are disjoint, we know that $x \neq y_n$. However, since x and y_n are both in $P_{nk(n)}$ and since the length of $P_{nk(n)}$ is $1/3^n$, we know that $\lim_{n\to\infty} y_n = x$. Therefore C is perfect. $\square$

Lemmas 2.1- 2.3 together yield the following theorem.

Theorem 2.17. C is a bounded, closed, totally disconnected and perfect subset of $[0, 1]$.

Sets of real numbers that are simultaneously closed, totally disconnected and perfect play a significant role in analysis, and are accorded a special name.

Definition 2.18. A set D of real numbers that is bounded, closed, totally disconnected and perfect is called a **Cantor set**.

By definition of a Cantor set, C is a Cantor set. Another property of C is that it has a "large" infinite number of points. To tell what we mean by a "large" infinite number of points, we will define countable and uncountable sets. We say that a subset D of real numbers is **countable** if we can list all of its members in the fashion $d_1, d_2, d_3, \ldots$, where the sequence is either finite or infinite. In mathematical language this means that we can find a one-to-one function F from D into the set of positive integers. It follows that finite sets, as well as the sequence $\{1/n\}_{n=1}^{\infty}$, are countable. Any set that is not countable is by definition **uncountable**. We mention that the uncountable sets are larger than countable ones, since uncountable sets are so large that they cannot be identified with the positive integers or subsets of the positive integers. As a result, if D is uncountable and $D \subseteq E$, then E is also uncountable. Below we will show that the collection of the numbers in the interval $[0, 1]$ is uncountable. First, we have a theorem and its corollary.

Theorem 2.19. The set of positive rational numbers is countable.

Proof. Let F be the function whose domain is the set of positive rational numbers and whose values are positive integers, defined in the following way. For any positive rational number p/q in reduced form, let

$$F(p/q) = 2^p \, 3^q$$

To show that F is one-to-one, we suppose that $F(p/q) = F(r/s)$, so that

$$2^p \, 3^q = 2^r \, 3^s, \quad \text{or equivalently,} \quad 2^{p-r} = 3^{s-q}$$

Let us assume without loss of generality that $p > r$, so that 2^{p-r} is an integer. The only way that a power of 2 can equal a power of 3 is for the powers to be 0. Therefore $p - r = 0 = s - q$, so that $p = r$ and $q = s$. Thus $p/q = r/s$. Consequently F is one-to-one, so that the set of positive rational numbers is a countable set. $\square$

Using Theorem 2.19, one can show that the collection of all rationals is also countable (Exercise 2(b)). In addition, since the set of rationals in $(0, 1)$ is a subset of the positive rationals, and since subsets of countable sets are again countable, we have the following corollary.

Corollary 2.20. The set of rational numbers in $(0, 1)$ is countable.

Next, we will prove that the set D of irrational numbers in $(0, 1)$ is uncountable by showing that there is *no* one-to-one correspondence between D and the set of positive integers. Before proving the result, we note that each number in $(0, 1)$ has a decimal expansion (which is not necessarily unique). The expansion is repeating (like $.23\overline{495}\cdots$) if and only if the number is rational (Exercise 2(a)).

Theorem 2.21. The set D of irrational numbers in $(0, 1)$ is uncountable.

Proof. We will prove the result by contradiction. Suppose that D were countable, so we could list its members $\{d_n\}_{n=1}^{\infty}$ as decimals:

$$d_1 = x_{11}x_{12}x_{13}x_{14}\cdots$$
$$d_2 = x_{21}x_{22}x_{23}x_{24}\cdots$$
$$d_3 = x_{31}x_{32}x_{33}x_{34}\cdots$$
$$d_4 = x_{41}x_{42}x_{43}x_{44}\cdots$$

and in general,

$$d_n = x_{n1}x_{n2}x_{n3}x_{n4} \cdots$$

Suppose that we were to define z to be the number in $(0, 1)$ with the decimal expansion $z_1z_2z_3 \cdots$ obtained by assigning $z_n = 1$ if $x_{nn} = 0$, and $z_n = 0$ if $x_{nn} \neq 0$. Then $z \neq d_n$ for each n, so that z would not appear in the listing of members of D. However, there is no assurance that z would be irrational! So we need to be more careful in defining z. Specifically, let z have the decimal expansion $z_1z_2z_3 \cdots$ obtained by letting $z_{2^n} = 1$ if $x_{n2^n} = 0$, and $z_{2^n} = 0$ if $x_{n2^n} \neq 0$, for each n. This ensures that $z \neq d_n$ for every n, so z is not in the listing for D. To guarantee that z is irrational, we will insert enough different sized strings of 0's and 1's into the expansion for z to make certain that the expansion is not repeating. We do this by letting $z_k = 0$ for all k strictly between 2^n and 2^{n+1} if n is even, and $z_k = 1$ for all k is strictly between 2^n and 2^{n+1} if n is odd. Thus the expansion of z is not repeating, so z is irrational and hence in D. However, z is not in the proposed listing of D, which means that D is uncountable. $\square$

> **Note:** The proof of Theorem 2.21 is called the Cantor Diagonalization argument. It first appeared in a paper by Cantor in 1891, though his proof that $\mathbb{R}$ was uncountable came more than 15 years earlier. A student interested in how this diagonalization argument is related to other logical paradoxes should refer to the article, "Paradoxes and Contemporary Logic" (2021) in the *Stanford Encyclopedia of Philosophy*, listed in the references.

Since the interval $[0, 1]$ contains the uncountable set of irrationals in $[0, 1]$, the following corollary is immediate.

Corollary 2.22. The set of all numbers in $[0, 1]$ is uncountable.

A slight alteration in the proof of Theorem 2.21 yields the following result.

Corollary 2.23. The set of all sequences of 0's and 1's is uncountable.

It is possible to show that between any two rational numbers in $(0, 1)$ there are infinitely many irrational numbers, and between any two irrational numbers in $(0, 1)$ there are infinitely many rational numbers. So how can the set of rational numbers in $(0, 1)$ be countable, whereas the set of irrational numbers in $(0, 1)$ is uncountable? There is nothing faulty in our mathematics; the paradox shows that our intuition can sometimes be faulty.

Our next goal is to show that C is uncountable. The uncountable set A of all sequences of 0's and 1's will come in handy. To start with, recall from the proof of Lemma 2.3 before Theorem 2.17 that if x is in C, then for each n, x

is in a subinterval $P_{nk(n)}$ of length $1/3^n$, where $k(n)$ is an appropriate integer such that $1 \leq k(n) \leq 2^n$. Let $h : C \rightarrow A$ be defined by

$$h(x) = x_1 x_2 x_3 \cdots, \quad \text{where } x_n = \begin{cases} 0 & \text{if } k(n) \text{ is an even integer} \\ 1 & \text{if } k(n) \text{ is an odd integer} \end{cases}$$

The function h and the set A are related to the function and set with similar names that appeared in Section 2.2. For convenience, we have altered the initial index of the sequence $h(x)$; the set A now includes all finite sequences as well as nonfinite sequences. We will prove that h is a **one-to-one correspondence** (or a **bijection**) between C and A, that is, $h : C \rightarrow A$ is one-to-one and onto.

Theorem 2.24. h is a one-to-one correspondence between C and A.

Proof. To show that h is one-to-one, we notice that if

$$x_1 x_2 x_3 \cdots = h(x) = h(y) = y_1 y_2 y_3 \cdots$$

then $x_1 = y_1$, $x_2 = y_2$, $x_3 = y_3$, ..., so that x and y both lie in the same subinterval $P_{nk(n)}$ for all n. Since the length of $P_{nk(n)}$ is $1/3^n$ and approaches 0 as n increases without bound, it follows that $h(x) = h(y)$, so that h is one-to-one. To show that h is onto, let $z_1 z_2 z_3 \cdots$ be an arbitrary sequence of 0's and 1's. Pick the subintervals $P_{1k(1)}, P_{2k(2)}, \ldots$ satisfying the following conditions:

$$x \text{ in } P_{1k(1)} \text{ implies that } h(x) \text{ begins } z_1$$
$$x \text{ in } P_{2k(2)} \text{ implies that } h(x) \text{ begins } z_1 z_2$$
$$x \text{ in } P_{3k(3)} \text{ implies that } h(x) \text{ begins } z_1 z_2 z_3$$

In general,

$$x \text{ in } P_{nk(n)} \text{ implies that } h(x) \text{ begins } z_1 z_2 \cdots z_n$$

Since

$$P_{1k(1)} \supseteq P_{2k(2)} \supseteq P_{3k(3)} \supseteq \cdots$$

and each of the sets is closed and bounded, the Heine-Borel Theorem (Theorem 2.7) implies that there is a point x^* that is simultaneously in $P_{nk(n)}$ for all n. By the definition of h, $h(x^*) = z_1 z_2 z_3 \cdots$, so that h is onto. Consequently h is a one-to-one correspondence between C and A. $\square$

Corollary 2.25. C is uncountable.

Proof. Since A is uncountable by Corollary 2.23, and since h is a one-to-one correspondence between C and A, it follows that C is also uncountable. $\square$

Now we have finished presenting the special features of C, including the fact that C is a closed, totally disconnected and perfect set (hence a Cantor set) and is uncountable. We will use the information we have accumulated about C as we discover properties of Q_μ for $\mu > 4$.

2.4.2   Strong Chaos of Functions in $\{Q_\mu\}$

What lies ahead in this section is to show that if $\mu > 4$, then the subset C_μ of $[0, 1]$ associated with Q_μ is a Cantor set on which Q_μ is strongly chaotic. First, we turn to the set C_μ itself. Recall that both C and C_μ were defined through nested sequences of disjoint, closed subintervals of $[0, 1]$. The proof that C is closed and uncountable carries over to show that C_μ also is a closed and uncountable subset of $[0, 1]$. If we knew that the lengths of the defining subintervals J_{nk} of C_μ shrank to 0 as n increased without bound, then we could also use the corresponding proofs for C to show that C_μ is totally disconnected and is perfect.

Lengths of nested bounded closed intervals need not converge to 0 (see Exercise 6). Although it is possible to prove that whenever $\mu > 4$, the length of J_{nk} converges to 0 as n increases without bound, the proof is much less complicated when $\mu > 2 + \sqrt{5} \approx 4.236$. The reason is that

$$\text{if } \mu > 2 + \sqrt{5}, \text{ then } |Q'_\mu(x)| > 1 \text{ for all } x \text{ in } J_{11} \cup J_{12} \qquad (2.14)$$

(Exercise 11). The proof that the length of J_{nk} shrinks to 0 will be based on the observation that if $0 \le f(x) \le 1$ and $|f'(x)| > s > 1$ for all x in $[a, b]$, then the length of $[a, b]$ is less than $1/s$. Indeed, since $0 \le f(x) \le 1$ for all x in $[a, b]$,

$$|f(b) - f(a)| \le 1 - 0 = 1 \qquad (2.15)$$

Since $|f'(x)| > s > 1$, the Mean Value Theorem implies that there is a w in (a, b) such that

$$|f(b) - f(a)| = |f'(w)||b - a| > s|b - a| \qquad (2.16)$$

Combining (2.15) and (2.16), we deduce that

$$1 \ge |f(b) - f(a)| > s|b - a|$$

so that $|b - a| < 1/s$, as asserted.

Theorem 2.26. Suppose that $\mu > 2 + \sqrt{5}$. Then there is a constant $r > 1$ such that J_{nk} has length less than $1/r^n$ for each n and k.

Proof. Since $J_{11} \cup J_{12}$ is closed and Q_μ is continuous, it follows from (2.14) that there is a constant $r > 1$ such that $|Q'_\mu(x)| > r$ for all x in $J_{11} \cup J_{12}$. Our first goal is to show that $|(Q_\mu^{[n]})'(x)| > r^n$ for all x in J_{nk}, for each k. To prove it, let x be in J_{nk}. By the definition of J_{nk}, $Q_\mu^{[j]}(x)$ is in $J_{11} \cup J_{12}$ for $j = 0, 1, 2, \ldots, n-1$. It follows that $|Q_\mu(Q_\mu^{[j]}(x))| > r$ for $j = 0, 1, 2, \ldots, n-1$. By the Chain Rule,

$$|(Q_\mu^{[n]})(x)| = |Q'_\mu(Q_\mu^{[n-1]}(x))||Q'_\mu(Q_\mu^{[n-2]}(x))| \cdots |Q'_\mu(x)| > r \cdot r \cdots r = r^n$$

Thus we have proved that $|(Q_\mu^{[n]})'| > r^n$ on J_{nk}. To complete the proof, we write J_{nk} as $[y, z]$. Since $0 \le Q_\mu^{[n]}(y) \le 1$ and $0 \le Q_\mu^{[n]}(z) \le 1$, the comment before the theorem implies that $|y - z| < 1/r^n$, so that the length of J_{nk} is less than $1/r^n$. □

Theorem 2.27. Let $\mu > 2 + \sqrt{5}$. Then C_μ is an uncountable Cantor set.

Proof. Theorem 2.26 is the missing link that allows us to deduce that C_μ is totally disconnected, perfect, and uncountable by mimicking the corresponding proofs for C (Lemmas 2.2 and 2.3 before Theorem 2.17, and Corollary 2.25, respectively). We have already noted that C_μ is closed. Therefore C_μ is a Cantor set. □

We know that Q_μ maps C_μ into C_μ, and that C_μ is an uncountable Cantor set. In what way can Q_μ be considered to be strongly chaotic on C_μ? In order to address this question we introduce the notion of distance between two sequences.

Let A denote, as before, the collection of sequences of 0's and 1's. To simplify notation we will write X for the sequence $x_1 x_2 x_3 \cdots$ and Y for the sequence $y_1 y_2 y_3 \cdots$. Then the **distance** between X and Y, written $||X - Y||$, is defined by the formula

$$||X - Y|| = \sum_{k=1}^{\infty} \frac{|x_k - y_k|}{2^k}$$

Since $|x_k - y_k| = 0$ or 1, it follows that

$$\sum_{k=1}^{\infty} \frac{|x_k - y_k|}{2^k} \leq \sum_{k=1}^{\infty} \frac{1}{2^k} = 1$$

This distance function has several important properties:

i. $0 \leq ||X - Y||$ for all X and Y, and $||X - Y|| = 0$ only if $X = Y$

ii. $||X - Y|| = ||Y - X||$ for all X and Y

iii. $||X - Z|| \leq ||X - Y|| + ||Y - Z||$ for all $X, Y,$ and Z

A distance function with properties (i)–(iii) is called a **metric**. The set on which the metric is defined, along with the metric, is a **metric space**. Thus A with its distance function is a metric space.

Next, we notice that

if the initial n terms of X and Y are identical, then $||X - Y|| \leq 1/2^n$ (2.17)

The reason is that in this case,

$$||X - Y|| = \sum_{k=n+1}^{\infty} \frac{|x_k - y_k|}{2^k} \leq \sum_{k=n+1}^{\infty} \frac{1}{2^k} = \frac{1}{2^n}$$

A similar argument shows that

if $||X - Y|| < 1/2^n$, then the initial n terms of X and Y are equal   (2.18)

(see Exercise 14). Consequently X and Y are close together precisely when their initial several terms are identical.

Now we define the **left shift map** $S : A \to A$ by

$$S(x_1 x_2 x_3 \cdots) = x_2 x_3 x_4 \cdots$$

The effect of S is to delete the first term of the given sequence, thereby shifting all terms to the left one place.   .

We would like to be able to show that S is a continuous function. However, the domain of S is a set of sequences, not a set of real numbers. As a result, we need to define what we mean by continuity for a function from one metric space to another. To that end, let $f : D \to E$, where D has metric $|| \cdot ||$ and E has metric $||| \cdot |||$. We say that f is **continuous at x** in D if for any arbitrary $\epsilon > 0$, there is a number $\delta > 0$ such that whenever z is in D and $||x - z|| < \delta$, it follows that $|||f(x) - f(z)||| < \epsilon$. Moreover, if f is continuous at each point of its domain, then we say that f is a **continuous function**.

Theorem 2.28. $S : A \to A$ is continuous.

Proof. Let X be in A, and let $\epsilon > 0$. Choose a positive integer n such that $1/2^n < \epsilon$, and let $\delta = 1/2^n$. It follows from (2.18) that if $||X - Y|| < \delta = 1/2^n$, then the initial n terms of X and Y are identical, so that

$$|S(X) - S(Y)|| = \sum_{k=n+1}^{\infty} \frac{|x_{k+1} - y_{k+1}|}{2^k} \leq \sum_{k=n+1}^{\infty} \frac{1}{2^k} = \frac{1}{2^n} < \epsilon$$

Therefore S is a continuous function.   □

The notions of sensitive dependence on initial conditions, dense set of periodic points, and dense orbit that were defined for functions whose domains are intervals can be extended to any metric space (Exercise 15). If D is endowed with a metric, then we will say that a function $f : D \to D$ is **strongly chaotic** on D if f has sensitive dependence on initial conditions, a dense set of periodic points, and a dense orbit.

Theorem 2.29. S is strongly chaotic on the set A.

Proof. To show that S has sensitive dependence on initial conditions, suppose that X and Y are distinct elements in A. This means that there is some initial n for which $x_n \neq y_n$. Since S shifts a given sequence left one term, $S^{[n-1]}$ shifts a given sequence left $n - 1$ terms, so that

$$||S^{[n-1]}(X) - S^{[n-1]}(Y)|| = ||x_n x_{n+1} x_{n+2} \cdots - y_n y_{n+1} y_{n+2} \cdots || \geq \frac{|x_n - y_n|}{2} = \frac{1}{2}$$

Thus S has sensitive dependence on initial conditions. Next, we will show that S admits a dense set of periodic points. Notice that since S shifts terms left, the periodic points for S are the repeating sequences of the form $\overline{x_1 x_2 x_3 \cdots x_n} \cdots$ for $n \geq 1$. Now suppose that $X = x_1 x_2 x_3 \cdots$ is any element of A, and n is any positive integer. Define Y by

$$Y = \text{the repeating sequence } \overline{x_1 x_2 x_3 \cdots x_n} \cdots$$

Then Y is periodic. Moreover, since the initial n terms of X and Y are identical, (2.17) implies that $\|X - Y\| \leq 1/2^n$. Consequently the periodic points are dense in A. Finally, the sequence

$$s = 0\,1 \quad 00\ 01\ 10\ 11 \quad 000\ 001\ 010\ 011\ 100\ 101\ 110\ 111 \cdots$$

(in which all n-tuples of 0's and 1's appear in order, for each positive integer n) has a dense orbit in A, since any partial sequence $x_1 x_2 x_3 \cdots x_n$ appears somewhere inside s. We conclude that S is strongly chaotic. $\qquad \square$

If D and E are metric spaces, then we say that a function $f : D \to E$ is a **homeomorphism** if f is one-to-one, onto, and continuous, and if f^{-1} is also continuous. In that case we say that D and E are **homeomorphic** to one another. For example, the function $f : [0, 2] \to [0, 4]$ defined by $f(x) = x^2$ is a homeomorphism if the metric on $[0, 2]$ and $[0, 4]$ is the usual distance function. If two metric spaces D and E are homeomorphic, then in many ways D and E are similar. In particular, they contain the same number of elements, and small distances in one space correspond to small distances in the other space. It is with this in mind that we will define a homeomorphism $H_\mu : C_\mu \to A$, so that the fact that S is strongly chaotic on A will yield corresponding information about Q_μ on C_μ.

Recall that x is in C_μ if and only if all the iterates $Q_\mu^{[n]}(x)$ are in $[0, 1]$, which means that for each n, $Q_\mu^{[n]}(x)$ is in either J_{11} or J_{12}. We will associate x in C_μ with the sequence in A whose nth term notes whether $Q_\mu^{[n]}(x)$ is in J_{11} or is in J_{12}. Specifically, let $H_\mu : C_\mu \to A$ be defined by

$$H_\mu(x) = x_0 x_1 x_2 \cdots, \text{ where } x_n = \begin{cases} 0 & \text{if } Q_\mu^{[n]}(x) \text{ is in } J_{11} \\ 1 & \text{if } Q_\mu^{[n]}(x) \text{ is in } J_{12} \end{cases}$$

Notice that the initial subsequence $x_0 x_1 x_2 \cdots$ identifies the nth stage subinterval J_{nk} in which x is located. The function H_μ is closely related to the function $h : C \to A$ that was proved in Theorem 2.24 to be a one-to-one correspondence.

Theorem 2.30. Let $\mu > 2 + \sqrt{5}$. Then the function H_μ is a homeomorphism from C_μ onto A.

Proof. To prove that H_μ is one-to-one and onto, one can make the needed modifications of the proof of Theorem 2.24 (Exercise 13). Therefore we need only show that H_μ and H_μ^{-1} are continuous. To prove that H_μ is continuous, let $\epsilon > 0$, and let x be an arbitrary point in C_μ. Furthermore, let n be so large that $1/2^n < \epsilon$. Next, pick $\delta > 0$ so small that if z is in C_μ and $|x - z| < \delta$, then x and z lie in the same $J_{jk(j)}$ for $j = 0, 1, 2, 3, \ldots, n$. (Such a δ exists by virtue of Theorem 2.26.) For such a z, the sequences $H_\mu(x)$ and $H_\mu(z)$ have the same initial $n + 1$ terms by the definition of H_μ. Then (2.17) implies that

$$\|H_\mu(x) - H_\mu(z)\| \leq \frac{1}{2^n} < \epsilon$$

We conclude that H_μ is continuous. That H_μ^{-1} is also continuous can be proved directly, or can be deduced from the fact that H_μ is a one-to-one, onto, continuous function whose domain is a closed, bounded subset of reals. Therefore H_μ is a homeomorphism. $\qquad\square$

Now we turn to conjugacy of two functions. Let D and E be two metric spaces, and let $f : D \to D$ and $g : E \to E$ be continuous functions. We say that f and g are **conjugate** provided that there is a homeomorphism $h : D \to E$ such that $h \circ f = g \circ h$. Our demonstration that Q_μ is strongly chaotic on C_μ will use this notion of conjugacy.

Theorem 2.31. Let $\mu > 2 + \sqrt{5}$. Then Q_μ (restricted to C_μ) and S are conjugate.

Proof. We observe that $Q_\mu : C_\mu \to C_\mu$, $S : A \to A$, $H_\mu : C_\mu \to A$. By Theorem 2.30, H_μ is a homeomorphism from C_μ onto A, so we need only show that $S \circ H_\mu = H_\mu \circ Q_\mu$. To that end, suppose that x is in C_μ, with $H_\mu(x) = x_0 x_1 x_2 \cdots$. Then

$$(S \circ H_\mu)(x) = S(H_\mu(x)) = x_1 x_2 x_3 \cdots$$

so that the nth term of $(S \circ H_\mu)(x)$ is x_n. By the definition of H_μ we know that $(H_\mu \circ Q_\mu)(x) = H_\mu(Q_\mu(x)) = z_1 z_2 z_3 \cdots$, where

$$z_n = \begin{cases} 0 & \text{if } Q_\mu^{[n]}(Q_\mu(x)) = Q_\mu^{[n+1]}(x) \text{ is in } J_{11} \\[2ex] 1 & \text{if } Q_\mu^{[n]}(Q_\mu(x)) = Q_\mu^{[n+1]}(x) \text{ is in } J_{12} \end{cases}$$

It follows that $z_n = x_n$, so that the nth terms of $(S \circ H_\mu)(x)$ and $(H_\mu \circ Q_\mu)(x)$ are x_n. Therefore $S \circ H_\mu = H_\mu \circ Q_\mu$, completing the proof. $\qquad\square$

On closed, bounded subsets of the real line, homeomorphisms automatically transfer all properties of separation (and hence sensitive dependence on initial conditions), periodicity, and density. Therefore, we are ready for our final result.

Theorem 2.32. If $\mu > 2+\sqrt{5}$, then Q_μ is strongly chaotic on the uncountable Cantor set C_μ.

Proof. Theorem 2.29 says that S is strongly chaotic on A. Since Q_μ and S are conjugate to one another by Theorem 2.31, it follows that Q_μ is strongly chaotic on C_μ. Finally, C_μ is an uncountable Cantor set by Theorem 2.27. $\square$

Since Q_μ is strongly chaotic on C_μ by Theorem 2.32 and hence is transitive, it follows that C_μ contains non-periodic points for Q_μ, as we indicated at the outset of the section.

We have now finished our study of the dynamics of functions of one variable. We have discussed many features of our two most illustrious parametrized families, $\{T_\mu\}$ and $\{Q_\mu\}$. Although the members of $\{T_\mu\}$ are polygonal lines and those of $\{Q_\mu\}$ are curved, nevertheless both families display similar dynamic behavior, and certain members of each family are not only chaotic but even strongly chaotic. In Chapter 3 we will turn to functions of more than one variable, and will use the information we have gained concerning functions of one variable.

2.4.3   Section 2.4 Exercises

Exercise 1.

(a) Show that any finite set is closed.

(b) Show that the union of a finite number of closed intervals is closed.

Exercise 2.

(a) Show that a number in $(0, 1)$ is rational if and only if its decimal expansion is repeating.

(b) Show that the set of all rational numbers is countable.

Exercise 3.   Show that between any two rationals there are infinitely many irrationals, and vice versa.

Exercise 4.

(a) Prove that any subset of a countable set is countable.

(b) Prove that if a set D is uncountable and if $E \supseteq D$, then E is also uncountable.

Exercise 5.   Determine whether the set of numbers in $(0, 1)$ whose decimal expansions contain no 1's or 5's is

(a) closed, open, or neither.

(b) totally disconnected.

Exercise 6. Find an example of a nested set of bounded closed intervals $\{B_n\}_{n=1}^{\infty}$ such that the length of B_n does not converge to 0 as n increases without bound.

Exercise 7. Let $D_n(C)$ denote the total length of the set L_n defined in (3) and employed in the definition of C. Find $D(C) = \lim_{n \to \infty} D_n(C)$. The number $D(C)$ is sometimes called the length of C. Is that reasonable?

Exercise 8. Let $C_{1/5}$ denote the set obtained the same way as is C, but with subintervals of length $1/5^n$ deleted at each stage. Let $D_n(C_{1/5})$ denote the total length of the remaining subintervals at the nth stage.

(a) Find $D_n(C_{1/5})$.

(b) Find $\lim_{n \to \infty} D_n(C_{1/5})$.

Exercise 9. Show that if $\mu > 4$, then C_μ contains all periodic points of Q_μ.

Exercise 10. Let $\mu = 5$. Show that J_{21} and J_{22} have different lengths.

Exercise 11. Prove that if $\mu > 2 + \sqrt{5}$, then there is an $r > 1$ such that $|Q_\mu'(x)| > r$ for all x in $J_{11} \cup J_{12}$.

Exercise 12. Let $\mu > 2 + \sqrt{5}$. Show that Q_μ has sensitive dependence on initial conditions on the interval $[0, 1]$.

Exercise 13. By modifying the proof of Theorem 2.24, prove that the function $H_\mu : C_\mu \to A$ is one-to-one and onto.

Exercise 14. Let $X = x_1 x_2 x_3 \cdots$ and $Y = y_1 y_2 y_3 \cdots$. Show that if $\|X - Y\| < 1/2^n$, then the initial n terms of X and Y are identical.

Exercise 15. Let D be a metric space, and let $f : D \to D$. Write down a definition, applicable to f and D, of

(a) sensitive dependence on initial conditions

(b) a dense set of periodic points

(c) a dense orbit

Exercise 16. Provide the details for the proofs of Corollary 2.22, 2.23, and 2.25

Exercise 17. Find μ such that C_μ is the middle-thirds Cantor set.

3

Two-Dimensional Chaos

In the first two chapters we discussed functions of one variable, with an eye toward those features of a chaotic nature. Now we turn to functions of two variables. Such functions can exhibit more varied behavior.

In the analysis of higher-dimensional functions, matrices serve as an indispensable tool. As a result, we give a brief review of matrices in Section 3.1, and in particular focus on features of similar matrices. While a course in linear algebra is not a prerequisite for this text, familiarity with the language is helpful. We will stick to two-dimensional systems, and so we won't go beyond 2×2 matrices. The theory of vector spaces is not needed here, but as with most subjects in mathematics, a strong foundation in linear algebra is recommended for those wishing to study any branch of mathematics.

In Section 3.2, with the help of matrices we study the dynamics of the simplest two-dimensional functions, the linear functions. Section 3.3 presents the basic ideas concerning dynamics of more general two-dimensional functions.

The remainder of Chapter 3 is devoted to well-known examples of two-dimensional functions: the two-dimensional baker's function, the Hénon map, and the Smale horseshoe map. We show that the first and third of these maps are chaotic, and that a Hénon map appears to be chaotic.

3.1 Review of Matrices

In the study of nonlinear dynamics in two (and higher) dimensions, the use of matrices cannot easily be avoided. As a result, we devote this section to the basic definitions and properties of matrices, and limit our discussion to 2×2 matrices because the chapter focuses on two-dimensional functions. The section begins with basic definitions relating to 2×2 matrices, and finishes with a longer discussion of similar matrices.

DOI: 10.1201/9781032678757-3

3.1.1   Brief Review of 2 × 2 Matrices

Let $A = \begin{pmatrix} a & b \\ c & d \end{pmatrix}$ and $B = \begin{pmatrix} e & f \\ g & h \end{pmatrix}$, and let r be a real number. Then

$$A + B = \begin{pmatrix} a+e & b+f \\ c+g & d+h \end{pmatrix}, \quad rA = \begin{pmatrix} ra & rb \\ rc & rd \end{pmatrix}, \quad \text{and}$$

$$AB = \begin{pmatrix} ae+bg & af+bh \\ ce+dg & cf+dh \end{pmatrix}$$

Example 3.1.1. Let a and d be any real numbers, and let $A = \begin{pmatrix} a & 0 \\ 0 & d \end{pmatrix}$. Find A^n for an arbitrary positive integer n.

Solution. First we have, for A given in the above example,

$$A^2 = A \cdot A = \begin{pmatrix} a & 0 \\ 0 & d \end{pmatrix} \begin{pmatrix} a & 0 \\ 0 & d \end{pmatrix} = \begin{pmatrix} a^2 & 0 \\ 0 & d^2 \end{pmatrix}$$

If $A^{n-1} = \begin{pmatrix} a^{n-1} & 0 \\ 0 & d^{n-1} \end{pmatrix}$, then

$$A^n = A^{n-1} \cdot A = \begin{pmatrix} a^{n-1} & 0 \\ 0 & d^{n-1} \end{pmatrix} \begin{pmatrix} a & 0 \\ 0 & d \end{pmatrix} = \begin{pmatrix} a^n & 0 \\ 0 & d^n \end{pmatrix}$$

By the Law of Induction, $A^n = \begin{pmatrix} a^n & 0 \\ 0 & d^n \end{pmatrix}$ for each positive integer n. $\square$

Matrices of the variety given in Example 3.1.1 are called **diagonal matrices** because the only nonzero entries appear along the major (upper left to lower right) diagonal. Example 3.1.1 shows that any power of a diagonal matrix is also a diagonal matrix. Equally easily we could show that any product of diagonal matrices is also a diagonal matrix.

Normally $AB \neq BA$. However, if A, B, and C are any 2×2 matrices, then $A(BC) = (AB)C$, so we can write ABC for the product of the three matrices.

The matrix $\begin{pmatrix} 1 & 0 \\ 0 & 1 \end{pmatrix}$ is called the **identity matrix** and is denoted by I. The matrix I has the property that $AI = IA = A$ for every 2×2 matrix A. Now suppose that A is an arbitrary 2×2 matrix. If there is a 2×2 matrix B such that $AB = I$ and $BA = I$, then B is the **inverse** of A, and we say that A is **invertible**. (The hypothesis that $BA = I$ is actually superfluous. See Exercise 9.) If such a matrix B exists for a given matrix A, then B is unique and is written A^{-1}.

Not every matrix has an inverse. In fact, we will show below that $\begin{pmatrix} a & b \\ c & d \end{pmatrix}$ has an inverse precisely when $ad - bc \neq 0$. The expression $ad - bc$ is called the **determinant** of A and is denoted $\det A$, or $|A|$. Thus

$$\det \begin{pmatrix} a & b \\ c & d \end{pmatrix} = ad - bc$$

Therefore if $A = \begin{pmatrix} 3 & 0 \\ 2 & -1 \end{pmatrix}$, then $\det A = (3)(-1) - (0)(2) = -3$. The determinant of the identity matrix I is given by

$$\det I = \det \begin{pmatrix} 1 & 0 \\ 0 & 1 \end{pmatrix} = (1)(1) - (0)(0) = 1$$

Next, assume that A and B are 2×2 matrices. A routine calculation establishes that

$$\det(AB) = (\det A)(\det B)$$

You'll prove this in Exercise 8. One consequence of this is that if A has an inverse, then

$$\det(A^{-1}) = \frac{1}{\det A}$$

These formulas also helps us to verify the following criterion concerning the existence of inverses.

Theorem 3.1. A 2×2 matrix A has an inverse if and only if $\det A \neq 0$.

Proof. Let $A = \begin{pmatrix} a & b \\ c & d \end{pmatrix}$. First, suppose that $\det A = ad - bc \neq 0$. Then you can check directly that

$$A^{-1} = \frac{1}{ad - bc} \begin{pmatrix} d & -b \\ -c & a \end{pmatrix} \tag{3.1}$$

Conversely, suppose that $\det A = 0$. Then by (3.1),

$$(\det AB) = (\det A)(\det B) = 0\,(\det B) = 0 \text{ for all } B$$

If there were a B such that $AB = I$, then by (3.1), $1 = \det I = \det(AB) = 0$. This contradiction shows that no such B exists. Therefore A has no inverse, and the proof is complete. $\square$

In the following sections we will consider each element of $\mathbb{R}^2$ either as a point, or an ordered pair of real numbers, or a two-dimensional vector. Thus the point whose x and y coordinates are -1 and 2, respectively, represents the ordered pair $(-1, 2)$ as well as the vector $\begin{pmatrix} -1 \\ 2 \end{pmatrix}$. As is customary, boldface letters such as $\mathbf{v}, \mathbf{w}, \mathbf{x}$, and $\mathbf{y}$ represent vectors. Three noteworthy, special vectors are

$$\mathbf{0} = \begin{pmatrix} 0 \\ 0 \end{pmatrix}, \mathbf{i} = \begin{pmatrix} 1 \\ 0 \end{pmatrix}, \text{and} \mathbf{j} = \begin{pmatrix} 0 \\ 1 \end{pmatrix}$$

By definition the sum and constant multiple of vectors are given by

$$\begin{pmatrix} x \\ y \end{pmatrix} + \begin{pmatrix} z \\ w \end{pmatrix} = \begin{pmatrix} x + z \\ y + w \end{pmatrix}$$

$$r \begin{pmatrix} x \\ y \end{pmatrix} = \begin{pmatrix} rx \\ ry \end{pmatrix} \quad \text{for every real number}$$

The product of the matrix $A = \begin{pmatrix} a & b \\ c & d \end{pmatrix}$ and the vector $\mathbf{v} = \begin{pmatrix} x \\ y \end{pmatrix}$ is the 2×1 matrix given by

$$A\mathbf{v} = \begin{pmatrix} a & b \\ c & d \end{pmatrix} \begin{pmatrix} x \\ y \end{pmatrix} = \begin{pmatrix} ax + by \\ cx + dy \end{pmatrix}$$

For example, if $A = \begin{pmatrix} 1 & 2 \\ 3 & 4 \end{pmatrix}$ and $\mathbf{v} = \begin{pmatrix} -2 \\ 1 \end{pmatrix}$, then

$$A\mathbf{v} = \begin{pmatrix} 1 & 2 \\ 3 & 4 \end{pmatrix} \begin{pmatrix} -2 \\ 1 \end{pmatrix} = \begin{pmatrix} 0 \\ -2 \end{pmatrix}$$

In the event that $\mathbf{v} = \mathbf{i}$ or $\mathbf{v} = \mathbf{j}$, we obtain

$$A\mathbf{i} = \begin{pmatrix} a & b \\ c & d \end{pmatrix} \begin{pmatrix} 1 \\ 0 \end{pmatrix} = \begin{pmatrix} a \\ c \end{pmatrix} \quad \text{and} \quad A\mathbf{j} = \begin{pmatrix} a & b \\ c & d \end{pmatrix} \begin{pmatrix} 0 \\ 1 \end{pmatrix} = \begin{pmatrix} b \\ d \end{pmatrix}$$

Consequently

$$A\mathbf{i} = \text{the first column of } A, \quad \text{and} \quad A\mathbf{j} = \text{the second column of } A \qquad (3.2)$$

Now we turn to eigenvalues and eigenvectors of 2×2 matrices.

Definition 3.2. Suppose that A is a 2×2 matrix. The real number λ is an **eigenvalue** of A provided that there is a nonzero vector $\mathbf{v}$ in $\mathbb{R}^2$ such that $A\mathbf{v} = \lambda\mathbf{v}$. In this case $\mathbf{v}$ is an **eigenvector** of A (relative to λ).

If $\mathbf{v}$ satisfies $A\mathbf{v} = \mathbf{v}$, then $\mathbf{v}$ looks very much like a "fixed point" of A. For later use we observe that the eigenvalues of any diagonal matrix $\begin{pmatrix} a & 0 \\ 0 & d \end{pmatrix}$ are the diagonal elements a and d (Exercise 14).

It is easy to demonstrate that the following conditions involving eigenvalues and eigenvectors are equivalent (see Exercise 13):

(i) λ is an eigenvalue of A.

(ii) $(A - \lambda I)\mathbf{v} = \mathbf{0}$ for some nonzero vector $\mathbf{v}$.

(iii) $\det(A - \lambda I) = 0$.

Example 3.1.2. Let $A = \begin{pmatrix} 1 & 1 \\ -2 & 4 \end{pmatrix}$. Find the eigenvalues of A, and corresponding eigenvectors.

Solution. By (iii), to find the (real) eigenvalues of A we need only determine the values of λ for which $\det(A - \lambda I) = 0$. To that end, we notice that

$$
\begin{aligned}
\det(A - \lambda I) &= \det \begin{pmatrix} 1 - \lambda & 1 \\ -2 & 4 - \lambda \end{pmatrix} \\
&= (1 - \lambda)(4 - \lambda) + 2 = \lambda^2 - 5\lambda + 6 = (\lambda - 2)(\lambda - 3)
\end{aligned}
$$

Thus 2 and 3 are the eigenvalues of A. To find an eigenvector $\mathbf{v}$ for $\lambda = 2$, we will solve $A\mathbf{v} = \lambda\mathbf{v}$ for $\mathbf{v}$. If $\mathbf{v} = \begin{pmatrix} x \\ y \end{pmatrix}$, this means we must find x and y such that

$$
\begin{pmatrix} 1 & 1 \\ -2 & 4 \end{pmatrix} \begin{pmatrix} x \\ y \end{pmatrix} = 2 \begin{pmatrix} x \\ y \end{pmatrix}, \text{ which is equivalent to } \begin{pmatrix} x + y \\ -2x + 4y \end{pmatrix}
$$
$$
= \begin{pmatrix} 2x \\ 2y \end{pmatrix}
$$

As a result, $x + y = 2x$, so that $y = x$. Therefore any nonzero vector of the form $\begin{pmatrix} x \\ x \end{pmatrix}$, like $\begin{pmatrix} 1 \\ 1 \end{pmatrix}$, is an eigenvector for $\lambda = 2$. For $\lambda = 3$, we need x and y to satisfy

$$
\begin{pmatrix} 1 & 1 \\ -2 & 4 \end{pmatrix} \begin{pmatrix} x \\ y \end{pmatrix} = 3 \begin{pmatrix} x \\ y \end{pmatrix}, \text{ which is equivalent to } \begin{pmatrix} x + y \\ -2x + 4y \end{pmatrix} = \begin{pmatrix} 3x \\ 3y \end{pmatrix}
$$

Thus $x + y = 3x$, so that $y = 2x$. Consequently any nonzero vector of the form $\begin{pmatrix} x \\ 2x \end{pmatrix}$, such as $\begin{pmatrix} 1 \\ 2 \end{pmatrix}$, is an eigenvector for $\lambda = 3$. $\square$

It is easy to find a general formula for the eigenvalues of any 2×2 matrix. To confirm this assertion, let $A = \begin{pmatrix} a & b \\ c & d \end{pmatrix}$. Then

$$
\det(A - \lambda I) = \det \begin{pmatrix} a - \lambda & b \\ c & d - \lambda \end{pmatrix} = (a - \lambda)(d - \lambda) - bc = \lambda^2 - (a + d)\lambda + (ad - bc)
$$

Thus $\det(A - \lambda I) = 0$ if and only if $\lambda^2 - (a + d)\lambda + (ad - bc) = 0$. The roots of this equation are

$$
\lambda = \frac{1}{2}(a + d) \pm \frac{1}{2}\sqrt{(a - d)^2 + 4bc}
$$

If the roots are real, they are eigenvalues of A. If the roots are complex, then we call them **complex eigenvalues**. Notice that if λ is a complex eigenvalue of the matrix A, then there are no nonzero vectors $\mathbf{v}$ in $\mathbb{R}^2$ such that $A\mathbf{v} = \lambda\mathbf{v}$, because in that case $A\mathbf{v}$ has real coordinates and $\lambda\mathbf{v}$ has complex coordinates.

3.1.2 Similar Matrices

The next topic we discuss is the notion of similarity, which will play an important role in the study of dynamics of two-dimensional functions.

Definition 3.3. Two 2×2 matrices A and B are **similar** if there is an invertible 2×2 matrix E such that $EA = BE$. In that case we write $A \approx B$, or we write $A \approx_E B$ if we wish to exhibit E, which is called a **similarity matrix** (for A and B).

Example 3.1.3. Let $A = \begin{pmatrix} 1 & 1 \\ -2 & 4 \end{pmatrix}$ and $B = \begin{pmatrix} 2 & 0 \\ 0 & 3 \end{pmatrix}$. Show that $A \approx B$.

Solution. Let $E = \begin{pmatrix} a & b \\ c & d \end{pmatrix}$. We will find numerical values of a, b, c, and d in order that $EA = BE$. Equivalently, we will solve

$$\begin{pmatrix} a & b \\ c & d \end{pmatrix} \begin{pmatrix} 1 & 1 \\ -2 & 4 \end{pmatrix} = EA = BE = \begin{pmatrix} 2 & 0 \\ 0 & 3 \end{pmatrix} \begin{pmatrix} a & b \\ c & d \end{pmatrix}$$

for a, b, c, and d. Since

$$\begin{pmatrix} a & b \\ c & d \end{pmatrix} \begin{pmatrix} 1 & 1 \\ -2 & 4 \end{pmatrix} = \begin{pmatrix} a - 2b & a + 4b \\ c - 2d & c + 4d \end{pmatrix} \quad \text{and} \quad \begin{pmatrix} 2 & 0 \\ 0 & 3 \end{pmatrix} \begin{pmatrix} a & b \\ c & d \end{pmatrix}$$

$$= \begin{pmatrix} 2a & 2b \\ 3c & 3d \end{pmatrix}$$

we must simultaneously solve the equations

$$\begin{aligned} a - 2b &= 2a \\ a + 4b &= 2b \\ c - 2d &= 3c \\ c + 4d &= 3d \end{aligned}$$

The first two equations yield $a = -2b$, and the last two equations yield $d = -c$. Therefore we can let $E = \begin{pmatrix} -2 & 1 \\ 1 & -1 \end{pmatrix}$, for example. Then $\det E = 1 \neq 0$, so that by Theorem 3.1, E is invertible. A straightforward computation yields $EA = BE$, so that $A \approx_E B$. $\square$

We remark that the invertible matrix E such that $EA = BE$ in Example 3.3.3 is not unique. There are infinitely many such matrices, including rE for any $r \neq 0$. In addition we observe that A is similar to B precisely when B is similar to A. After all, $EA = BE$ if and only if $AE^{-1} = E^{-1}B$.

Next, suppose that $A \approx_E B$. Then $EA = BE$, or equivalently, $A = E^{-1}BE$. It follows that for any positive integer n,

$$\begin{aligned} A^n &= (E^{-1}BE)^n = (E^{-1}BE)(E^{-1}BE) \cdots (E^{-1}BE) = E^{-1}BIBI \cdots IBE \\ &= E^{-1}B^n E \end{aligned}$$

where the expression between the second and third equal signs has n expressions $E^{-1}BE$. Consequently $A^n \approx_E B^n$, so that

$$\text{if } A \approx_E B, \text{ then } A^n \approx_E B^n \tag{3.3}$$

This formula will be useful when we analyze iterates of linear functions on $\mathbb{R}^2$.

Now we will show that any 2×2 matrix A is similar to a matrix of one of three simple forms.

Theorem 3.4. Let $A = \begin{pmatrix} a & b \\ c & d \end{pmatrix}$. For appropriate real numbers $\lambda, \mu, \beta,$ and γ, the matrix A is similar to one of the normal forms

$\begin{pmatrix} \lambda & 0 \\ 0 & \mu \end{pmatrix}$, if A has two distinct real eigenvalues λ, μ

$\begin{pmatrix} \lambda & \beta \\ 0 & \lambda \end{pmatrix}$, if A has one real eigenvalue λ

$\begin{pmatrix} \beta & -\gamma \\ \gamma & \beta \end{pmatrix}$, if A has two complex eigenvalues, $\beta + i\gamma$ and $\beta - i\gamma$

Proof. Suppose that $A = \begin{pmatrix} a & b \\ c & d \end{pmatrix}$ has distinct real eigenvalues, λ and μ. We will show that $B \approx A$, where $B = \begin{pmatrix} \lambda & 0 \\ 0 & \mu \end{pmatrix}$. To that end, let $\mathbf{v} = \begin{pmatrix} r \\ t \end{pmatrix}$ and $\mathbf{w} = \begin{pmatrix} s \\ u \end{pmatrix}$ be nonzero eigenvectors of A corresponding to the eigenvalues λ and μ, respectively, and let $E = \begin{pmatrix} r & s \\ t & u \end{pmatrix}$. We will verify that $EB = AE$ and that E is invertible. First, we notice that

$$EB = \begin{pmatrix} r & s \\ t & u \end{pmatrix} \begin{pmatrix} \lambda & 0 \\ 0 & \mu \end{pmatrix} = \begin{pmatrix} r\lambda & s\mu \\ t\lambda & u\mu \end{pmatrix}$$

and

$$AE = \begin{pmatrix} a & b \\ c & d \end{pmatrix} \begin{pmatrix} r & s \\ t & u \end{pmatrix} = \begin{pmatrix} ar + bt & as + bu \\ cr + dt & cs + du \end{pmatrix}$$

Next, we observe that $A\mathbf{v} = \lambda\mathbf{v}$ and $A\mathbf{w} = \mu\mathbf{w}$, because $\mathbf{v}$ and $\mathbf{w}$ are eigenvectors of A corresponding to the eigenvalues λ and μ, respectively. It follows that

$$\begin{pmatrix} ar + bt \\ cr + dt \end{pmatrix} = \begin{pmatrix} a & b \\ c & d \end{pmatrix} \begin{pmatrix} r \\ t \end{pmatrix} = A\mathbf{v} = \lambda\mathbf{v} = \lambda \begin{pmatrix} r \\ t \end{pmatrix} = \begin{pmatrix} \lambda r \\ \lambda t \end{pmatrix}$$

and

$$\begin{pmatrix} as + bu \\ cs + du \end{pmatrix} = \begin{pmatrix} a & b \\ c & d \end{pmatrix} \begin{pmatrix} s \\ u \end{pmatrix} = A\mathbf{w} = \mu\mathbf{w} = \mu \begin{pmatrix} s \\ u \end{pmatrix} = \begin{pmatrix} \mu s \\ \mu u \end{pmatrix}$$

Therefore we can substitute in the columns of AE to obtain

$$\left(\begin{array}{cc} ar+bt & as+bu \\ cr+dt & cs+du \end{array}\right) = \left(\begin{array}{cc} \lambda r & \mu s \\ \lambda t & \mu u \end{array}\right)$$

Consequently

$$EB = \left(\begin{array}{cc} r\lambda & s\mu \\ t\lambda & u\mu \end{array}\right) = \left(\begin{array}{cc} ar+bt & as+bu \\ cr+dt & cs+du \end{array}\right) = AE$$

To finish the proof we need to certify that E is invertible, which by Theorem 3.1 is equivalent to showing that $\det E \neq 0$. However, if $\det E = 0$, then $ru - st = 0$. Since $\mathbf{v} \neq \mathbf{0}$ and $\mathbf{w} \neq \mathbf{0}$, this means that either

$$\frac{r}{t} = \frac{s}{u} \quad \text{or} \quad \frac{t}{r} = \frac{u}{s}$$

Either way, $\mathbf{v}$ would be a constant multiple of $\mathbf{w}$, say, $\mathbf{v} = q\mathbf{w}$. But then

$$\lambda \mathbf{v} = A\mathbf{v} = A(q\mathbf{w}) = qA\mathbf{w} = q\mu(\mathbf{w}) = \mu(q\mathbf{w}) = \mu\mathbf{v}$$

so that $\lambda = \mu$. But this contradicts the hypothesis that λ and μ are distinct eigenvalues of A. Consequently $\det E \neq 0$, so that E is invertible.

We defer to the exercises the proofs involving one real eigenvalue and complex eigenvalues (see Exercises 21 and 22). $\qquad\square$

What makes the normal forms of matrices appearing in Theorem 3.4 important for the study in dynamics is the following result.

Theorem 3.5. If $A \approx B$, then A and B have identical eigenvalues.

Proof. Suppose that $A \approx_E B$, so that $EA = BE$, or equivalently, $A = E^{-1}BE$. Then by (3.1) and (3.2),

$$\begin{aligned}
\det(A - \lambda I) &= \det(E^{-1}BE - \lambda I) \\
&= \det[E^{-1}BE - E^{-1}(\lambda I)E] \\
&= \det(E^{-1}(B - \lambda I)E] \\
&\overset{(1)}{=} [\det(E^{-1})][\det(B - \lambda I)][\det E] \\
&\overset{(2)}{=} \frac{1}{\det E}[\det(B - \lambda I)][\det E] \\
&= \det(B - \lambda I)
\end{aligned}$$

Therefore λ is an eigenvalue of A if and only if λ is an eigenvalue of B. $\square$

3.1.3   Section 3.1 Exercises

In Exercises 1–3, find the eigenvalues, and the normal form, of the given matrix. Also find an eigenvector for each eigenvalue that is a real number.

Exercise 1. $\left(\begin{array}{cc} 0 & 1 \\ 1 & 0 \end{array}\right)$

Exercise 2. $\begin{pmatrix} 3 & 1 \\ -1 & 1 \end{pmatrix}$

Exercise 3. $\begin{pmatrix} 1 & 1 \\ -1 & 1 \end{pmatrix}$

Exercise 4. Let $A = \begin{pmatrix} 1 & -2 \\ 0 & 3 \end{pmatrix}$. Find $\mathbf{v} = \begin{pmatrix} x \\ y \end{pmatrix}$ such that $A\mathbf{v} = \begin{pmatrix} 0 \\ 1 \end{pmatrix}$.

Exercise 5. Let a and b be real numbers, and $A = \begin{pmatrix} -2a & 1 \\ b & 0 \end{pmatrix}$. Determine the relationship that must hold between a and b in order for A to have a real eigenvalue.

Exercise 6. Let θ denote an angle in radians, and let $A = \begin{pmatrix} \cos\theta & \sin\theta \\ -\sin\theta & \cos\theta \end{pmatrix}$.

(a) Show that A is invertible, and find A^{-1}.

(b) Show that A has a real eigenvalue if and only if θ is a multiple of π.

Exercise 7. Let $A = \begin{pmatrix} a & b \\ c & d \end{pmatrix}$. Describe the relationships that must hold between a, b, c, and d in order for $AB = BA$ for every 2×2 matrix B.

Exercise 8. Let A and B be 2×2 matrices. Prove that $\det(AB) = (\det A)(\det B)$.

Exercise 9. Let A be a 2×2 matrix, and suppose that there is a 2×2 matrix B such that $AB = I$. Show that A has an inverse. (Thus A has an inverse if and only if there is a B such that $AB = I$.)

Exercise 10. Let A be a 2×2 matrix. Show that if B and C are both inverses of A, then $B = C$. (This means that the inverse of A is unique.)

Exercise 11. Let A and B be invertible 2×2 matrices. Show that AB is also invertible.

Exercise 12. Suppose that A, B, and C are 2×2 matrices.

(a) Assume that A is invertible, and that $AB = AC$. Show that $B = C$.

(b) Show that if A is *not* invertible, then there are numbers a, b, and r such that A has the form $\begin{pmatrix} a & b \\ ra & rb \end{pmatrix}$.

(c) Suppose that A is *not* invertible. Show that there are two distinct matrices B and C such that $AB = AC$.

Exercise 13. Let A be a 2×2 matrix. Show that the following three conditions are equivalent:

(a) λ is an eigenvalue of A.

(b) There is a nonzero vector $\mathbf{v}$ such that $(A - \lambda I)\mathbf{v} = \mathbf{0}$.

(c) $\det(A - \lambda I) = 0$.

Exercise 14. Let $A = \begin{pmatrix} a & 0 \\ 0 & d \end{pmatrix}$. Show that the eigenvalues of A are the diagonal elements a and d.

Exercise 15. Show that 0 is an eigenvalue of the matrix A if and only if A is not invertible.

Exercise 16. Suppose that λ is a real eigenvalue of A.

(a) Show that λ^n is an eigenvalue of A^n, for each integer $n \geq 1$.

(b) Show that A and A^n have the same eigenvectors.

Exercise 17. Let A and B be 2×2 matrices. Show that AB and BA have the same eigenvalues.

Exercise 18. Let A be a 2×2 matrix. Show that if $A \approx I$ then $A = I$.

Exercise 19. Let $A = \begin{pmatrix} \lambda & a \\ 0 & \lambda \end{pmatrix}$ and $B = \begin{pmatrix} \lambda & b \\ 0 & \lambda \end{pmatrix}$, where a, b, and λ are real numbers, and a and b are nonzero. Show that $A \approx_E B$, where E is a diagonal matrix.

Exercise 20. Let b and c be nonzero real numbers. Show that $\begin{pmatrix} 0 & b \\ 0 & 0 \end{pmatrix} \approx \begin{pmatrix} 0 & 0 \\ c & 0 \end{pmatrix}$.

Exercise 21. Suppose that the 2×2 matrix A has the single real eigenvalue λ. Show that $A \approx B$ for some matrix $B = \begin{pmatrix} \lambda & \beta \\ 0 & \lambda \end{pmatrix}$.

Exercise 22. Suppose that the 2×2 matrix A has complex eigenvalues $\beta + i\gamma$ and $\beta - i\gamma$, where β and γ are real numbers. Show that $A \approx B$ for the matrix $B = \begin{pmatrix} \beta & -\gamma \\ \gamma & \beta \end{pmatrix}$.

Exercise 23. Consider the square in $\mathbb{R}^2$ with corners at $(0,0), (1,0), (1,1)$, and $(0,1)$. We can think of this square as being bounded by the vectors in the standard basis for $\mathbb{R}^2$.

- Draw the square above in the plane.

- Use the matrix $\begin{pmatrix} 3 & -2 \\ 1 & 0 \end{pmatrix}$ as a linear transformation on the vectors that represent the four corners of your square. Draw the new region in the same plane as your original square.

- Compute the eigenvalues and eigenvectors of this above matrix.

- Interpret the eigenvalues and eigenvectors in terms of how your region in the plane was transformed. Did it stretch or shrink? In what directions did it change shape? Explain any connections you see.

Exercise 24. Repeat Exercise 23, but using the matrix

$$\begin{pmatrix} \sqrt{3}/2 & -1/2 \\ 1/2 & \sqrt{3}/2 \end{pmatrix}.$$

Demonstrate that your square has been rotated by an angle of $\pi/6$. Make a conjecture about the role of the sine and cosine function in a matrix with complex eigenvalues.

Exercise 25. Suppose you have the matrix

$$\begin{pmatrix} a & b \\ c & d \end{pmatrix}$$

with all values integers and $a + b = c + d$. Prove that the eigenvalues of this matrix are $a + b$ and $a - c$.

3.2   Dynamics of Linear Functions

Functions of the form $f(x) = ax$, where a is a constant and x is a real number, are linear functions because their graphs are lines. The dynamics of such functions are simple to analyze (see Exercise 13 of Section 1.2). For linear functions in two dimensions, the analysis is also reasonable. It is the dynamics of such functions that we study in this section.

Note: We will use the terms "fixed point," "attracting," "repelling," and a host of other familiar terms, throughout these next few sections. To clarify, we redefine them each time a new context is introduced. While we've done things in a single dimension before, we will proceed into linear and non-linear two-dimensional dynamics. We encourage students to look at the most general version of the definition when it is introduced, and to verify for themselves that all early, simplified versions of the definition remain valid.

3.2.1   Linear Functions

We begin with the definition of a linear function defined on $\mathbb{R}^2$.

Definition 3.6. The function $L : \mathbb{R}^2 \to \mathbb{R}^2$ is **linear** if

$$L(b\mathbf{v} + c\mathbf{w}) = bL(\mathbf{v}) + cL(\mathbf{w})$$

for all $\mathbf{v}$ and $\mathbf{w}$ in $\mathbb{R}^2$, and all real numbers b and c. A linear function is also called a **linear map**.

The reason that such functions are called linear is that the image of any line is a line or, in special cases, a point. Before we show this in Example 3.5.1, we recall that a line in the plane can be represented parametrically as the collection of all vectors of the form $t\mathbf{v}+\mathbf{w}$, where $\mathbf{v}$ and $\mathbf{w}$ are fixed vectors with $\mathbf{v} \neq \mathbf{0}$, and where t is any real number. The line passes through $\mathbf{w}$, and if $\mathbf{w} = \mathbf{0}$, then the line passes through the origin. The vector $\mathbf{v}$ tells the slope of the line. Specifically, let $\mathbf{v} = \begin{pmatrix} r \\ s \end{pmatrix}$. If $r \neq 0$, then the slope of the line is s/r; if $r = 0$, then the line is vertical.

Example 3.2.1. Let $L : \mathbb{R}^2 \to \mathbb{R}^2$ be linear. Show that the image of any line γ is a line or a point.

Solution. Let γ be the collection of all vectors of the form $t\mathbf{v} + \mathbf{w}$, for all real t. By the definition of linearity,

$$L(t\mathbf{v} + \mathbf{w}) = tL(\mathbf{v}) + L(\mathbf{w})$$

On the one hand, if $L(\mathbf{v}) \neq \mathbf{0}$, then the image of γ is the line passing through $L(\mathbf{w})$ with slope determined by the coordinates of $L(\mathbf{v})$. On the other hand, if $L(\mathbf{v}) = \mathbf{0}$, then the image of γ is the point $L(\mathbf{w})$. $\square$

Any vector $\mathbf{v} = \begin{pmatrix} x \\ y \end{pmatrix}$ in $\mathbb{R}^2$ can be written in terms of the coordinate vectors $\mathbf{i} = \begin{pmatrix} 1 \\ 0 \end{pmatrix}$ and $\mathbf{j} = \begin{pmatrix} 0 \\ 1 \end{pmatrix}$. Indeed,

$$\mathbf{v} = \begin{pmatrix} x \\ y \end{pmatrix} = x \begin{pmatrix} 1 \\ 0 \end{pmatrix} + y \begin{pmatrix} 0 \\ 1 \end{pmatrix} = x\mathbf{i} + y\mathbf{j}$$

As a result, if $L : \mathbb{R}^2 \to \mathbb{R}^2$ is any linear function, and if $\mathbf{v} = \begin{pmatrix} x \\ y \end{pmatrix}$, then by (3.4),

$$L(\mathbf{v}) = L(x\mathbf{i} + y\mathbf{j}) = xL(\mathbf{i}) + y$$

Therefore $L(\mathbf{i})$ and $L(\mathbf{j})$ uniquely define the linear function L.

Assume once again that $L : \mathbb{R}^2 \to \mathbb{R}^2$ is a linear function, and suppose that $L(\mathbf{i}) = \begin{pmatrix} a \\ c \end{pmatrix}$ and $L(\mathbf{j}) = \begin{pmatrix} b \\ d \end{pmatrix}$. Then we define the 2×2 **associated matrix** A_L by

$$A_L = \begin{pmatrix} a & b \\ c & d \end{pmatrix} \tag{3.4}$$

Example 3.2.2. Let $L \begin{pmatrix} x \\ y \end{pmatrix} = \begin{pmatrix} x + y \\ -2x + 4y \end{pmatrix}$ for all x and y in R. Find A_L.

Solution. Since

$$L(\mathbf{i}) = L \begin{pmatrix} 1 \\ 0 \end{pmatrix} = \begin{pmatrix} 1 \\ -2 \end{pmatrix} \quad \text{and} \quad L(\mathbf{j}) = \begin{pmatrix} 0 \\ 1 \end{pmatrix} = \begin{pmatrix} 1 \\ 4 \end{pmatrix}$$

it follows from (3.6) that $A_L = \begin{pmatrix} 1 & 1 \\ -2 & 4 \end{pmatrix}$. $\square$

A natural relationship between a linear function L on $\mathbb{R}^2$ and its associated matrix A_L is given in the next theorem.

Theorem 3.7. Let $L : \mathbb{R}^2 \to \mathbb{R}^2$ be an arbitrary linear function, and A_L the associated matrix defined in 3.4. Then

$$L(\mathbf{v}) = A_L\mathbf{v} \text{ for all } \mathbf{v} \text{ in } \mathbb{R}^2$$

Proof. If $L(\mathbf{i}) = \begin{pmatrix} a \\ c \end{pmatrix}$ and $L(\mathbf{j}) = \begin{pmatrix} b \\ d \end{pmatrix}$ then by (3.6), $A_L = \begin{pmatrix} a & b \\ c & d \end{pmatrix}$. Thus for any $\mathbf{v} = \begin{pmatrix} x \\ y \end{pmatrix}$ we have

$$A_L\mathbf{v} = \begin{pmatrix} a & b \\ c & d \end{pmatrix} \begin{pmatrix} x \\ y \end{pmatrix} = \begin{pmatrix} ax + by \\ cx + dy \end{pmatrix}$$

Next, the formula in (3.5) tells us that

$$L(\mathbf{v}) = L(x\mathbf{i} + y\mathbf{j}) \overset{(2)}{=} xL(\mathbf{i}) + yL(\mathbf{j}) = x \begin{pmatrix} a \\ c \end{pmatrix} + y \begin{pmatrix} b \\ d \end{pmatrix} = \begin{pmatrix} xa + yb \\ xc + yd \end{pmatrix}$$

Consequently $L(\mathbf{v}) = A_L\mathbf{v}$, which completes the proof. $\square$

The following result is a nearly immediate consequence of the theorem.

Corollary 3.8. If L is any linear function, then the matrix A_L is unique. Conversely, if A is any matrix, then there is a unique linear function L such that $A_L = A$.

Proof. Let L be an arbitrary linear function. Since

$$A_L \mathbf{i} = \text{first column of } A_L \quad \text{and} \quad A_L \mathbf{j} = \text{second column of } A_L$$

it follows that the matrix A_L is unique. Conversely, if A is any given matrix, then the function L defined by $L(\mathbf{v}) = A\mathbf{v}$ for all $\mathbf{v}$ in $\mathbb{R}^2$ is linear, and has the property that $A_L = A$ (Exercise 12). $\square$

We know from Example 3.5.1 that if L is a linear function, then the image of a line is a line or a point. One of our goals is to determine conditions under which the image of a line is contained in the same line. We say that a line γ is **invariant** under L if $L(\gamma) \subseteq \gamma$. Thus we seek lines that are invariant for a given linear function.

Suppose that L is a linear function. If λ is a real eigenvalue of A_L and corresponds to the eigenvector $\mathbf{v}$, then the line $\gamma_{\mathbf{v}}$ through the origin and parallel to $\mathbf{v}$ is invariant under L. The reason is that points on $\gamma_{\mathbf{v}}$ have the form $t\mathbf{v}$, and by Theorem 3.7,

$$L(t\mathbf{v}) = A_L(t\mathbf{v}) = \lambda t\mathbf{v} \text{ for any real number } t \tag{3.5}$$

Therefore the image of $\gamma_{\mathbf{v}}$ is contained in the line $\gamma_{\mathbf{v}}$.

Example 3.2.3. Let $L\begin{pmatrix} x \\ y \end{pmatrix} = \begin{pmatrix} x + y \\ -2x + 4y \end{pmatrix}$. Find lines through the origin that are invariant under L.

Solution. By Example 3.2.2, $A_L = \begin{pmatrix} 1 & 1 \\ -2 & 4 \end{pmatrix}$. In Example 3.2.2 of Section 3.1 we found that $\mathbf{v} = \begin{pmatrix} 1 \\ 1 \end{pmatrix}$ and $\mathbf{w} = \begin{pmatrix} 1 \\ 2 \end{pmatrix}$ are eigenvectors for the eigenvalues (respectively 2 and 3) of A_L. Consequently by (3.7) the lines $\gamma_{\mathbf{v}}$ and $\gamma_{\mathbf{w}}$ determined by $\mathbf{v}$ and $\mathbf{w}$ and passing through the origin are invariant under L. $\square$

The linear function L in Example 3.2.3 doubles distances between points in $\gamma_{\mathbf{v}}$ because $\mathbf{v}$ corresponds to the eigenvalue 2 of A_L. Similarly, L triples distances between points on $\gamma_{\mathbf{w}}$, since the corresponding eigenvalue of A_L is 3. Despite these results, we have not completed our analysis of the dynamics of L, because we do not as yet know the behavior of iterates of points not on $\gamma_{\mathbf{v}}$ or $\gamma_{\mathbf{w}}$. Later we will be able to resolve this issue.

Eigenvalues of A_L play an integral part in the analysis of L. For convenience, we will refer to λ as an **eigenvalue** of L if λ is an eigenvalue of the corresponding matrix A_L.

The next theorem shows that composition of linear functions corresponds to multiplication of matrices.

Theorem 3.9. Let $L : \mathbb{R}^2 \to \mathbb{R}^2$ and $M : \mathbb{R}^2 \to \mathbb{R}^2$ be linear functions. Then $A_{L \circ M} = A_L A_M$.

Proof. If $\mathbf{v}$ is arbitrary, then by repeated applications of Theorem 3.7 we have

$$A_{L \circ M} \mathbf{v} = (L \circ M)\mathbf{v} = L(M(\mathbf{v})) = L(A_M \mathbf{v}) = A_L(A_M \mathbf{v}) = A_L A_M \mathbf{v}$$

Since

1st column of $A_{L \circ M} = A_{L \circ M} \mathbf{i} = A_L A_M \mathbf{i} =$ 1st column of $A_L A_M$

and similarly,

2nd column of $A_{L \circ M} =$ 2nd column of $A_L A_M$

we conclude that $A_{L \circ M} = A_L A_M$. □

In the unlikely event that $A_L A_M = I$, Theorems 3.7 and tell us that

$$(L \circ M)\mathbf{v} = A_{L \circ M} \mathbf{v} = (A_L A_M)\mathbf{v} = I\mathbf{v} = \mathbf{v} \text{ for all } \mathbf{v}$$

Consequently $L \circ M = I$. In such a case, by analogy with functions of one variable, we say that L is **invertible**, and that M is the **inverse** of L. We write L^{-1} for the inverse of L. Notice that the linear functions L and M are inverses of one another if their associated matrices A_L and A_M are inverses of each other (see Exercise 13).

Example 3.2.4. Let $A_L = \begin{pmatrix} 3 & 0 \\ 2 & -1 \end{pmatrix}$ and $A_M = \begin{pmatrix} 1/3 & 0 \\ 2/3 & -1 \end{pmatrix}$. Show that L and M are inverses of one another.

Solution. By Theorem 3.9,

$$A_{L \circ M} = A_L A_M = \begin{pmatrix} 3 & 0 \\ 2 & -1 \end{pmatrix} \begin{pmatrix} 1/3 & 0 \\ 2/3 & -1 \end{pmatrix} = \begin{pmatrix} 1 & 0 \\ 0 & 1 \end{pmatrix}$$

By the remarks preceding the example, L and M are inverses of one another. □

If $L = M$, then Theorem 3.9 implies that $A_{L^{[2]}} = (A_L)^2$. More generally,

$$A_{L^{[n]}} = (A_L)^n \text{ for any } n \geq 2 \tag{3.6}$$

To illustrate (3.6), let $L \begin{pmatrix} x \\ y \end{pmatrix} = \begin{pmatrix} x + y \\ -2x + 4y \end{pmatrix}$, so that $A_L = \begin{pmatrix} 1 & 1 \\ -2 & 4 \end{pmatrix}$. Therefore

$$A_{L^{[2]}} = \begin{pmatrix} 1 & 1 \\ -2 & 4 \end{pmatrix}^2 = \begin{pmatrix} -1 & 5 \\ -10 & 14 \end{pmatrix} \text{ and } A_{L^{[3]}} = \begin{pmatrix} 1 & 1 \\ -2 & 4 \end{pmatrix}^3$$

$$= \begin{pmatrix} -11 & 19 \\ -38 & -46 \end{pmatrix}$$

Calculating the entries of $A_{L^{[n]}}$ by using (3.8) becomes cumbersome as n increases. The notion of conjugacy will allow us to evaluate $A_{L^{[n]}}$ with a minimum of effort.

Definition 3.10. Let $L : \mathbb{R}^2 \to \mathbb{R}^2$ and $M : \mathbb{R}^2 \to \mathbb{R}^2$ be linear functions. Then L and M are **linearly conjugate** if there exists an invertible linear function $P : \mathbb{R}^2 \to \mathbb{R}^2$ such that $P \circ L = M \circ P$. In that case we write $L \approx M$, or if we wish to emphasize the role of P, we write $L \approx_P M$.

Linear conjugacy of functions corresponds to similarity of the associated matrices, as Theorem 3.11 suggests.

Theorem 3.11. Let $L : \mathbb{R}^2 \to \mathbb{R}^2$ and $M : \mathbb{R}^2 \to \mathbb{R}^2$ be linear functions. Then $L \approx_P M$ if and only if $A_L \approx_{A_P} A_M$.

Proof. By definition, $L \approx_P M$ if and only if $P \circ L = M \circ P$, which is equivalent to $A_{P \circ L} = A_{M \circ P}$ by Corollary 3.8. By Theorem 3.9, this is tantamount to $A_P A_L = A_M A_P$, which means that $A_L \approx_{A_P} A_M$. $\qquad\square$

Theorem 3.11 and (3.8) help us to solve the following example expeditiously.

Example 3.2.5. Let $A_L = \begin{pmatrix} 1 & 1 \\ -2 & 4 \end{pmatrix}$. Find the entries of $A_{L^{[n]}}$.

Solution. By Example 3.3.3 of Section 3.1, $A_L \approx_E A_M$, where $A_M = \begin{pmatrix} 2 & 0 \\ 0 & 3 \end{pmatrix}$ and $E = \begin{pmatrix} -2 & 1 \\ 1 & -1 \end{pmatrix}$. You can check that $E^{-1} = \begin{pmatrix} -1 & -1 \\ -1 & -2 \end{pmatrix}$. Then (3.8) in the present section and (3.8) in Section 3.1 together yield

$$A_{L^{[n]}} = (A_L)^n = E^{-1}(A_M)^n E = \begin{pmatrix} -1 & -1 \\ -1 & -2 \end{pmatrix} \begin{pmatrix} 2 & 0 \\ 0 & 3 \end{pmatrix}^n \begin{pmatrix} -2 & 1 \\ 1 & -1 \end{pmatrix}$$

$$= \begin{pmatrix} -1 & -1 \\ -1 & -2 \end{pmatrix} \begin{pmatrix} 2^n & 0 \\ 0 & 3^n \end{pmatrix} \begin{pmatrix} -2 & 1 \\ 1 & -1 \end{pmatrix}$$

$$= \begin{pmatrix} 2^{n+1} - 3^n & -2^n + 3^n \\ 2^{n+1} - 2(3^n) & -2^n + 2(3^n) \end{pmatrix}$$

$\square$

Now we are ready to discuss the dynamics of linear functions defined on $\mathbb{R}^2$.

3.2.2 Dynamics of Linear Functions

Every linear function L has the fixed point $\mathbf{0}$, since $L(\mathbf{0}) = \mathbf{0}$. Before we can say whether $\mathbf{0}$ is attracting or repelling (or neither), we need to indicate a distance on $\mathbb{R}^2$. To that end, let $\mathbf{v} = \begin{pmatrix} x \\ y \end{pmatrix}$ and $\mathbf{w} = \begin{pmatrix} r \\ s \end{pmatrix}$. As is usual, we let the **distance** $||\mathbf{v} - \mathbf{w}||$ between $\mathbf{v}$ and $\mathbf{w}$ be the distance between the corresponding points in $\mathbb{R}^2$, that is,

$$||\mathbf{v} - \mathbf{w}|| = \sqrt{(r - x)^2 + (s - y)^2}$$

If $\mathbf{w} = \mathbf{0}$, then we find that $||\mathbf{v}|| = \sqrt{x^2 + y^2}$, which is the distance between the point corresponding to $\mathbf{v}$ and the origin. We observe that $|| \cdot ||$ is a metric on $\mathbb{R}^2$, because for any $\mathbf{v}$ and $\mathbf{w}$,

$$||\mathbf{v}|| \geq 0, \text{ and } ||\mathbf{v}|| = 0 \text{ if and only if } \mathbf{v} = \mathbf{0}$$

$$||r\mathbf{v}|| = |r| \, ||\mathbf{v}|| \text{ for any real number } r$$

$$||\mathbf{v} + \mathbf{w}|| \leq ||\mathbf{v}|| + ||\mathbf{w}|| \text{(triangle inequality)} \tag{3.7}$$

In addition, if $\{\mathbf{v}_n\}_{n=0}^{\infty}$ is a sequence of elements of $\mathbb{R}^2$ and $\mathbf{w}$ is in $\mathbb{R}^2$, then $\{\mathbf{v}_n\}_{n=0}^{\infty}$ **converges to** $\mathbf{w}$ if $||\mathbf{v}_n - \mathbf{w}|| \to 0$ as n increases without bound. In this case we write $\mathbf{v}_n \to \mathbf{w}$.

We say that $L : \mathbb{R}^2 \to \mathbb{R}^2$ is **continuous at 0** if

for all $\epsilon > 0$ there is a $\delta > 0$ such that if $||\mathbf{v}|| < \delta$, then $||L(\mathbf{v})|| < \epsilon$

This corresponds to the notion of continuity for functions of one variable and is equivalent to the following:

$$\text{if } \mathbf{v}_n \to \mathbf{0}, \text{ then } L(\mathbf{v}_n) \to \mathbf{0} \tag{3.8}$$

In order to prove that every linear function L on $\mathbb{R}^2$ is continuous at $\mathbf{0}$, we need the following preliminary result.

Theorem 3.12. Let $A = \begin{pmatrix} a & b \\ c & d \end{pmatrix}$ be any 2×2 matrix. Suppose that $\mathbf{v}_n = \begin{pmatrix} x_n \\ y_n \end{pmatrix} = x_n\mathbf{i} + y_n\mathbf{j}$ for $n = 1, 2, \ldots$, and that $\mathbf{v}_n \to \mathbf{0}$. Then $A\mathbf{v}_n \to \mathbf{0}$.

Proof. Let $L : \mathbb{R}^2 \to \mathbb{R}^2$ be the linear function guaranteed by Corollary 3.8 to exist such that $L(\mathbf{v}) = A\mathbf{v}$ for all $\mathbf{v}$. Also, let $\mathbf{v}_n = x_n\mathbf{i} + y_n\mathbf{j}$. Since $||\mathbf{v}_n|| = \sqrt{x_n^2 + y_n^2}$, it follows that if $\mathbf{v}_n \to \mathbf{0}$, then $x_n \to 0$ and $y_n \to 0$. Consequently

$$A\mathbf{v}_n = L(\mathbf{v}_n) = L(x_n\mathbf{i} + y_n\mathbf{j}) = x_nL(\mathbf{i}) + y_nL(\mathbf{j}) \to \mathbf{0}$$

$\square$

Corollary 3.13. Let $L : \mathbb{R}^2 \to \mathbb{R}^2$ be a linear function. Then L is continuous at $\mathbf{0}$.

Proof. Let $\mathbf{v}_n \to \mathbf{0}$. Then $L(\mathbf{v}_n) = A_L\mathbf{v}_n \to \mathbf{0}$ by Theorem 3.12. Therefore (3.10) implies that L is continuous at $\mathbf{0}$. □

If L is linear on $\mathbb{R}^2$ and $\mathbf{v}$ is any element of $\mathbb{R}^2$, and if $\mathbf{v}_n \to \mathbf{0}$, then by Corollary 3.13,

$$L(\mathbf{v} + \mathbf{v}_n) = L(\mathbf{v}) + L(\mathbf{v}_n) \to L(\mathbf{v}) + \mathbf{0} = L(\mathbf{v})$$

so that L is continuous at $\mathbf{v}$. Thus if L is linear on $\mathbb{R}^2$, then L is a continuous function on $\mathbb{R}^2$.

Our next goal is to show that if L has two real eigenvalues λ and μ with $|\lambda| < 1$ and $|\mu| < 1$, then for each $\mathbf{v}$ in $\mathbb{R}^2$, $L^{[n]}(\mathbf{v}) \to \mathbf{0}$. If $L^{[n]}(\mathbf{v}) \to \mathbf{0}$ for all $\mathbf{v}$ in $\mathbb{R}^2$, we call $\mathbf{0}$ an **attracting fixed point** of L.

Theorem 3.14. Let $L : \mathbb{R}^2 \to \mathbb{R}^2$ have the property that A_L has distinct real eigenvalues λ and μ, with $|\lambda| < 1$ and $|\mu| < 1$. Then for every $\mathbf{v}$ in $\mathbb{R}^2$, $L^{[n]}(\mathbf{v}) \to \mathbf{0}$ as $n \to \infty$. Therefore $\mathbf{0}$ is an attracting fixed point of L, and $\mathbb{R}^2$ is the basin of attraction of $\mathbf{0}$.

Proof. Let $\mathbf{v}$ be arbitrary in $\mathbb{R}^2$. Since $L^{[n]}(\mathbf{v}) = (A_L)^n(\mathbf{v})$ by (3.8), we need only show that $(A_L)^n\mathbf{v} \to \mathbf{0}$ as $n \to \infty$. Now let $B = \begin{pmatrix} \lambda & 0 \\ 0 & \mu \end{pmatrix}$. By Theorem 3.4, $A_L \approx B$, so that there is an invertible matrix E such that $A_L \approx_E B$. Then (3.8) in Section 3.1 implies that $(A_L)^n \approx_E B^n$, so that $(A_L)^n = E^{-1}B^nE$, and hence

$$(A_L)^n\mathbf{v} = E^{-1}B^nE\mathbf{v} = E^{-1}(B^nE\mathbf{v}) \tag{3.9}$$

The proof will be complete if we show that $E^{-1}(B^nE\mathbf{v}) \to \mathbf{0}$.

From Example 3.5.1 in Section 3.1 we know that $B^n = \begin{pmatrix} \lambda^n & 0 \\ 0 & \mu^n \end{pmatrix}$.

Next, let $E\mathbf{v} = \begin{pmatrix} x \\ y \end{pmatrix}$. Then

$$B^nE\mathbf{v} = \begin{pmatrix} \lambda^n & 0 \\ 0 & \mu^n \end{pmatrix}\begin{pmatrix} x \\ y \end{pmatrix} = \begin{pmatrix} \lambda^n x \\ \mu^n y \end{pmatrix}$$

Since $|\lambda| < 1$ and $|\mu| < 1$ by hypothesis, we find that

$$\|B^nE\mathbf{v}\| = \left\|\begin{pmatrix} \lambda^n x \\ \mu^n y \end{pmatrix}\right\| = \sqrt{\lambda^{2n}x^2 + \mu^{2n}y^2} \to 0$$

as n increases without bound. Therefore (3.11) and Theorem 3.12, with $\mathbf{v}_n$ replaced by $B^nE\mathbf{v}$, tell us that

$$L^{[n]}(\mathbf{v}) = (A_L)^n\mathbf{v} = E^{-1}(B^nE\mathbf{v}) \to \mathbf{0}$$

Since $\mathbf{v}$ is an arbitrary vector in $\mathbb{R}^2$, it follows that not only is $\mathbf{0}$ an attracting fixed point of L, but also the basin of attraction of $\mathbf{0}$ is $\mathbb{R}^2$ itself. This completes the proof. □

The conclusions of Theorem 3.14 remain valid if the eigenvalues are not necessarily real, as the next result states.

Corollary 3.15. Let $L : \mathbb{R}^2 \to \mathbb{R}^2$ be a linear function with a single eigenvalue λ with $|\lambda| < 1$, or two complex eigenvalues $\beta + i\lambda$ and $\beta - i\lambda$, where β and λ are real numbers and $\beta^2 + \lambda^2 < 1$. Then $L^{[n]}(\mathbf{v}) \to \mathbf{0}$ for each $\mathbf{v}$ in $\mathbb{R}^2$. Therefore $\mathbf{0}$ is an attracting fixed point whose basin of attraction is $\mathbb{R}^2$.

Proof. Theorem 3.14 proves the result if λ and μ are real and distinct. The cases in which L has one real eigenvalue, or complex eigenvalues, are left as exercises (Exercises 19 and 17, respectively). □

We observe that if $|\lambda| > 1$ and $|\mu| > 1$, then the proof of Theorem 3.14 shows that $\|L^{[n]}(\mathbf{v})\| \to \infty$ as n increases without bound. In this case we say that $\mathbf{0}$ is a **repelling fixed point** of L. For example, if $L\begin{pmatrix} x \\ y \end{pmatrix} = \begin{pmatrix} x + y \\ -2x + 4y \end{pmatrix}$, then the eigenvalues are 2 and 3, by Example 3.3.2 of Section 3.1. It follows that $\mathbf{0}$ is a repelling fixed point, and iterates for L of all points except $\mathbf{0}$ recede from $\mathbf{0}$.

Lines determined by eigenvectors are invariant under a linear function L, as we have already seen. We can extend the notion of invariance to other sets. A set C in $\mathbb{R}^2$ is **invariant** under a linear function $L : \mathbb{R}^2 \to \mathbb{R}^2$ if $L(C) \subseteq C$. Thus C is invariant under L if C is mapped into itself by L.

Using invariant parabolas, we can determine the dynamics of the linear function in the following example.

Example 3.2.6. Let $L\begin{pmatrix} x \\ y \end{pmatrix} = \begin{pmatrix} x/2 \\ y/4 \end{pmatrix}$. Show that $\mathbf{0}$ is an attracting fixed point whose basin of attraction is $\mathbb{R}^2$. Also show that for any number $r \neq 0$, the parabola $y = rx^2$ is invariant under L.

Solution. We find that

$$L\begin{pmatrix} x \\ y \end{pmatrix} = \begin{pmatrix} 1/2 & 0 \\ 0 & 1/4 \end{pmatrix}\begin{pmatrix} x \\ y \end{pmatrix}$$

so that the eigenvalues of L are $1/2$ and $1/4$. Theorem 3.14 implies that $\mathbf{0}$ is an attracting fixed point whose basin of attraction is $\mathbb{R}^2$. If $y = rx^2$ with $r \neq 0$, then

$$L\begin{pmatrix} x \\ y \end{pmatrix} = L\begin{pmatrix} x \\ rx^2 \end{pmatrix} = \begin{pmatrix} x/2 \\ rx^2/4 \end{pmatrix}\begin{pmatrix} x/2 \\ r(x/2)^2 \end{pmatrix}$$

The latter point lies on the parabola $y = rx^2$, so it is invariant under L. □

In addition to the invariant parabolas, the axes are also invariant under L in Example 3.2.6. Since every point in $\mathbb{R}^2$ lies on some parabola $y = rx^2$ or on a coordinate axis, it follows that the iterates of every vector $\mathbf{v}$ converge to $\mathbf{0}$ along a suitable parabola or along a coordinate axis. With these observations, we are ready to draw a figure, called a **portrait** or **phase portrait**, that

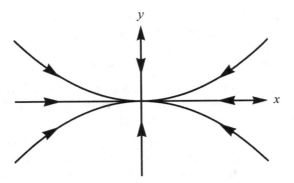

FIGURE 3.1
Phase portrait for a linear function

describes the dynamics of L. Figure 3.1 is a portrait of the linear function L in Example 3.2.6.

More generally, when a linear function L has two real eigenvalues whose absolute values are less than 1, the portrait of L can have the same general form as that in Figure 3.1, the invariant lines and curves adjusted appropriately. To support this claim, let $L\begin{pmatrix} x \\ y \end{pmatrix} = \begin{pmatrix} ax \\ dy \end{pmatrix}$, so that $A_L = \begin{pmatrix} a & 0 \\ 0 & d \end{pmatrix}$. Suppose $a > 0$ and $d > 0$. We will determine a value of α for which the graph C of $y = rx^\alpha$ is invariant under L for every real number r and all positive x. If $\begin{pmatrix} x \\ y \end{pmatrix}$ lies on C, then $L\begin{pmatrix} x \\ y \end{pmatrix} = \begin{pmatrix} ax \\ drx^\alpha \end{pmatrix}$. This latter point lies on C only if

$$drx^\alpha = r(ax)^\alpha, \text{ so that } \alpha = \frac{\ln d}{\ln a} \qquad (3.10)$$

As a result, curves of the form $y = rx^\alpha$ are invariant under L.

Now suppose that the two real eigenvalues of L satisfy $|\lambda| < 1$ and $|\mu| < 1$, but the associated matrix is not necessarily diagonal. This is where similar matrices play a decisive role. The reason is that if M corresponds to a diagonal matrix such that $L \approx_P M$, and if C is a set that is invariant under M, then

$$(PLP^{-1})(C) = M(C) \subseteq C, \text{ so that } L(P^{-1}(C)) \subseteq P^{-1}(C)$$

This means that $P^{-1}(C)$ is the adjusted curve and is invariant under L.

Example 3.2.7. Let $A_L = \begin{pmatrix} 1/2 & 1/8 \\ 1/2 & 1/2 \end{pmatrix}$. Analyze the dynamics of L.

Solution. We observe that the eigenvalues of A_L are $1/4$ and $3/4$, since $\det(A_L - \lambda I) = \lambda^2 - \lambda + \dfrac{3}{16} = 0$ if $\lambda = 1/4$ or $3/4$. You can calculate that corresponding eigenvectors for A_L are $\begin{pmatrix} 1 \\ -2 \end{pmatrix}$ and $\begin{pmatrix} 1 \\ 2 \end{pmatrix}$. Next, if

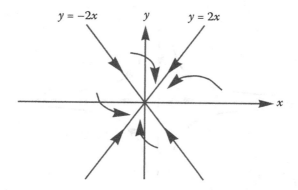

FIGURE 3.2
Phase portrait for Example 3.2.7

$A_M = \begin{pmatrix} 1/4 & 0 \\ 0 & 3/4 \end{pmatrix}$, then by Theorems 3.4 and , $A_L \approx A_M$, and also A_L and A_M have the same eigenvalues. Using (9), we calculate that with respect to M, iterates of points converge to **0** along curves C of the form $y = rx^\alpha$, where $x > 0$ and

$$\alpha = \frac{\ln(3/4)}{\ln(1/4)} = \frac{\ln 3 - \ln 4}{\ln 1 - \ln 4} = 1 - \frac{\ln 3}{\ln 4} \approx .2075$$

If P is a linear function such that $L \approx_P M$, then from the remarks preceding the example, $P^{-1}(C)$ is invariant under L. As a result, we can draw a portrait of L in Figure 3.2. $\square$

Next, we will discover what happens when L has real eigenvalues λ and μ with $|\lambda| > 1$ and $|\mu| < 1$.

Example 3.2.8. Let $L\begin{pmatrix} x \\ y \end{pmatrix} = \begin{pmatrix} 2x \\ y/2 \end{pmatrix}$. Show that for each $r \neq 0$, the hyperbola $y = r/x$ is invariant under L. Then analyze the dynamics of L and draw a portrait of L.

Solution. Let $r \neq 0$. If $y = r/x$, then

$$L\begin{pmatrix} x \\ y \end{pmatrix} = L\begin{pmatrix} x \\ r/x \end{pmatrix} = \begin{pmatrix} 2x \\ r/(2x) \end{pmatrix}$$

Since the latter point lies on the hyperbola $y = r/x$, it follows that the hyperbola is invariant under L. Next, we notice that if $y = r/x$, then

$$L^{[n]}\begin{pmatrix} x \\ y \end{pmatrix} = \begin{pmatrix} 2^n x \\ r/(2^n x) \end{pmatrix}$$

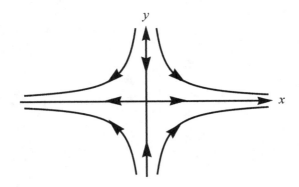

FIGURE 3.3
Phase portrait for Example 3.2.8

As n increases without bound, $L^{[n]}\begin{pmatrix} x \\ y \end{pmatrix}$ approaches the x-axis (since $r/(2^n x) \to 0$) and recedes from the y-axis (because $|2^n x| \to \infty$). Every point not on the x- or y-axis is on a hyperbola $y = r/x$ for an appropriate value of r. We conclude that the iterates of all points not on either the x- or y-axis eventually recede from the origin (Figure 3.3). Since

$$L^{[n]}\begin{pmatrix} x \\ 0 \end{pmatrix} = \begin{pmatrix} 2^n x \\ 0 \end{pmatrix} \quad \text{and} \quad L^{[n]}\begin{pmatrix} 0 \\ y \end{pmatrix} \to \mathbf{0}$$

iterates of all points on the x-axis except for $\mathbf{0}$ also recede from $\mathbf{0}$, whereas iterates of all points on the y-axis except for $\mathbf{0}$ converge to $\mathbf{0}$. These conclusions are registered in the portrait of L given in 3.3. $\square$

For L, if λ and μ are real, and $|\lambda| > 1$ and $|\mu| < 1$, then $\mathbf{0}$ is a **saddle point**. Also, if A_L is a diagonal matrix, with $\lambda > 1$ and $0 < \mu < 1$, then there is an $\alpha < 0$ such that the curve $y = rx^\alpha$ is invariant, for each $r \neq 0$ (see Exercise 20). The portrait of such an L has the same general appearance as that in Figure 3.3, with appropriate distortions.

Next, we turn to linear functions whose associated eigenvalues are not real.

Example 3.2.9. Let $L\begin{pmatrix} x \\ y \end{pmatrix} = \begin{pmatrix} -y \\ x \end{pmatrix}$. Find the eigenvalues of L, and discuss the dynamics of L.

Solution. Notice that

$$L\begin{pmatrix} x \\ y \end{pmatrix} = \begin{pmatrix} 0 & -1 \\ 1 & 0 \end{pmatrix}\begin{pmatrix} x \\ y \end{pmatrix}$$

Since

$$\det\begin{pmatrix} 0 - \lambda & -1 \\ 1 & 0 - \lambda \end{pmatrix} = \lambda^2 + 1 = 0 \text{ if and only if } \lambda = \pm i$$

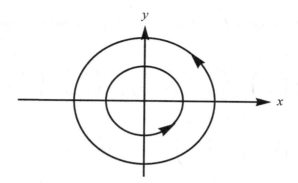

FIGURE 3.4
Phase portrait for Example 3.2.9

it follows that the eigenvalues of L are the complex numbers i and $-i$. Next, we use polar coordinates for x and y:

$$x = r \cos \theta \quad \text{and} \quad y = r \sin \theta$$

From trigonometry we know that

$$-\sin \theta = \cos(\theta + \pi/2) \quad \text{and} \quad \cos \theta = \sin(\theta + \pi/2)$$

so that

$$L \left(\begin{array}{c} r \cos \theta \\ r \sin \theta \end{array} \right) = \left(\begin{array}{c} -r \sin \theta \\ r \cos \theta \end{array} \right) = \left(\begin{array}{c} r \cos \left(\theta + \pi/2 \right) \\ r \sin \left(\theta + \pi/2 \right) \end{array} \right) \tag{3.11}$$

The formula in (10) tells us that if $\mathbf{v}$ has polar coordinates (r, θ), then $L(\mathbf{v})$ has polar coordinates $(r, \theta + \pi/2)$. In other words, $L(\mathbf{v})$ lies on the same circle of radius r centered at $\mathbf{0}$, but is rotated counterclockwise $\pi/2$ radians (3.4). Evidently each such circle is invariant under L. $\square$

In general, if $L : \mathbb{R}^2 \to \mathbb{R}^2$ is linear, and if

$$A_L = \left(\begin{array}{cc} a & -b \\ b & a \end{array} \right) \quad \text{with } a^2 + b^2 = 1 \text{ and } a \neq 0$$

then it can be shown that L acts as a rotation through an angle of $\arctan(b/a)$. Equivalently, if

$$A_L = \left(\begin{array}{cc} \cos \theta & -\sin \theta \\ \sin \theta & \cos \theta \end{array} \right)$$

then L acts as a rotation about the origin through an angle of θ in the counterclockwise direction (Exercise 7). For such L, there are *no* real-valued eigenvectors.

In our final example we discuss what can happen if L has but one real eigenvalue.

Example 3.2.10. Let $L\begin{pmatrix} x \\ y \end{pmatrix} = \begin{pmatrix} 2x + y \\ 2y \end{pmatrix}$. Discuss the dynamics of L.

Solution. Here we have

$$L\begin{pmatrix} x \\ y \end{pmatrix} = \begin{pmatrix} 2 & 1 \\ 0 & 2 \end{pmatrix}\begin{pmatrix} x \\ y \end{pmatrix} = 2\begin{pmatrix} x \\ y \end{pmatrix} + \begin{pmatrix} y \\ 0 \end{pmatrix}$$

Letting

$$L_1\begin{pmatrix} x \\ y \end{pmatrix} = 2\begin{pmatrix} x \\ y \end{pmatrix} \quad \text{and} \quad L_2\begin{pmatrix} x \\ y \end{pmatrix} = \begin{pmatrix} y \\ 0 \end{pmatrix}$$

we notice that L_1 multiplies the distance between $\begin{pmatrix} x \\ y \end{pmatrix}$ and $\mathbf{0}$ by a factor of 2, and L_2 projects $\begin{pmatrix} x \\ y \end{pmatrix}$ onto the y-axis and then rotates $90°$ clockwise. □

As you can see, the values of the eigenvalues λ and μ of the linear function L determine the behavior of the iterates of L. We summarize some of the behaviors below:

(i) Suppose that λ and μ are real. If $|\lambda| < 1$ and $|\mu| < 1$, we have $L^{[n]}(\mathbf{v}) \to \mathbf{0}$ for all $\mathbf{v}$ in $\mathbb{R}^2$, so that $\mathbf{0}$ is an attracting fixed point of L. If $|\lambda| > 1$ and $|\mu| \geq 1$, then $||L^{[n]}(\mathbf{v})|| \to \infty$ for all nonzero $\mathbf{v}$ in $\mathbb{R}^2$, so that $\mathbf{0}$ is a repelling fixed point of L.

(ii) Suppose that λ and μ are real. If $|\lambda| > 1$ and $|\mu| < 1$, then $\mathbf{0}$ is a saddle point.

(iii) If the eigenvalues λ and μ are complex, then L has a rotation component.

3.2.3   Section 3.2 Exercises

Exercise 1. Let $L\begin{pmatrix} x \\ y \end{pmatrix} = \begin{pmatrix} x/2 + y/8 \\ x/2 + y/2 \end{pmatrix}$, so that L is the linear function in Example 3.2.7.

(a) Confirm the eigenvalues and corresponding eigenvectors of L.

(b) Confirm the value of α such that the graph of $y = rx^\alpha$ is invariant under L for each $r \neq 0$.

(c) Confirm that Figure 3.2 is a portrait of L.

In Exercises 2–6, let $L : \mathbb{R}^2 \to \mathbb{R}^2$. Find the eigenvalues of L, and the eigenvectors where they exist. Draw a portrait of L.

Exercise 2. $L\begin{pmatrix} x \\ y \end{pmatrix} = \begin{pmatrix} x/2 \\ y/3 \end{pmatrix}$

Exercise 3. $L\begin{pmatrix} x \\ y \end{pmatrix} = \begin{pmatrix} x/3 \\ 3y \end{pmatrix}$

Exercise 4. $L\begin{pmatrix} x \\ y \end{pmatrix} = \begin{pmatrix} 4y \\ -x \end{pmatrix}$

Exercise 5. $L\begin{pmatrix} x \\ y \end{pmatrix} = \begin{pmatrix} x + y \\ -2x + 4y \end{pmatrix}$

Exercise 6. $L\begin{pmatrix} x \\ y \end{pmatrix} = \begin{pmatrix} -x/2 \\ 9x/2 + 4y \end{pmatrix}$

Exercise 7. Let $L : \mathbb{R}^2 \to \mathbb{R}^2$ be linear, with

$$A_L = \begin{pmatrix} \cos\theta & -\sin\theta \\ \sin\theta & \cos\theta \end{pmatrix}$$

Show that L represents a rotation of angle θ in the counterclockwise direction about the origin.

Exercise 8. Find a linear function $L : \mathbb{R}^2 \to \mathbb{R}^2$ such that $L^{[3]} = I$ and $L \neq I$. (*Hint*: Use the result of Exercise 7.)

Exercise 9. Let $L\begin{pmatrix} x \\ y \end{pmatrix} = \begin{pmatrix} ax + by \\ dy \end{pmatrix}$. Show that if $L^{[3]} = I$, then $L = I$.

Exercise 10. Let $L\begin{pmatrix} x \\ y \end{pmatrix} = \begin{pmatrix} ax + by \\ dy \end{pmatrix}$. Find nonzero values of a, b, and d such that $L^{[4]} = I$.

Exercise 11. Let $a \neq 0$ and $b \neq 0$, and suppose that $L\begin{pmatrix} x \\ y \end{pmatrix} = \begin{pmatrix} ax \\ by \end{pmatrix}$ for all x and y in R. Show that the image of any circle centered at the origin is an ellipse centered at the origin.

Exercise 12. Let A be an arbitrary 2×2 matrix. Define $L : \mathbb{R}^2 \to \mathbb{R}^2$ by $L(\mathbf{v}) = A\mathbf{v}$, for all $\mathbf{v}$ in $\mathbb{R}^2$. Show that L is a linear function and that $A_L = A$.

Exercise 13. Let $L : \mathbb{R}^2 \to \mathbb{R}^2$ and $M : \mathbb{R}^2 \to \mathbb{R}^2$. Show that $M = L^{-1}$ if and only if $A_M = (A_L)^{-1}$.

Exercise 14. Let $L : \mathbb{R}^2 \to \mathbb{R}^2$. Show that L is linear if and only if there are real numbers a, b, c, and d such that $L\begin{pmatrix} x \\ y \end{pmatrix} = \begin{pmatrix} ax + by \\ cx + dy \end{pmatrix}$ for all real x and y.

Exercise 15. Let $L : \mathbb{R}^2 \to \mathbb{R}^2$ be a linear function.

(a) Suppose that there is a fixed point $\mathbf{v}_0$ that is not the origin. Show that L may *not* be the identity function I, where $I(\mathbf{v}) = \mathbf{v}$ for all $\mathbf{v}$ in $\mathbb{R}^2$.

(b) Suppose that there are nonzero fixed points $\mathbf{v}_0$ and $\mathbf{w}_0$ such that $\mathbf{w}_0$ is not a multiple of $\mathbf{v}_0$. Determine whether or not L must be I. Explain your answer.

Exercise 16. Use (3.9) to show that $\|\mathbf{v} - \mathbf{w}\| \geq |\,\|\mathbf{v}\| - \|\mathbf{w}\|\,|$ for all $\mathbf{v}$ and $\mathbf{w}$ in $\mathbb{R}^2$.

Exercise 17. Let $L : \mathbb{R}^2 \to \mathbb{R}^2$, and suppose that A_L has complex eigenvalues $\beta + i\gamma$ and $\beta - i\gamma$, with $\beta^2 + \gamma^2 < 1$. Show that $\lim_{n \to \infty} L^{[n]}(\mathbf{v}) = \mathbf{0}$ for all $\mathbf{v}$ in $\mathbb{R}^2$.

Exercise 18. Suppose that $|\lambda| < 1$, and let $\epsilon > 0$ be so small that $|\lambda| + \epsilon < 1$. Also, let $A = \begin{pmatrix} \lambda & \epsilon \\ 0 & \lambda \end{pmatrix}$. Show that if $c = |\lambda| + \epsilon$, then $\|A^n \mathbf{v}\| \leq c^n \|\mathbf{v}\|$ for every integer $n \geq 1$, and hence that $\lim_{n \to \infty} L^{[n]}(\mathbf{v}) = \mathbf{0}$ for all $\mathbf{v}$ in $\mathbb{R}^2$.

Exercise 19. Let $L : \mathbb{R}^2 \to \mathbb{R}^2$, and suppose that A_L has but one real eigenvalue λ, with $|\lambda| < 1$. Show that $\lim_{n \to \infty} L^{[n]}(\mathbf{v}) = \mathbf{0}$ for all $\mathbf{v}$ in $\mathbb{R}^2$.

Exercise 20. Let the linear function $L : \mathbb{R}^2 \to \mathbb{R}^2$ have real eigenvalues λ and μ such that $\lambda > 1$ and $0 < \mu < 1$. Assume that $A_L \approx A_M$, with A_M a diagonal matrix, and $A_L = E^{-1} A_m E$. Then show that for $\alpha = (\ln \mu)/(\ln \lambda)$, the curves $y = E \begin{pmatrix} x \\ rx^\alpha \end{pmatrix}$ are invariant under M, for all $r \neq 0$.

Exercise 21. Prove or disprove: The determinant is a linear function on matrices.

Exercise 22. Assume T is a matrix, and T^4 is the matrix with all entries zero. Show that $(I - T)^{-1} = I + T + T^2 + T^3$. Could you generalize this to the case where T^n is the zero matrix?

3.3 Nonlinear Maps

Section 3.2 was devoted to linear functions defined on $\mathbb{R}^2$. In the remainder of the chapter we will discuss functions defined in $\mathbb{R}^2$ that are not necessarily linear. In contrast to linear functions, whose dynamics are relatively tame, nonlinear functions can have very rich dynamics. The present section is preparatory, formulating concepts that will play a role in the ensuing discussions of nonlinear functions.

Let V be a subset of $\mathbb{R}^2$, and let $F : V \to \mathbb{R}^2$. Frequently such a function is called a **map**. The function F can always be represented in the form

$$F(\mathbf{v}) = \begin{pmatrix} f(\mathbf{v}) \\ g(\mathbf{v}) \end{pmatrix} \quad \text{for all } \mathbf{v} \text{ in } V$$

where f and g are real-valued **coordinate functions** of F. For example, if

$$F \begin{pmatrix} x \\ y \end{pmatrix} = \begin{pmatrix} y \\ a \sin x + by \end{pmatrix} \tag{3.12}$$

then $V = \mathbb{R}^2$, $f \begin{pmatrix} x \\ y \end{pmatrix} = y$, and $g \begin{pmatrix} x \\ y \end{pmatrix} = a \sin x + by$. If the constants a and b in (1) are negative real numbers, then F could represent the motion of a damped, unforced pendulum, which we will study in more detail in Chapter 4.

Just as the derivative is important in the analysis of a function of one variable, the differential plays a key role in the study of functions of several variables.

Definition 3.16. Let V be a subset of $\mathbb{R}^2$, and consider $F : V \to \mathbb{R}^2$. Assume that the first partials of the coordinate functions f and g of F exist at $\mathbf{v}_0$. The **differential** of F at $\mathbf{v}_0$ is the linear function $DF(\mathbf{v}_0)$ defined on $\mathbb{R}^2$ by

$$[DF(\mathbf{v}_0)](\mathbf{v}) = \begin{pmatrix} \dfrac{\partial f}{\partial x}(\mathbf{v}_0) & \dfrac{\partial f}{\partial y}(\mathbf{v}_0) \\ \dfrac{\partial g}{\partial x}(\mathbf{v}_0) & \dfrac{\partial g}{\partial y}(\mathbf{v}_0) \end{pmatrix} \mathbf{v} \quad \text{for all } \mathbf{v} \text{ in } \mathbb{R}^2$$

Notice that $DF(\mathbf{v}_0)$ is a linear function, by Corollary 3.8. We will normally identify $DF(\mathbf{v}_0)$ with the associated **Jacobian matrix**

$$\begin{pmatrix} \dfrac{\partial f}{\partial x}(\mathbf{v}_0) & \dfrac{\partial f}{\partial y}(\mathbf{v}_0) \\ \dfrac{\partial g}{\partial x}(\mathbf{v}_0) & \dfrac{\partial g}{\partial y}(\mathbf{v}_0) \end{pmatrix}$$

The one-dimensional version of $DF(\mathbf{v}_0)$ is the derivative $f'(x_0)$ of the linear function L defined by $L(x) = [f'(x_0)]x$ for all real numbers x. Analogously, $DF(\mathbf{v}_0)$ can be considered as a two-dimensional derivative.

Example 3.3.1. Let $F \begin{pmatrix} x \\ y \end{pmatrix} = \begin{pmatrix} y \\ a \sin x + by \end{pmatrix}$. Find $DF \begin{pmatrix} x_0 \\ y_0 \end{pmatrix}$.

Solution. We have $f\begin{pmatrix} x \\ y \end{pmatrix} = y$ and $g\begin{pmatrix} x \\ y \end{pmatrix} = a\sin x + by,$ so that

$$\frac{\partial f}{\partial x} = 0, \quad \frac{\partial f}{\partial y} = 1, \quad \frac{\partial g}{\partial x} = a\cos x, \quad \frac{\partial g}{\partial y} = b$$

Therefore at $\begin{pmatrix} x_0 \\ y_0 \end{pmatrix}$ the partials are

$$\frac{\partial f}{\partial x} = 0, \quad \frac{\partial f}{\partial y} = 1, \quad \frac{\partial g}{\partial x} = a\cos x_0, \quad \frac{\partial g}{\partial y} = b$$

Consequently

$$DF\begin{pmatrix} x_0 \\ y_0 \end{pmatrix}(\mathbf{v}) = \begin{pmatrix} 0 & 1 \\ a\cos x_0 & b \end{pmatrix}\mathbf{v} \text{ for all } \mathbf{v} \text{ in } \mathbb{R}^2$$

or in the Jacobian matrix form,

$$DF\begin{pmatrix} x_0 \\ y_0 \end{pmatrix} = \begin{pmatrix} 0 & 1 \\ a\cos x_0 & b \end{pmatrix}$$

□

Example 3.3.2. Let $L\begin{pmatrix} x \\ y \end{pmatrix} = \begin{pmatrix} ax + by \\ cx + dy \end{pmatrix} = \begin{pmatrix} a & b \\ c & d \end{pmatrix}\begin{pmatrix} x \\ y \end{pmatrix}$. Find $DL\begin{pmatrix} 0 \\ 0 \end{pmatrix}$.

Solution. In this case we have $f\begin{pmatrix} x \\ y \end{pmatrix} = ax + by$ and $g\begin{pmatrix} x \\ y \end{pmatrix} = cx + dy.$ Therefore

$$\frac{\partial f}{\partial x} = a, \quad \frac{\partial f}{\partial y} = b, \quad \frac{\partial g}{\partial x} = c, \quad \frac{\partial g}{\partial y} = d$$

It follows that

$$DL\begin{pmatrix} 0 \\ 0 \end{pmatrix} = \begin{pmatrix} a & b \\ c & d \end{pmatrix}$$

so that $DL\begin{pmatrix} 0 \\ 0 \end{pmatrix} = A_L.$ □

Example 3.3.2 tells us that for a linear function L on $\mathbb{R}^2$, the differential at the origin $\mathbf{0}$ is the same linear function L, because the associated matrix is A_L. Is the same true for the differential of L at any $\mathbf{v}_0$?

If the partials of the coordinate functions f and g are continuous in a neighborhood of $\mathbf{v}_0$, then it is possible to prove that

$$\frac{F(\mathbf{v}) - F(\mathbf{v}_0) - [DF(\mathbf{v}_0)](\mathbf{v} - \mathbf{v}_0)}{\|\mathbf{v} - \mathbf{v}_0\|} \text{ approaches } \mathbf{0} \text{ as } \mathbf{v} \text{ approaches } \mathbf{v}_0 \quad (3.13)$$

Nonlinear Maps

Therefore $F(\mathbf{v_0}) + [DF(\mathbf{v_0})](\mathbf{v} - \mathbf{v_0})$ is a good approximation to $F(\mathbf{v})$ if $\mathbf{v}$ is near to $\mathbf{v_0}$. Thus the differential $DF(\mathbf{v_0})$ can tell us something about the behavior of $F(\mathbf{v})$ when $\mathbf{v}$ is near to $\mathbf{v_0}$.

Moreover, $DF(\mathbf{v_0})$ indicates whether F significantly contracts or expands areas of regions near to $\mathbf{v_0}$. To be more specific, we consider a function $F :$ $V \to \mathbb{R}^2$ with coordinate functions f and g whose partials exist at $\mathbf{v_0}$. The determinant of $DF(\mathbf{v_0})$ is called the **Jacobian** of F at $\mathbf{v_0}$ and is given by

$$\det DF(\mathbf{v_0}) = \det \begin{pmatrix} \dfrac{\partial f}{\partial x}(\mathbf{v_0}) & \dfrac{\partial f}{\partial y}(\mathbf{v_0}) \\ \dfrac{\partial g}{\partial x}(\mathbf{v_0}) & \dfrac{\partial g}{\partial y}(\mathbf{v_0}) \end{pmatrix} \tag{3.14}$$

If $|\det DF(\mathbf{v_0})| < 1$, then F is **area-contracting** at $\mathbf{v_0}$, in the sense that F shrinks small regions containing $\mathbf{v_0}$. Similarly, if $|\det DF(\mathbf{v_0})| > 1$, then F is **area-expanding** at $\mathbf{v_0}$.

Example 3.3.3. Let $F\begin{pmatrix} x \\ y \end{pmatrix} = \begin{pmatrix} y \\ a\sin x + by \end{pmatrix}$. Find the Jacobian of the function F at $\begin{pmatrix} \pi/3 \\ 4 \end{pmatrix}$, and determine conditions on a and b that imply that F is area-contracting at $\begin{pmatrix} \pi/3 \\ 4 \end{pmatrix}$.

Solution. From Example 3.1.1 with $x_0 = \pi/3$ and $y_0 = 4$, we have

$$DF\begin{pmatrix} \pi/3 \\ 4 \end{pmatrix} = \begin{pmatrix} 0 & 1 \\ a/2 & b \end{pmatrix}$$

This means that the Jacobian is given by $\det DF\begin{pmatrix} \pi/3 \\ 4 \end{pmatrix} = -\dfrac{a}{2}$. Consequently F is area-contracting at $\begin{pmatrix} \pi/3 \\ 4 \end{pmatrix}$ provided that $|a| < 2$. $\square$

Now we turn to fixed points of F.

Definition 3.17. Let $\mathbf{p}$ be a fixed point of F. Then $\mathbf{p}$ is **attracting** if and only if there is a disk centered at $\mathbf{p}$ such that $F^{[n]}(\mathbf{v}) \to \mathbf{p}$ for every point $\mathbf{v}$ in the disk. By contrast, $\mathbf{p}$ is **repelling** if and only if there is a disk centered at $\mathbf{p}$ such that $||F(\mathbf{v}) - F(\mathbf{p})|| > ||\mathbf{v} - \mathbf{p}||$ for every $\mathbf{v}$ in the disk for which $\mathbf{v} \neq \mathbf{p}$.

These definitions extend to two dimensions the notions of attracting and repelling fixed points presented in Section 1.2. Often attracting fixed points of multi-variable functions are called **sinks**, and repelling fixed points are called **sources**.

Recall that a fixed point p of a function f defined in R is attracting provided that $|f'(p)| < 1$. Below we will prove a two-dimensional version of this criterion. Before we do it, we need to recall the Chain Rule from Section 1.3:

$$(f^{[n]})'(x) = [f'(f^{[n-1]}(x))][f'(f^{[n-2]}(x))] \cdots [f'(f(x))][f'(x)]$$

If x is a fixed point, say $x = p$, then the formula reduces to

$$(f^{[n]})'(p) = [f'(p)]^n$$

(3.15)

The two-dimensional analogue of (3.14), applied to the function F defined in $\mathbb{R}^2$, is

$$(DF^{[n]})(\mathbf{p}) = [DF(\mathbf{p})]^n$$

(3.16)

You can find a proof of (3.15), as well as the following theorem and its proof, in multivariable analysis textbooks. You may also wish to prove it yourself, and this is included in the exercises.

Theorem 3.18. Let $\mathbf{p}$ be a fixed point of F. Assume that $DF(\mathbf{p})$ exists, with real eigenvalues λ and μ. If $|\lambda| < 1$ and $|\mu| < 1$, then $\mathbf{p}$ is attracting. By contrast, if $|\lambda| > 1$ and $|\mu| \geq 1$, then $\mathbf{p}$ is repelling.

If $\mathbf{p}$ is a fixed point of F and $DF(\mathbf{p})$ has real eigenvalues λ and μ such that $|\lambda| < 1$ and $|\mu| > 1$, then $\mathbf{p}$ is called a **saddle point**. This corresponds to the definition of saddle point given in Section 3.4 for linear functions. In order to understand the behavior of F near a saddle point, let $\mathbf{v}_\lambda$ and $\mathbf{v}_\mu$ be eigenvectors for λ and μ, respectively. If $\mathbf{v}$ is near to $\mathbf{p}$, then

$$\|F(\mathbf{v}) - \mathbf{p}\| < \|\mathbf{v} - \mathbf{p}\| \text{ if } \mathbf{v} \text{ lies in the direction of } \mathbf{v}_\lambda$$

$$\|F(\mathbf{v}) - \mathbf{p}\| > \|\mathbf{v} - \mathbf{p}\| \text{ if } \mathbf{v} \text{ lies in the direction of } \mathbf{v}_\mu$$

If F is a linear function, then the iterates of points on the line through $\mathbf{p}$ in the direction of $\mathbf{v}_\lambda$ converge to $\mathbf{p}$, and the iterates of points on the line through $\mathbf{p}$ in the direction of $\mathbf{v}_\mu$ separate from $\mathbf{p}$. However, if F is not linear, then there is no assurance that these conclusions hold. Nevertheless there is a famous theorem, with the imposing name **Stable and Unstable Manifold Theorem**, which states that there are differentiable curves C_λ and C_μ through $\mathbf{p}$ such that

C_λ is tangent to $\mathbf{v}_\lambda$ at $\mathbf{p}$, and $F^{[n]}(\mathbf{v}) \to \mathbf{p}$ for all $\mathbf{v}$ on C_λ

C_μ is tangent to $\mathbf{v}_\mu$ at $\mathbf{p}$, and $F^{-[n]}(\mathbf{v}) \to \mathbf{p}$ for all $\mathbf{v}$ on C_μ

The expression $F^{-[n]}(\mathbf{v}) \to \mathbf{p}$ signifies that the pre-images of $\mathbf{v}$ on C_μ approach $\mathbf{p}$, and suggests that the iterates of points on C_μ recede from $\mathbf{p}$. The curve C_λ is often called a **local stable manifold** for $\mathbf{p}$ and is denoted $W_{loc}^s(\mathbf{p})$. Similarly, the curve C_μ is often called a **local unstable manifold** for $\mathbf{p}$ and is denoted $W_{loc}^u(\mathbf{p})$. (For more details, see the books by Devaney, 2003, or Guckenheimer and Holmes, 1983.) The geometrical interpretation of these ideas is given in Figure 3.5.

We return once again to the function F defined in Example 3.5.1, with b assigned the value -1.

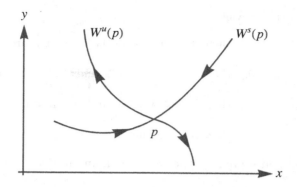

FIGURE 3.5
Illustration of stable and unstable manifolds

Example 3.3.4. Let $F\begin{pmatrix} x \\ y \end{pmatrix} = \begin{pmatrix} y \\ a\sin x - y \end{pmatrix}$. Determine the values of a for which $DF(\mathbf{0})$ has real eigenvalues and the fixed point $\mathbf{0}$ is

(a) attracting

(b) repelling

(c) a saddle point

Solution. By the result of Example 3.5.1 with DF at $\mathbf{0}$, we have

$$DF(\mathbf{0}) = \begin{pmatrix} 0 & 1 \\ a\cos 0 & -1 \end{pmatrix} = \begin{pmatrix} 0 & 1 \\ a & -1 \end{pmatrix}$$

so that

$$\det[DF(\mathbf{0}) - \lambda I] = \det\begin{pmatrix} -\lambda & 1 \\ a & -1-\lambda \end{pmatrix} = \lambda^2 + \lambda - a$$

Now $\lambda^2 + \lambda - a = 0$ if and only if

$$\lambda = \frac{-1 \pm \sqrt{1+4a}}{2}$$

Next, let

$$\lambda = \frac{-1 - \sqrt{1+4a}}{2} \quad \text{and} \quad \mu = \frac{-1 + \sqrt{1+4a}}{2}$$

A little calculation yields

if $a < -1/4$,   then $1 + 4a < 0$, so that λ and μ are not real

if $-1/4 < a < 0$,   then $-1 < \lambda < -1/2$ and $-1/2 < \mu < 0$

if $0 < a < 2$,   then $-2 < \lambda < -1$ and $0 < \mu < 1$

if $a > 2$,   then $\lambda < -2$ and $\mu > 1$

Therefore λ and μ are real when $a > -1/4$, and in that case,

0 is an attracting fixed point if $-1/4 < a < 0$
0 is a saddle point if $0 < a < 2$
0 is a repelling fixed point if $a > 2$

This completes the solution. □

You may have observed that our discussion has revolved around those fixed points **p** whose eigenvalues satisfy $|\lambda| \neq 1$ and $|\mu| \neq 1$. Such fixed points are called **hyperbolic**; they are either attracting fixed points or repelling fixed points or saddle points, and the behavior of the function near such a point is amenable to analysis. If the eigenvalues satisfy $|\lambda| = 1$ or $|\mu| = 1$, then **p** is **non–hyperbolic**, and the analysis of the function near such a point is much more difficult.

3.3.1   Baker's Functions

We complete this section with a discussion of a two-dimensional version of the baker's function described in Section 1.3. We define the function $B_0 : \mathbb{R}^2 \to \mathbb{R}^2$ by the two formulas

$$B_0 \begin{pmatrix} x \\ y \end{pmatrix} = \begin{pmatrix} x/4 \\ 3y \end{pmatrix} \text{ for } 0 \leq x \leq 1 \text{ and } 0 \leq y < \frac{1}{3}$$

and

$$B_0 \begin{pmatrix} x \\ y \end{pmatrix} = \begin{pmatrix} \frac{1}{2} + \frac{1}{3}x \\ \frac{3}{2}(y - \frac{1}{3}) \end{pmatrix} \text{ for } 0 \leq x \leq 1 \text{ and } \frac{1}{3} \leq y \leq 1$$

Then B_0 is called a **baker's function** because it is linear on two portions of the domain the way the one-dimensional baker's function is. The domain of B_0 consists of the unit square $S = [0, 1] \times [0, 1]$. The effect of B_0 on S can be seen graphically in Figure 3.6, where the rectangles in the right graph are the images of the rectangles with similar shading in the left graph.

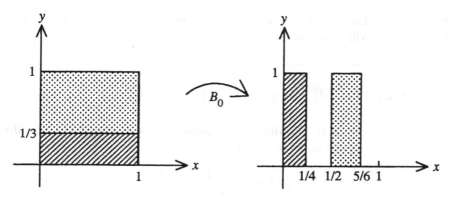

FIGURE 3.6

Illustration of kneading in the baker's function

The image $B_0^{[2]}(S)$ is depicted on the right in Figure 3.7, and suggests that

$$S \supseteq B_0(S) \supseteq B_0^{[2]}(S) \supseteq \cdots$$

One can prove that $\{B_0^{[2]}(S)\}_{n=0}^{\infty}$ is, in fact, a nested sequence of strips in S. Next, let

$$A_{B_0} = \text{the collection of } \mathbf{v} \text{ that are in } B_0^{[n]}(S) \text{ for all}$$
$$n \geq 0, \text{ including boundary}$$

Then A_{B_0} is composed of a set of vertical line segments of length 1. The intersection of A_{B_0} and the x-axis is a one-dimensional Cantor set (though *not* the Cantor ternary set). The set A_{B_0} is called an **attractor** because the iterates of every point in S approach A_{B_0}. Notice that $\mathbf{0}$ is a fixed point of B_0.

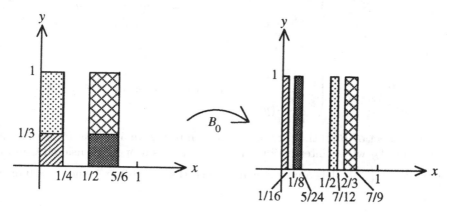

FIGURE 3.7

Squaring the baker's function

Example 3.3.5. Determine whether $\mathbf{0}$ is an attracting or repelling fixed point, or a saddle point, of B_0.

Solution. By the first formula that defines B_0,

$$B_0(\mathbf{v}) = \left(\begin{array}{c} f(\mathbf{v}) \\ g(\mathbf{v}) \end{array} \right), \text{ where } f\left(\begin{array}{c} x \\ y \end{array} \right) = \frac{1}{4}x \text{ and } g\left(\begin{array}{c} x \\ y \end{array} \right) = 3y$$

Because $\mathbf{0}$ is on the border of the domain, we can only take one-sided partial derivatives of B_0 at $\mathbf{0}$. We obtain

$$DB_0(\mathbf{0}) = \left(\begin{array}{cc} 1/4 & 0 \\ 0 & 3 \end{array} \right)$$

Therefore $\mathbf{0}$ is a saddle point, contracting by a factor of $1/4$ along the x-axis and expanding by a factor of 3 along the y-axis. $\square$

It follows from the solution to Example 3.3.5 that the interval $[0, 1]$ along the x-axis is a local stable manifold $W_{loc}^s(\mathbf{0})$. By contrast, the interval $[0, 1/3]$ along the y-axis is a local unstable manifold $W_{loc}^u(\mathbf{0})$.

The point $\mathbf{0}$ is not the only periodic point of B_0. There is another fixed point, as well as periodic points that are not fixed points (Exercises 13–14). In addition, B_0 has sensitive dependence on initial conditions because B_0 separates nearby points in the domain by at least a factor of $3/2$ (Exercise 15).

The function B_0 described above is but one of a whole family of functions. The full family is defined in the following way. Let a, b, and c be constants, with $0 < a \leq 1/2$, and $0 < b < c < 1/2$. Then a (generalized) **baker's function**, which we designate by B rather than the more accurate but cumbersome B_{abc}, is defined on the unit square S by the two formulas:

$$B\left(\begin{array}{c} x \\ y \end{array} \right) = \left(\begin{array}{c} bx \\ \dfrac{1}{a}y \end{array} \right) \text{ for } 0 \leq x \leq 1 \text{ and } 0 \leq y < a$$

and

$$B\left(\begin{array}{c} x \\ y \end{array} \right) = \left(\begin{array}{c} \dfrac{1}{2} + cx \\ \dfrac{1}{1-a}(y - a) \end{array} \right) \text{ for } 0 \leq x \leq 1 \text{ and } a \leq y \leq 1$$

The effect of B on the domain S is shown in Figure 3.8. Like B_0, B has an attractor A_B whose intersection with the x-axis is a one-dimensional Cantor set. If $\mathbf{v}_0 = \left(\begin{array}{c} x_0 \\ y_0 \end{array} \right)$ is any vector in S such that $x_0 \neq 0$ or 1 and $y_0 \neq 0$ or a

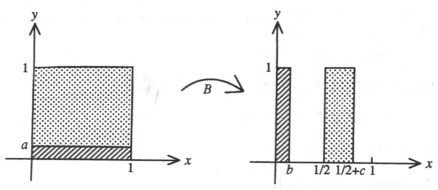

FIGURE 3.8
Effect of the baker's function on a subdomain

or 1, then $DB(\mathbf{v}_0)$ exists, and moreover,

$$DB(\mathbf{v}_0) = \begin{pmatrix} b & 0 \\ 0 & \dfrac{1}{a} \end{pmatrix} \quad \text{if } 0 < y_0 < a \quad \text{and} \quad DB(\mathbf{v}_0)$$

$$= \begin{pmatrix} c & 0 \\ 0 & \dfrac{1}{1-a} \end{pmatrix} \quad \text{if } a < y_0 < 1$$

Exercises 17–19 are devoted to features of B.

3.3.2   Section 3.3 Exercises

In Exercises 1–3, find the fixed points of F, and determine whether F is area-expanding, area-contracting, or neither at these points.

Exercise 1. $F\begin{pmatrix} x \\ y \end{pmatrix} = \begin{pmatrix} x^2 + \dfrac{1}{4} \\ 4x - y^2 \end{pmatrix}$

Exercise 2. $F\begin{pmatrix} x \\ y \end{pmatrix} = \begin{pmatrix} 1 - x^2 + y \\ x \end{pmatrix}$

Exercise 3. $F\begin{pmatrix} x \\ y \end{pmatrix} = \begin{pmatrix} y^2 - \dfrac{1}{2}x \\ \dfrac{1}{4}x + \dfrac{1}{2}y \end{pmatrix}$ 4. $F\begin{pmatrix} x \\ y \end{pmatrix} = \begin{pmatrix} 2xy + y \\ 3y - x \end{pmatrix}$

In Exercises 4–7, determine whether the fixed points of F are attracting, repelling, or saddle points.

Exercise 4. F in Exercise 1

Exercise 5. F in Exercise 2

Exercise 6. F in Exercise 3

Exercise 7. F in Exercise 4

Exercise 8. Let $L \begin{pmatrix} x \\ y \end{pmatrix} = \begin{pmatrix} 2y \\ 0 \end{pmatrix}$.

(a) Show that 0 is the only eigenvalue of L, but $||L(\mathbf{v})|| > ||\mathbf{v}||$ for all nonzero $\mathbf{v}$ on the y-axis.

(b) Does the result of part (a) contradict Corollary 3.15? Explain your answer.

Exercise 9. Let $L \begin{pmatrix} x \\ y \end{pmatrix} = \begin{pmatrix} \frac{1}{2}x \\ 10x + \frac{1}{4}y \end{pmatrix}$.

(a) Show that the eigenvalues of L are less than 1 in absolute value.

(b) Find a vector $\mathbf{v}$ such that $||L(\mathbf{v})|| > ||\mathbf{v}||$.

(c) Do (a) and (b) contradict Theorem 3.18? Explain your answer.

Exercise 10. Let $L \begin{pmatrix} x \\ y \end{pmatrix} = \begin{pmatrix} \sin x \\ y^2 \end{pmatrix}$. Show that $\mathbf{0}$ is an attracting fixed point of F, although $DF(\mathbf{0})$ has an eigenvalue that is not less than 1 in absolute value.

Exercise 11. Find an example of a function $F : \mathbb{R}^2 \to \mathbb{R}^2$ that has a saddle point $\mathbf{p}$, such that F is

(a) area-expanding at $\mathbf{p}$

(b) area-contracting at $\mathbf{p}$

Exercise 12. Determine a fixed point of B_0 that is not $\mathbf{0}$.

Exercise 13. Find a 2-cycle for B_0.

Exercise 14. Show that B_0 has sensitive dependence on initial conditions.

Nonlinear Maps

Exercise 15. Use the formulas that define B_0 to show explicitly that $S \supseteq B_0(S)$.

...8 are devoted to the (generalized) baker's function B.

Exercise. Determine the fixed points of B.

Exercise 17.

(a) Use the formulas that define B to show explicitly that $S \supseteq B(S)$.

(b) Find the areas of $B(S)$ and $B^{[2]}(S)$.

Exercise 1... Show that 0 is a saddle point for B, for any choice of a, b, and c.

...se 19. Let $L : \mathbb{R}^2 \to \mathbb{R}^2$ be a *linear function*, and let $\mathbf{v}_0$ be an arbitrary element of $\mathbb{R}^2$. Is $DL(\mathbf{v}_0) = L$? Explain your answer.

The cat map C, introduced by the Russian mathematician V. I. Arnold, is defined on the half-open unit square $[0, 1) \times [0, 1)$, using the "mod" notation. By "$z \bmod 1$" we mean $z - n$, where n is the integer such that $0 \le z - n < 1$. Then

$$C\begin{pmatrix} x \\ y \end{pmatrix} = \begin{pmatrix} (x+y) \bmod 1 \\ (x+2y) \bmod 1 \end{pmatrix}$$

Exercises 20–21 explore features of C.

Exercise 20. Show that 0 is the only fixed point of C, and find two 2-cycles for C.

Exercise 21. Let $\mathbf{v}$ be any vector at which C is continuous. Note that

$$DC(\mathbf{v}) = \begin{pmatrix} 1 & 1 \\ 1 & 2 \end{pmatrix}.$$

(a) Find the eigenvalues of $DC(0)$, and show that 0 is a saddle point.

(b) Let $c_1 = 1 = c_2$, and recursively let $c_n = c_{n-1} + c_{n-2}$ for $n \ge 3$. Then $c_3 = 2, c_4 = 3, c_5 = 5$, etc., and the sequence of c_n's is the **Fibonacci sequence**. Use induction to show that

$$DC^{[n]}(\mathbf{v}) = \begin{pmatrix} c_{2n-1} & c_{2n} \\ c_{2n} & c_{2n+1} \end{pmatrix} \quad \text{for } n \ge 2$$

Exercise 22. Construct a matrix that maps $M : \mathbb{R}^2 \to \mathbb{R}^2$, with Jacobian 1 for all points in $\mathbb{R}^2$.

Exercise 23. Construct a rectangular region in $\mathbb{R}^2$. Using the matrix in Exercise 23, transform that region. Compare the area of the rectangle before and after the transformation.

Exercise 24. Repeat Exercise 24 *with two new matrices. One should have a Jabobian less than 1, and the other a Jacobian greater u... Make a conjecture about the relationship between the Jacobian of a linea... and how that transformation affects the area of the region being ...formation ...ed.*

Exercise 25. Prove

$$(DF^{[n]})(\mathbf{p}) = [DF(\mathbf{p})]^n.$$

3.4 The Hénon Map

In the 1970s the French astronomer-mathematician Michel Hénon was ...ing for a simple two-dimensional function possessing special properties of m... complicated systems. The result was a family of functions denoted by H_{ab} and given by

$$H_{ab}\begin{pmatrix} x \\ y \end{pmatrix} = \begin{pmatrix} 1 - ax^2 + y \\ bx \end{pmatrix}, \text{ where } a \text{ and } b \text{ are real numbers} \quad (3.17)$$

(See Hénon, 1976.) The maps defined in (3.17) are called **Hénon maps**. Many authors refer to the Hénon maps as H, and just call them the Hénon map.

Notice that if $b = 1, x = t$ and $y = 0$, then (3.17) becomes

$$H_{ab}\begin{pmatrix} t \\ 0 \end{pmatrix} = \begin{pmatrix} 1 - at^2 \\ t \end{pmatrix}$$

Thus the image of the real line is the parabola given parametrically by $x = 1 - at^2$ and $y = t$ (Figure 3.9). As a result, the Hénon maps constitute a

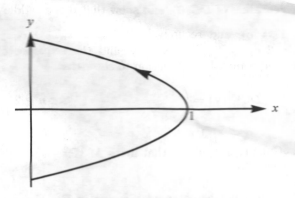

FIGURE 3.9
The Hénon map with $b = 1, x = t$, and $y = 0$

two-dimensional generalization of the family $\{F_C\}$ appearing in Exercise 3 of Section 2.3, where $F_C(x) = 1 - Cx^2$. Next, we will find the Jacobian of H_{ab}.

Theorem 3.19. Let a and b be any fixed real numbers. Then $\det DH_{ab}\begin{pmatrix} x \\ y \end{pmatrix} = -b$ for all x, y in $\mathbb{R}^2$. If $a^2x^2 + b \geq 0$, then the eigenvalues of $DH_{ab}\begin{pmatrix} x \\ y \end{pmatrix}$ are the real numbers $-ax \pm \sqrt{a^2x^2 + b}$.

Proof. Since the coordinate functions of H_{ab} are given by

$$f\begin{pmatrix} x \\ y \end{pmatrix} = 1 - ax^2 + y \quad \text{and} \quad g\begin{pmatrix} x \\ y \end{pmatrix} = bx$$

we find that

$$DH_{ab}\begin{pmatrix} x \\ y \end{pmatrix} = \begin{pmatrix} -2ax & 1 \\ b & 0 \end{pmatrix}$$

so that the Jacobian is given by

$$\det DH_{ab}\begin{pmatrix} x \\ y \end{pmatrix} = \det\begin{pmatrix} -2ax & 1 \\ b & 0 \end{pmatrix} = -b$$

To determine the eigenvalues of $DH_{ab}\begin{pmatrix} x \\ y \end{pmatrix}$ we observe that

$$\det\left(DH_{ab}\begin{pmatrix} x \\ y \end{pmatrix} - \lambda I\right) = \det\begin{pmatrix} -2ax - \lambda & 1 \\ b & -\lambda \end{pmatrix} = \lambda^2 + 2ax\lambda - b$$

Therefore λ is an eigenvalue of $DH_{ab}\begin{pmatrix} x \\ y \end{pmatrix}$ if $\lambda^2 + 2ax\lambda - b = 0$. This means that

$$\lambda = \frac{-2ax \pm \sqrt{4a^2x^2 + 4b}}{2} = -ax \pm \sqrt{a^2x^2 + b}$$

Thus the eigenvalues are real if $a^2x^2 + b \geq 0$. $\qquad\square$

The map H_{ab} has a constant Jacobian because $\det DH_{ab} = -b$. Hénon noted in his original paper (1976) that H_{ab} is the "most general quadratic mapping [on $\mathbb{R}^2$] with constant Jacobian." Next, recall that the Jacobian of H_{ab} determines whether H_{ab} is area-expanding or area-contracting (or neither). By Theorem 3.19, the map H_{ab} is area-contracting if $0 \leq b < 1$; it is genuinely a two-dimensional map if $b \neq 0$. Thus we will henceforth assume $0 < b < 1$. For such values of b, $H_{ab}(\mathbf{v})$ has distinct real eigenvalues for every value of the parameter a, and all $\mathbf{v}$.

It is straightforward to show that H_{ab} is one-to-one.

Theorem 3.20. H_{ab} is one-to-one.

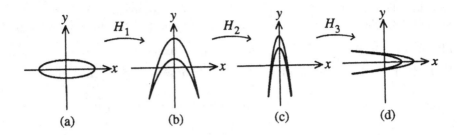

FIGURE 3.10
A composition of Hénon maps

Proof. Let x, y, z, and w be real numbers. Then

$$H_{ab}\begin{pmatrix} x \\ y \end{pmatrix} = H_{ab}\begin{pmatrix} z \\ w \end{pmatrix} \quad \text{if and only if} \quad \begin{pmatrix} 1 - ax^2 + y \\ bx \end{pmatrix}$$

$$= \begin{pmatrix} 1 - az^2 + w \\ bz \end{pmatrix}$$

that is,

$$1 - ax^2 + y = 1 - az^2 + w \quad \text{and} \quad bx = bz$$

Since $b \neq 0$, it follows that $x = z$. Therefore $y = w$ as well, so that $\begin{pmatrix} x \\ y \end{pmatrix} = \begin{pmatrix} z \\ w \end{pmatrix}$. Consequently H_{ab} is one-to-one.   □

The map H_{ab} is composed of the three functions H_1, H_2, and H_3, where

$$H_1\begin{pmatrix} x \\ y \end{pmatrix} = \begin{pmatrix} x \\ 1 - ax^2 + y \end{pmatrix}, \quad H_2\begin{pmatrix} x \\ y \end{pmatrix} = \begin{pmatrix} bx \\ y \end{pmatrix}, \quad \text{and} \quad H_3\begin{pmatrix} x \\ y \end{pmatrix}$$

$$= \begin{pmatrix} y \\ x \end{pmatrix}$$

More precisely, $H_{ab} = H_3 \circ H_2 \circ H_1$ (where for convenience we suppress the subscript ab on H_1, H_2, and H_3). To interpret H_1, H_2, and H_3 geometrically, suppose that $a > 1$. Then H_1 begins the folding process. The effect of H_1 on the ellipse in Figure 3.10(a) is shown in Figure 3.10(b). Next, H_2 contracts curves in the x direction, since $0 < b < 1$ by hypothesis (Figure 3.10(c)). The folding started by H_1 is enhanced by H_2. Finally, H_3 flips shapes across the line $y = x$. The total effect of H_1, H_2, and H_3 (that is, of H_{ab}) on an ellipse is shown in Figure 3.10(d).

Now we will prove that H_{ab} is invertible.

Theorem 3.21. H_{ab} is invertible, and $H_{ab}^{-1}\begin{pmatrix} x \\ y \end{pmatrix} = \begin{pmatrix} \dfrac{1}{b}y \\ -1 + \dfrac{a}{b^2}y^2 + x \end{pmatrix}$.

Proof. We could show that $H_1, H_2,$ and H_3 are invertible, and then that

$$H_{ab}^{-1} = (H_3 \circ H_2 \circ H_1)^{-1} = H_1^{-1} \circ H_2^{-1} \circ H_3^{-1}$$

It is easier to show that by the formula given in the theorem's statement, we have $H_{ab} \circ H_{ab}^{-1} = I$:

$$(H_{ab} \circ H_{ab}^{-1}) \begin{pmatrix} x \\ y \end{pmatrix} = H_{ab} \begin{pmatrix} \frac{1}{b} y \\ -1 + \frac{a}{b^2} y^2 + x \end{pmatrix} = \begin{pmatrix} x \\ y \end{pmatrix} \quad \text{for all } x \text{ and } y$$

$\square$

Next, we determine the values of a and b for which H_{ab} has fixed points.

Theorem 3.22. Let $a \neq 0$. Then H_{ab} has a fixed point if $a \geq -\frac{1}{4}(1-b)^2$.

Proof. The point $\begin{pmatrix} x \\ y \end{pmatrix}$ is a fixed point of H_{ab} provided that

$$\begin{pmatrix} x \\ y \end{pmatrix} = H_{ab} \begin{pmatrix} x \\ y \end{pmatrix} = \begin{pmatrix} 1 - ax^2 + y \\ bx \end{pmatrix}$$

The left-hand and right-hand vectors in the equation are equal if $x = 1 - ax^2 + y$ and $y = bx$, which implies that $x = 1 - ax^2 + bx$. This is equivalent to $ax^2 + (1-b)x - 1 = 0$, which by the quadratic formula yields

$$x = \frac{1}{2a} \left(b - 1 \pm \sqrt{(1-b)^2 + 4a} \right)$$

Such an x exists if $(1-b)^2 + 4a \geq 0$, that is, if $a \geq -\frac{1}{4}(1-b)^2$. $\square$

In the event that H_{ab} has two fixed points $\mathbf{p}$ and $\mathbf{q}$, they are given by

$$\mathbf{p} = \begin{pmatrix} \frac{1}{2a}(b - 1 + \sqrt{(1-b)^2 + 4a}) \\ \frac{b}{2a}(b - 1 + \sqrt{(1-b)^2 + 4a}) \end{pmatrix}, \quad \mathbf{q} = \begin{pmatrix} \frac{1}{2a}(b - 1 - \sqrt{(1-b)^2 + 4a}) \\ \frac{b}{2a}(b - 1 - \sqrt{(1-b)^2 + 4a}) \end{pmatrix}$$

$$(3.18)$$

Since we know the fixed points of H_{ab} and the eigenvalues of $DH_{ab} \begin{pmatrix} x \\ y \end{pmatrix}$ for all x and y, we can determine conditions under which the fixed point $\mathbf{p}$ is attracting.

Theorem 3.23. The fixed point $\mathbf{p}$ is attracting provided that a is a nonzero number lying in the interval

$$J = \left(-\frac{1}{4}(1-b)^2, \frac{3}{4}(1-b)^2 \right)$$

Proof. Theorem 3.18 tells us that $\mathbf{p}$ is attracting if the eigenvalues of $DH_{ab}(\mathbf{p})$ are less than 1 in absolute value. Letting $\mathbf{p} = \begin{pmatrix} p_1 \\ p_2 \end{pmatrix}$, we know from (3.18) that

$$p_1 = \frac{1}{2a}\left(b - 1 + \sqrt{(1-b)^2 + 4a}\right) \tag{3.19}$$

so that $2ap_1 = b - 1 + \sqrt{(1-b)^2 + 4a}$. Therefore

$$2ap_1 > b - 1, \text{ or equivalently, } 2ap_1 + 1 > b \tag{3.20}$$

By Theorem 3.19 the eigenvalues of $DH_{ab}(\mathbf{p})$ are less than 1 in absolute value if $|-ap_1 \pm \sqrt{a^2p_1^2 + b}| < 1$. We will show that if a is in the interval J, then

$$0 \le -ap_1 + \sqrt{a^2p_1^2 + b} < 1 \tag{3.21}$$

So let a be in J. On the one hand, since we assumed that $0 < b < 1$, in particular $b > 0$, so that

$$-ap_1 + \sqrt{a^2p_1^2 + b} \ge -ap_1 + \sqrt{a^2p_1^2} = -ap_1 + |ap_1| \ge 0$$

On the other hand, $a > -(1-b)^2/4$ by hypothesis, so that $(1-b)^2 + 4a > 0$. Since p_1 is a real number, we use (3.20) to show that

$$(ap_1 + 1)^2 = a^2p_1^2 + 2ap_1 + 1 > a^2p_1^2 + b > 0$$

It follows that $ap_1 + 1 > \sqrt{a^2p_1^2 + b}$, so that $-ap_1 + \sqrt{a^2p_1^2 + b} < 1$, and thus $0 \le -ap_1 + \sqrt{a^2p_1^2 + b} < 1$. Therefore (3.21) is proved. An analogous argument proves that if a is in J, then

$$-1 < -ap_1 - \sqrt{a^2p_1^2 + b} < 0$$

(see Exercise 9). Consequently the eigenvalues of $DH_{ab}(\mathbf{p})$ are less than 1 in absolute value, so that $\mathbf{p}$ is an attracting fixed point. $\square$

From Theorem 3.23, the fixed point $\mathbf{p}$ is attracting for certain values of a. By contrast the fixed point $\mathbf{q}$ given in (3.18) is a saddle point (Exercise 8). Thus for given values of b in $(0, 1)$ we have the following:

(i) If $a < -\dfrac{1}{4}(1-b)^2$,   then H_{ab} has no fixed points.

(ii) If $-\dfrac{1}{4}(1-b)^2 < a < \dfrac{3}{4}(1-b)^2$ and $a \ne 0$. then H_{ab} has two fixed points, $\mathbf{p}$ and $\mathbf{q}$, of which $\mathbf{p}$ is attracting and $\mathbf{q}$ is a saddle point.

For the present, let b be fixed in the interval $(0, 1)$, and let the parameter a increase. In addition to the bifurcation at $-(1-b)^2/4$, H_{ab} has a bifurcation at $a = 3(1-b)^2/4$, because one of the two eigenvalues of $H_{ab}(\mathbf{p})$ descends through -1 (Exercise 10), so that **p** is transformed from an attracting fixed point to a saddle point. Recollect that an attracting 2-cycle for the quadratic family $\{Q_\mu\}$ emerges as μ increases and passes through 3. Thus we might suspect that as a passes through $3(1-b)^2/4$, an attracting 2-cycle for H_{ab} would be born. This is the case. In order to prove it, one would have to solve the equation

$$\begin{pmatrix} x \\ y \end{pmatrix} = H_{ab}^{[2]} \begin{pmatrix} x \\ y \end{pmatrix} = \begin{pmatrix} 1 - a(1 - ax^2 + y)^2 + bx \\ b(1 - ax^2 + y) \end{pmatrix}$$

for x and y. Of course this entails solving a fourth-degree equation in x, which is possible because two roots are known from the two fixed points of H_{ab}. The result is that H_{ab} has a period-doubling bifurcation at $a = 3(1-b)^2/4$.

As a increases further, H_{ab} undergoes a period-doubling cascade. For certain special values of b the bifurcation values of a, as well as the Feigenbaum number, are known. In particular, Derrida, Gervois, and Pomeau (1979) have calculated the following bifurcation values of a for $b = 0.3$:

bifurcation point a	period-n cycle appears
a	n
-0.1225	1
0.3675	2
0.9125	4
$1.0260\cdots$	8
$1.0510\cdots$	16
$1.0565\cdots$	32

The cascade terminates at approximately 1.0580459, which is the "Feigenbaum number" for the Hénon map. How does $H_{a(.3)}$ behave when $a > 1.06$? One might imagine that for an arbitrary $a > 1.06$, the iterates of virtually any initial point would be sprinkled unpredictably throughout a region in the plane. However, that does not happen. For example, let $a = 1.4$, and designate $H_{(1.4)(.3)}$ by H. If we neglect the first few iterates of **0** and plot the next 10,000 iterates, then we obtain the shape A_H appearing in Figure 3.11. The set A_H is called the **Hénon attractor** of the map, because the iterates of every point in a certain quadrilateral Q surrounding A_H approach the attractor.

Although A_H may appear to consist of a few fairly simple curves, when we zoom in on a small rectangle containing the fixed point **p**, we see that there are several strands (Figure 3.12(a)). No matter how much we magnify the region, nearly identical new sets of strands appear (Figures 3.12(b) and (c)). It turns out that there are in reality an infinite number of such strands that make the region near to **p** look like a product of a line and a Cantor set.

The iterates of nearly all points in a rectangular region Q containing A_H not only converge to A_H but seem to trace out a dense subset of A_H. Thus whether the initial point is **0** or another point, after a few initial iterates the next several thousand iterates yield a virtually identical shape.

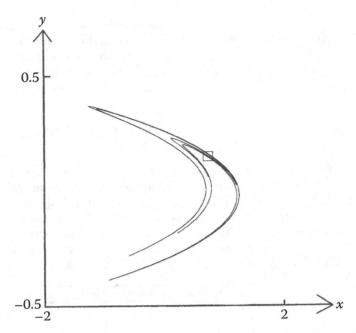

FIGURE 3.11
Hénon attractor

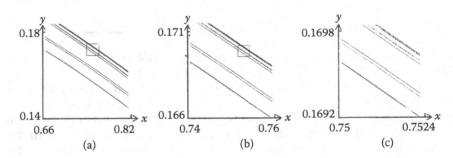

FIGURE 3.12
Hénon attractor in the neighborhood of a fixed point

Two significant features of the Hénon attractor are evident from Figure 3.13, which displays iterates 34 through 40 of the origin (denoted by an isolated asterisk). First, the figure shows that iterates bounce around the attractor erratically. Second, the figure suggests that the Hénon attractor has sensitive dependence on initial conditions. Iterates of the origin identified by asterisks were computed by a CRAY supercomputer with single precision accuracy, whereas iterates identified by little squares were computed by the same

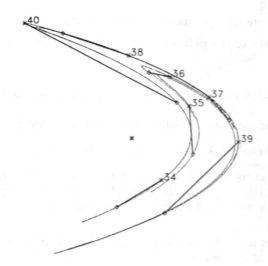

FIGURE 3.13
Hénon attractor with double precision

CRAY supercomputer, but with double precision. The difference in accuracy between single and double precision is approximately 10^{-14} units. However, after 40 iterates the results are completely unrelated. Because of its sensitive dependence, A_H is called a **chaotic attractor**.

In his famous article of 1976, Hénon featured the attractor A_H with $a = 1.4$ and $b = 0.3$. Why did he select these parameters? On the one hand, he noticed that if a lies somewhere near 1.4, then as b increases from 0 to 0.3 the attractor A_H grows from nothing to a robust set in $\mathbb{R}^2$; moreover, if b is much larger than 0.3, then the iterates of various points near $\mathbf{0}$ are unbounded. On the other hand, he found that if $b = 0.3$, and if $a <$ the Feigenbaum number ≈ 1.06, then H_{ab} has an attracting cycle, so the attractor is a finite point set; however, if $a > 1.55$ then iterates of all points are unbounded. These observations led Hénon to select $a = 1.4$ and $b = 0.3$ as parameters that provide an attractor that is full and has interesting characteristics. Nevertheless, other parameters near 1.4 and 0.3 yield interesting attractors. In fact, some colleagues use $a = 1.42$ instead of $a = 1.4$. You might be interested to see what differences there are between the attractors corresponding to these two values of a.

3.4.1 Section 3.4 Exercises

Exercise 1. Using the program HENON, determine whether it takes 100, 500, 1000, 2000, or 5000 iterates of $\mathbf{0}$ to fill out A_H so that it appears as a collection of (relatively) complete curves.

Exercise 2. Using the program HENON, show that after the first few iterates, the iterates of the points $\mathbf{0}$ and $\begin{pmatrix} 0.1 \\ 0.2 \end{pmatrix}$ produce virtually the same shape for A_H.

Exercise 3. Recall that H has parameters $a = 1.4$ and $b = 0.3$.

(a) Find the (approximate) coordinates of the fixed point $\mathbf{p}$ for H.

(b) Find the eigenvalues λ and μ of $H(\mathbf{p})$, with $\lambda < \mu$.

(c) Find eigenvectors $\mathbf{v}_\lambda$ and $\mathbf{v}_\mu$ of $DH(\mathbf{p})$. Confirm that H stretches distances along (that is, in the direction of) the attractor at $\mathbf{p}$, and contracts distances in a direction oblique to the direction of the attractor at $\mathbf{p}$.

Exercise 4. Let $a = 1.4$. Use the program HENON to discuss what happens to the attractor of H_{ab} when

(a) b increases from 0.3 to 0.5

(b) b decreases from 0.3 to 0.1

Exercise 5. Let $b = 0.3$. Use the program HENON to discuss what happens to the attractor of H_{ab} when

(a) a increases from 1.4

(b) a decreases from 1.4

Exercise 6. Let $b = 0.3$. Use the program HENON to find a value of a such that H_{ab} has an attracting n-cycle, and determine an n-cycle.

(a) $n = 4$

(b) $n = 8$

(c) $n = 7$

(d) $n = 3$

Exercise 7. Let $a = 0$.

(a) Show that for each b in the interval $(0, 1)$, H_{ab} has a unique fixed point $\mathbf{p}$, and find $\mathbf{p}$.

(b) Determine whether $\mathbf{p}$ is attracting, repelling, or a saddle point.

(c) Find an eigenvector corresponding to each eigenvalue of H_{ab}.

Exercise 8.

(a) Let $a = -(1-b)^2/4$ and $0 < b < 1$. Find the eigenvalues of $H_{ab}(\mathbf{q})$.

(b) Let $a > -(1-b)^2/4$ and $a \neq 0$. Show that the fixed point $\mathbf{q}$ in (2) is a saddle point of H_{ab} for each b in the interval $(0, 1)$.

(c) Use the results of (a) and (b) to discuss the type of bifurcation of H_{ab} that occurs at $a = -0.1225$.

Exercise 9. Assume that $0 < b < 1$ and $-(1-b)^2/4 < a < 3(1-b)^2/4$. Let

$$p_1 = \frac{1}{2a}(b - 1 + \sqrt{(1-b)^2 + 4a})$$

Show that $-1 < -ap_1 - \sqrt{a^2 p_1^2 + b} < 0$.

Exercise 10. Let b be in $(0, 1)$ and $a = 3(1-b)^2/4$. Find the eigenvalues of H_{ab}, and convince yourself that H_{ab} has a bifurcation at this value of a.

Exercise 11. Let $H^* \begin{pmatrix} x \\ y \end{pmatrix} = \begin{pmatrix} a - x^2 + by \\ x \end{pmatrix}$.

(a) Show that $H_{ab} \approx_E H^*$, where E is an appropriate linear function.

(b) What does the result of (a) tell you about the attractor of H^*? (H^* is often used as an alternative to H_{ab}.)

Exercise 12. Use the bifurcation table for $H_{a(.3)}$ that appears in this section to compute an approximate Hénon version of the Feigenbaum constant (see Section 1.5). Compare your answer with the Feigenbaum constant for the quadratic family.

3.5 The Horseshoe Map

One of the earliest examples of a function defined on $\mathbb{R}^2$ that exhibits interesting dynamics is the horseshoe map described in the 1960s by the American mathematician Stephen Smale (1967). The horseshoe map will be denoted by M. Its domain is the set S in $\mathbb{R}^2$ composed of the unit square $T = [0,1] \times [0,1]$, bounded on the left and right by semicircles B and E (Figure 3.14(a)). We assume that S contains its boundary. The function M shrinks S vertically by a factor of $a < 1/3$, and expands S horizontally by a factor of $b = 3$, with the semicircles B and E altered so as to continue to be semicircular. The resulting

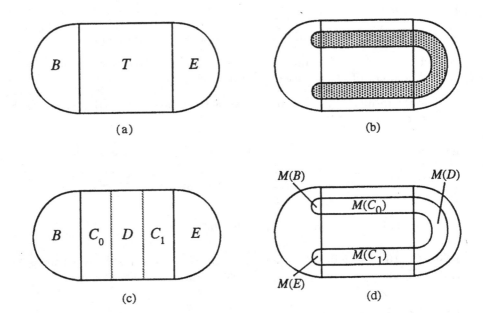

FIGURE 3.14
Illustration of the horseshoe map

figure is folded by M so that it fits again inside S, with only the semicircles protruding to the left of T (Figure 3.14(b)). Thus the range of M looks like a horseshoe. When S is partitioned as in Figure 3.14(c), we can see the effect of M on each member of the partition (Figure 3.14(d)). Specifically, M sends semicircles B and E into B, and sends the square T into two strips inside T plus a curved strip inside E.

Even though we have not defined M by a formula or a series of formulas, we are able to systematically analyze it. To start with, we note that M is well-defined and the range is contained in S. Next, we will show that M is a homeomorphism.

Theorem 3.24. $M : S \to M(S)$ is a homeomorphism.

Proof. By definition, M maps onto $M(S)$. That M is one-to-one follows from the fact that stretching and contracting are one-to-one operations, and M folds in a nonoverlapping manner. To prove that M is continuous, we compare subregions in Figure 3.14(c) with their corresponding images in Figure 3.14(d). The map M expands distances in C_0 and C_1 by a factor of 3, and shrinks distances in B and E. The largest expansion of distances for points in D occurs at the top boundary, which maps onto the exterior boundary γ of $M(D)$. Since the length of the top boundary of D is $1/3$ and the length of γ is less than $\pi/2$, it follows that M expands distances in D by no more than a factor of 6. Consequently if $\mathbf{v}$ and $\mathbf{w}$ are in S and if $||\mathbf{v} - \mathbf{w}|| < \epsilon$, then $||M(\mathbf{v}) - M(\mathbf{w})|| < 6\epsilon$. Therefore M is continuous. The inverse map M^{-1} is continuous by the same kind of argument. Therefore M is a homeomorphism. □

Assuming that B in Figure 3.14(c) contains its boundary, we show next that M has a fixed point that lies in B and on that boundary.

Theorem 3.25. M has a unique fixed point in B, to which the iterates of all points in B and E converge.

Proof. Since M shrinks the domain vertically by a factor of $a < 1/3$, this means that for each n, $M^{[n]}(B)$ is closed and semicircular with diameter a^n. In addition,

$$B \supseteq M(B) \supseteq M^{[2]}(B) \supseteq \cdots \supseteq M^{[n]}(B)$$

The two-dimensional version of the Heine-Borel Theorem (Theorem 3.7) implies that a nested sequence of closed, bounded sets in $\mathbb{R}^2$ has a common point. Thus the intersection B_∞ of the sets $M^{[n]}(B)$ for $n \geq 1$ has at least one element. Notice that the diameter of $M^{[n]}(B)$ is a^n, and $\lim_{n \to \infty} a^n = 0$. Thus B_∞ contains exactly one element, which we denote by $\mathbf{p}$. Now $\mathbf{p}$ lies on the boundary of B and C_0. Since $\mathbf{p}$ and $M(\mathbf{p})$ are in $M^{[n]}(B)$ for all n, it follows that $||\mathbf{p} - M(\mathbf{p}))|| \leq a^n \to 0$ as n grows without bound. Consequently $\mathbf{p} = M(\mathbf{p})$, which means that $\mathbf{p}$ is a fixed point. Finally, because $B \supseteq M(E)$ and all elements of B converge to $\mathbf{p}$, we conclude that all elements of E also converge to $\mathbf{p}$. □

Although $\mathbf{p}$ attracts all points in B and E, $\mathbf{p}$ is not an attracting fixed point because the iterates of points in the interior of C_0 are drawn away from $\mathbf{p}$.

Figure 3.15 shows the image of $M^{[2]}$, which is composed of two connected horseshoes. Similarly, for each positive integer n, the image of $M^{[n]}$ contains 2^{n-1} connected horseshoes whose width is approximately a^n.

From the definition of M, if $\mathbf{v}$ is in T, then $M(\mathbf{v})$ is also in T only if $\mathbf{v}$ is in $C_0 \cup C_1$. Let C_+ denote the collection of points in $C_0 \cup C_1$ *all* of whose iterates lie in $C_0 \cup C_1$. Thus

$$C_+ = \{\mathbf{v} \text{ in } C_0 \cup C_1 : M^{[n]}(\mathbf{v}) \text{ is in } C_0 \cup C_1 \text{ for } n = 0, 1, 2, 3, \ldots\}$$

All points in the domain of M either migrate toward $\mathbf{p}$, which is the most illustrious member of C_+, or start out in C_+ and stay there. Consequently C_+ serves as the attractor A_M of M.

What does C_+ look like geometrically? To obtain an answer, let us note that the set C_+ is an intersection of a nested sequence of ever thinner vertical strips. To show this, we observe first that the collection of all points $\mathbf{v}$ such that $M(\mathbf{v})$ is in $C_0 \cup C_1$ consists of the four strips in Figure 3.16(a), where C_{ik} is the set of all $\mathbf{v}$ such that $\mathbf{v}$ is in C_i and $M(\mathbf{v})$ is in C_k, for $i, k = 0, 1$. (For

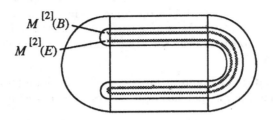

FIGURE 3.15
Squaring the horseshoe map

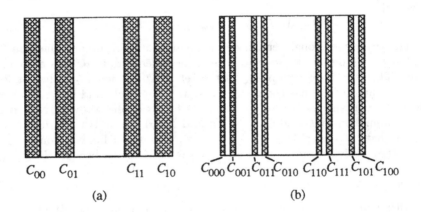

FIGURE 3.16
Tracking the points in the horseshoe

example, $\mathbf{v}$ is in C_{01} if and only if $\mathbf{v}$ is in C_0 and $M(\mathbf{v})$ is in C_1.) Similarly, the collection of all $\mathbf{v}$ such that $M^{[2]}(\mathbf{v})$ is in $C_0 \cup C_1$ consists of the eight strips in Figure 3.16(b), and so forth. Therefore C_+, and hence the attractor A_M, is a collection of vertical lines in the square $T = [0, 1] \times [0, 1]$ whose intersection with the x-axis is a Cantor-like set.

If $\mathbf{v}$ is in C_+, then each iterate of $\mathbf{v}$ lies either in C_0 or in C_1, so we can associate with $\mathbf{v}$ the **forward sequence** $z = z_0 z_1 z_2 \cdots$, where

$$z_n = \begin{cases} 0 & \text{if } M^{[n]}(\mathbf{v}) \text{ is in } C_0 \\ 1 & \text{if } M^{[n]}(\mathbf{v}) \text{ is in } C_1 \end{cases}$$

Notice that the sequence $z = z_0 z_1 z_2 \cdots$ identifies the forward iterates of $\mathbf{v}$. This is reminiscent of the sequences identified with the numbers in $[0, 1]$ for the tent function described in Section 2.2. However, since C_0 and C_1 are

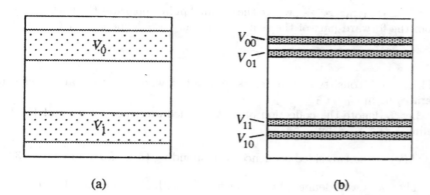

FIGURE 3.17
Horizonal stripes in the horseshoe map

separated by a rectangle of width $1/3$, *every* sequence of 0's and 1's is the image of an element of C_+. (The corresponding result is false for T!)

In contrast to the tent function, each sequence of 0's and 1's corresponds to a whole vertical line, not just an individual point. To identify sequences of 0's and 1's with unique points in S, we need to examine the pre-images of points in S.

Figure 3.14(d) indicates that $M(C_0)$ and $M(C_1)$ are horizontal strips in T that we denote by V_0 and V_1, respectively (Figure 3.17(a)). Similarly, $M^{[2]}(C_{00})$, $M^{[2]}(C_{01})$, $M^{[2]}(C_{11})$, and $M^{[2]}(C_{10})$ are horizontal strips in Figure 3.15, and we designate these strips as V_{00}, V_{01}, V_{11}, and V_{10}, respectively (Figure 3.17(b)). Letting $M^{-[m]}(P)$ denote the set Q such that $M^{[m]}(Q) = P$, we find that

$$M^{-[1]}(V_i) = C_i \quad \text{and} \quad M^{-[2]}(V_{ik}) = C_{ik} \quad \text{for } i = 0, 1 \text{ and } k = 0, 1$$

Continuing in this fashion, we obtain a nested sequence of ever thinner horizontal strips whose intersection we denote by C_-. Thus

$$C_- = \{\mathbf{v} \text{ in } S : M^{-[n]}(\mathbf{v}) \text{ is in } C_0 \cup C_1 \text{ for } n = 1, 2, 3, \dots\}$$

In the same way as we identified a sequence of 0's and 1's for each element of C_+, we now assign the **backward sequence** $\cdots z_{-3}z_{-2}z_{-1}$ to each $\mathbf{v}$ in C_-, where

$$z_{-n} = \begin{cases} 0 & \text{if } M^{-[n]}(\mathbf{v}) \text{ is in } C_0 \\ 1 & \text{if } M^{-[n]}(\mathbf{v}) \text{ is in } C_1 \end{cases}$$

Each such sequence corresponds to a horizontal line in T, and C_- is a collection of horizontal lines in T whose intersection with the y-axis is a Cantor-like set.

Combining the forward sequence $z_0z_1z_2\cdots$ and the backward sequence $\cdots z_{-3}z_{-2}z_{-1}$, we obtain the **two-sided sequence** (or **bi-infinite sequence**)

$\cdots z_{-3}z_{-2}z_{-1}.z_0z_1z_2\cdots$, where the decimal point separates the backward part from the forward part of the sequence. Finally, we define the set C^* by

$$C^* = C_+ \cap C_-$$

The set C^* consists of the points in T all of whose forward and backward iterates lie in $C_0 \cup C_1$.

Let $\mathbb{Z}$ denote the collection of all two-sided sequences of 0's and 1's, and define $h : C^* \to \mathbb{Z}$ by

$$h(\mathbf{v}) = \text{the two-sided sequence corresponding to } \mathbf{v}, \text{ for } \mathbf{v} \text{ in } C^* \qquad (3.22)$$

Then h is well-defined, and identifies C^* with $\mathbb{Z}$, as Lemma 1 asserts.

Lemma 3.1. $h : C^* \to Z$ is one-to-one and onto.

Proof. If $\mathbf{v}$ and $\mathbf{w}$ are in C^* and $h(\mathbf{v}) = h(\mathbf{w})$, then because $h(\mathbf{v})$ and $h(\mathbf{w})$ have the same forward (backward) sequence, they lie on the same vertical (horizontal) line in T. Therefore $\mathbf{v} = \mathbf{w}$, so that h is one-to-one. To show that h is onto, assume that $x = \cdots x_{-3}x_{-2}x_{-1}.x_0x_1x_2\cdots$ is in $\mathbb{Z}$. For $n \geq 0$, let

$$J_n = \{\mathbf{v} \text{ in } C_0 \cup C_1 : h(\mathbf{v}) = \cdots z_{-3}z_{-2}z_{-1}.z_0z_1z_2\cdots \text{ and } z_0z_1z_2\cdots z_n$$
$$= x_0x_1x_2\cdots x_n\}$$

and

$$J_{-n} = \{\mathbf{v} \text{ in } C_0 \cup C_1 : h(\mathbf{v}) = \cdots z_{-3}z_{-2}z_{-1}.z_0z_1z_2\cdots \text{ and}$$
$$z_{-n}\cdots z_{-3}z_{-2}z_{-1} = x_{-n}\cdots x_{-3}x_{-2}x_{-1}\}$$

Then J_n and J_{-n} are closed for all n. Because $\bigcap_{n\geq 0} J_n$ is a single vertical line and $\bigcap_{n<0} J_n$ is a single horizontal line in T, it follows that $\bigcap_{-\infty<n<\infty} J_n$ is a unique point $\mathbf{v}^*$. By construction, $h(\mathbf{v}^*) = x$, so that h is onto. $\qquad \square$

The function h has the property that

if $h(\mathbf{v}) = \cdots z_{-3}z_{-2}z_{-1}.z_0z_1z_2\cdots$, then $h(M(\mathbf{v})) = \cdots z_{-2}z_{-1}z_0.z_1z_2z_3\cdots$

This means that the sequence associated with $M(\mathbf{v})$ is the sequence associated with $\mathbf{v}$, shifted to the left one place with respect to the decimal point. This fact, along with the association between points of C^* and two-sided sequences of 0's and 1's, bears immediate fruit. In particular, the two doubly repeated sequences

$$\cdots \overline{0.0} \cdots \quad \text{and} \quad \cdots \overline{1.1} \cdots$$

correspond to fixed points of M. You can check that the fixed point $\mathbf{p}$ on the border between B and C_0 corresponds to $\cdots \overline{0.0} \cdots$. Can you locate the other fixed point in T? Next, the sequences $\cdots \overline{10.10} \cdots$ and $\cdots \overline{01.01} \cdots$ comprise a 2-cycle for M. Using these sequences as models, one can exhibit two-sided sequences corresponding to n-cycles for any positive integer n. From the definition of $\cdots z_{-3}z_{-2}z_{-1}.z_0z_1z_2\cdots$, one can even indicate where in T members of such a cycle lie.

Next, we will introduce a distance on the set $\mathbb{Z}$ of two-sided sequences. This distance will make it possible for us to show that C^* and $\mathbb{Z}$ are homeomorphic. Let $x = \cdots x_{-3}x_{-2}x_{-1}.x_0x_1x_2\cdots$ and $z = \cdots z_{-3}z_{-2}z_{-1}.z_0z_1z_2\cdots$. Then we define the **distance** $||x-z||$ between x and z by the formula

$$||x-z|| = \sum_{k=-\infty}^{\infty} \frac{|x_k - z_k|}{2^{|k|}} \tag{3.23}$$

The distance is a metric on the space of two-sided sequences (Exercise 6). If $x_k = z_k$ for $|k| \le n$, then $||x-z|| \le 1/2^{n-1}$. Moreover, if $||x-z|| \le 1/2^n$, then $x_k = z_k$ for $|k| \le n+1$. Thus the distance between x and z is small provided that the central blocks of x and z are identical.

Theorem 3.26. $h : C^* \to Z$ is a homeomorphism.

Proof. By Lemma 1, h is one-to-one and onto. Therefore we need only show that h and h^{-1} are continuous. Let $\epsilon > 0$, and choose n so large that $1/2^{n-1} < \epsilon$. Next, let $\mathbf{v}$ and $\mathbf{w}$ be in C^*, with

$$h(\mathbf{v}) = x = \cdots x_{-3}x_{-2}x_{-1}.x_0x_1x_2\cdots \text{ and } h(\mathbf{w}) = z = \cdots z_{-3}z_{-2}z_{-1}.z_0z_1z_2\cdots$$

If $||\mathbf{v}-\mathbf{w}|| < 1/3^{n+1}$, then $\mathbf{v}$ and $\mathbf{w}$ lie in the same vertical strip of width $1/3^{n+1}$, so that $x_k = z_k$ for $k = 0,1,2,\ldots,n$. Similarly, there is a $\delta_1 > 0$ such that if $||\mathbf{v}-\mathbf{w}|| < \delta_1$, then $\mathbf{v}$ and $\mathbf{w}$ lie in the same horizontal strip at the nth stage, which means that $x_k = z_k$ for $k = -1,-2,\ldots,-n$. Now choose $\delta > 0$ such that $\delta < 1/3^{n+1}$ and $\delta < \delta_1$. It follows that if $||\mathbf{v}-\mathbf{w}|| < \delta$, then $x_k = z_k$ for $|k| \le n$, and thus

$$||x-z|| = \sum_{|k|=n+1}^{\infty} \frac{|x_k - z_k|}{2^{|k|}} \le \frac{1}{2^{n-1}} < \epsilon$$

Consequently h is continuous. The proof that h^{-1} is continuous follows by a similar argument (but with the roles of vertical contraction and horizontal expansion interchanged). $\square$

The **left shift map** $\sigma : Z \to Z$ is defined as the name suggests:

if $z = \cdots z_{-3}z_{-2}z_{-1}.z_0z_1z_2\cdots$, then $\sigma(z) = \cdots z_{-3}z_{-2}z_{-1}z_0.z_1z_2\cdots$

Thus σ shifts the entries to the left one place with respect to the decimal point, or equivalently, shifts the decimal point one place to the right. That σ is a homeomorphism on $\mathbb{Z}$ can be proved in a straightforward manner (Exercise 7). Moreover, we can show that σ is **strongly chaotic**, which by definition means that

    i.   its domain has a dense set of periodic points

    ii.   it has sensitive dependence on initial conditions

    iii.   it is transitive (that is, there is an element with dense orbit)

Theorem 3.27. σ is strongly chaotic on $\mathbb{Z}$.

Proof. First, we will show that the set of periodic points of σ is dense in $\mathbb{Z}$. To that end, let $z = \cdots z_{-3}z_{-2}z_{-1}.z_0z_1z_2\cdots$ be an arbitrary element of $\mathbb{Z}$, and let n be an arbitrary positive integer. If x is the doubly-repeating two-sided sequence $\overline{z_{-n}\cdots z_{-3}z_{-2}z_{-1}.z_0z_1z_2\cdots z_n}$, then it follows that x is periodic (with period $2n+1$). Moreover, $z_k = x_k$ for $|k| \leq n$, so that $||x-z|| \leq 1/2^{n-1}$. Thus the periodic points are dense in $\mathbb{Z}$. To show that σ has sensitive dependence on initial conditions, let z be in $\mathbb{Z}$, $\epsilon = 1/2$, $\delta > 0$, and n so large that $1/2^{n-1} < \delta$. If x is chosen with $x_k = z_k$ for all k such that $|k| \leq n$ but $x_{n+1} \neq z_{n+1}$, then $||x - z|| \leq 1/2^{n-1} < \delta$. However,

$$\sigma^{[n+1]}(x) = \cdots x_n.x_{n+1}\cdots \quad \text{and} \quad \sigma^{[n+1]}(z) = \cdots z_n.z_{n+1}\cdots$$

so that $||\sigma^{[n+1]}(x) - \sigma^{[n+1]}(z)|| \geq 1 > \epsilon$. Therefore σ has sensitive dependence on initial conditions. Finally, we will show that σ is transitive. To see this, let the forward portion of the two-sided sequence z^* have the form

0 1 000 001 010 011 100 101 110 111 00000 00001 $\cdots$

(where for each positive odd integer n, all possible n-tuples appear in order), and let the backward portion of z^* have the form

$\cdots$0100 0011 0010 0001 0000 11 10 01 00

(where for each positive even integer n, all possible n-tuples appear in backward order). Then it is possible to show that the orbit of z^* is dense in $\mathbb{Z}$ (Exercise 4). With this, we have completed the proof that σ is strongly chaotic on $\mathbb{Z}$.   □

The functions $M : C^* \to C^*$ and $\sigma : Z \to Z$ are **conjugate**, in the sense that there is a homeomorphism $h : C^* \to Z$ such that $h \circ M = \sigma \circ h$. The homeomorphism we have in mind, of course, is the one defined in (3.22). You should check that indeed $h(M(\mathbf{v})) = \sigma(h(\mathbf{v}))$ for all $\mathbf{v}$ in C^* (Exercise 8). The strong chaotic nature of one function is inherited by another function conjugate to it (as we saw in Section 2.3). Thus we obtain the following consequence of Theorem 3.27.

Corollary 3.28. M is strongly chaotic on C^*.

Note: The Smale horseshoe map is justifiably famous for several reasons. First, it has a very simple geometric definition and is easy to visualize. Second, it has all the characteristics of a strongly chaotic map: sensitive dependence, a dense set of periodic points, and an element with dense orbit. It also has, in the most obvious way, the telltale signs of a chaotic map: stretching and folding. The stretching yields sensitive dependence, and the folding allows the map to be bounded. It is conjectured that in some sense every map that is strongly chaotic has a subset of its domain on which the map acts like the horseshoe map M.

Properties of this map have been extended to entire classifications of dynamical systems. The interested reader is directed to Ruelle (2004) for the definition of a Smale Space, or Putnam (2014) for a deeper look into the dynamics of Smale Spaces.

3.5.1 Homoclinic Points

Let f be a function, and p a fixed point of f. It can happen that there is a point q in the domain of f whose forward iterates converge to p and such that a sequence of backward iterates of q also converges to p. The following example illustrates such behavior.

Example 3.5.1. Let T be the tent function, defined by

$$T(x) = \begin{cases} 2x & \text{for } 0 \le x \le 1/2 \\ 2 - 2x & \text{for } 1/2 < x \le 1 \end{cases}$$

Show that the forward and backward iterates of $1/4$ converge to the fixed point 0.

Solution. First, we notice that

$$T(\frac{1}{4}) = \frac{1}{2}, \quad T^{[2]}(\frac{1}{4}) = 1, \quad \text{and} \quad T^{[3]}(\frac{1}{4}) = 0$$

so that the forward iterates of $1/4$ are eventually 0, and hence converge to 0. By contrast, $T^{-1}(1/4)$ contains $1/8$, $T^{-[2]}(1/4)$ contains $1/16$, and in general, $T^{-[n]}(1/4)$ contains the number $1/2^{n+2}$. Thus there are backward iterates of $1/4$ that converge to 0. Figure 3.18 shows both forward and backward iterates of $1/4$. $\square$

Since the forward iterates of $1/4$ for T converge to 0, $1/4$ is in the local stable manifold $W^s_{loc}(0)$ of 0. In the same manner, $1/4$ is in the local unstable manifold $W^u_{loc}(0)$ of 0, because there are backward iterates of $1/4$ that converge to 0. This means that $1/4$ is in both $W^s_{loc}(0)$ and $W^u_{loc}(0)$. Points with this property are called "homoclinic points."

Definition 3.29. Let F be defined on a subset of R^n, where $n = 1$ or 2, and let **p** be a fixed point of F. A point **q** is **homoclinic** to **p**, or is a **homoclinic point**, if $\mathbf{p} \ne \mathbf{q}$ but **q** is in both $W^s_{loc}(\mathbf{p})$ and $W^u_{loc}(\mathbf{p})$.

From Example 3.5.1 we know that $1/4$ is homoclinic to 0 for T. In fact, there are infinitely many numbers in $(0, 1)$ that are homoclinic to 0 for T

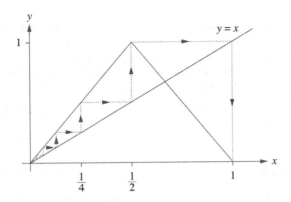

FIGURE 3.18
Forward and backward iterates of 1/4

(Exercise 11). Analogously, there are numbers in (0, 1) that are homoclinic to 0 for the quadratic function Q_4 (Exercise 12). By contrast, there are no numbers in (0, 1) that are homoclinic to 0 for the quadratic function Q_μ if $0 < \mu < 4$ (also Exercise 12).

The horseshoe map M has points homoclinic to the fixed point **p** in B. Since M is conjugate to the left shift σ, we can verify this by showing, equivalently, that the left shift σ has points homoclinic to the fixed point $z^{**} = \cdots \overline{0}.\overline{0} \cdots$ of σ. We will indicate how one can determine such homoclinic points of σ, but will leave the details to be completed in Exercise 10. On the one hand, $W^s_{loc}(z^{**})$ consists of all two-sided sequences all of whose entries to the right of some entry are identical to those of z^{**}. On the other hand, $W^u_{loc}(z^{**})$ consists of those two-sided sequences all of whose entries to the left of some entry are identical to those of z^{**}. It follows from these facts that σ has points homoclinic to z^{**}. Furthermore, there are infinitely many points homoclinic to z^{**}.

Homoclinic points play a significant role in the study of the dynamics of higher-dimensional functions.

3.5.2   The Williams Solenoid

A fascinating follow up to the work of Smale came in Williams (1974). This introduced the idea of a solenoid, now commonly called the **Smale-Williams attractor**. We discuss it briefly here to illustrate how the homoclinic maps we have introduced could be extended to higher dimensions.

Begin with a solid torus embedded in $\mathbb{R}^3$. If you are not familiar with the notion of a torus, we mean a donut. To be precise, let $T = S^1 \times D$, the Cartesian product of a unit circle and a filled disc. We give the circle the usual coordinate in radians (modulus 2π), and the disc gets a complex coordinate $z = x + iy$.

FIGURE 3.19
Multiple applications of the map f to generate a solenoid

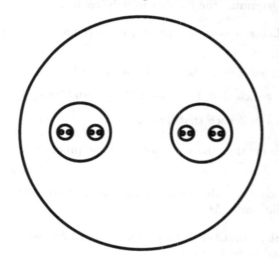

FIGURE 3.20
The Williams Solenoid

Define a map

$$f(t, z) = (2t, \frac{z}{4} + \frac{e^{it}}{2}).$$

The angle on the circle is doubled (mod 2π), and the point on the disc has its distance from the origin reduced by a factor of $1/4$, and then moved left or right along the real axis depending on the angle t. This essentially shrinks the torus to $1/4$ its usual thickness, and wraps it twice around itself. We can illustrate this in Figure 3.19, taken from Isaeva et al. (2015).

Something interesting occurs when we apply this map over and over again. If we choose some angle t on our original circle, we can cut a slice through the torus, and look at where the torus crosses that slice under multiple iterations of the map. In Figure 3.20, we see this illustrated. The first circle is unchanged, but it contains two smaller circles, each $1/4$ the diameter of the original circle.

Those in turn contain two circles each 1/4 the diameter of the previous circles. This process can be repeated indefinitely. We're essentially stretching and folding the torus, just like we did with the Horseshoe. This should, as with the Horseshoe, remind you of the Cantor set.

3.5.3   Section 3.5 Exercises

Exercise 1. Find the approximate location of the point in S corresponding to the sequence $\cdots \bar{1} . \bar{1} \cdots$ in $\mathbb{Z}$.

Exercise 2. Determine the number of n-cycles in C^*.

Exercise 3. Let $x = \cdots x_{-3} x_{-2} x_{-1} . x_0 x_1 x_2 \cdots$ and $z = \cdots z_{-3} z_{-2} z_{-1} . z_0 z_1 z_2 \cdots$.

(a) Show that if $x_k = z_k$ for $|k| \leq n$, then $||x - z|| \leq 1/2^{n-1}$.

(b) Find x and z such that $x_k = z_k$ for $|k| \leq n$ and $||x - z|| = 1/2^{n-1}$.

(c) Find x and z in $\mathbb{Z}$ such that $||x - z|| = 3$.

Exercise 4. Show that the sequence z^* in the proof of Theorem 3.27 has a dense orbit.

Exercise 5. Show that there is an element of C^* that neither is periodic nor has a dense orbit under M.

Exercise 6. Show that the distance defined in (3.23) has the properties of a metric as defined in Section 2.4.

Exercise 7. Show that the left shift $\sigma : \mathbb{Z} \to \mathbb{Z}$ is a homeomorphism.

Exercise 8. Show that $h(M(\mathbf{v})) = \sigma(h(\mathbf{v}))$ for all $\mathbf{v}$ in C^*.

Exercise 9. Show that there is a one-to-one function whose domain is C^* and whose range is the Cantor ternary set C.

Exercise 10. Let $z^{**} = \cdots \bar{0} . \bar{0} \cdots$, which is one of the two fixed points of σ.

(a) Show that $W^s_{loc}(z^{**})$ consists of all two-sided sequences all of whose entries to the right of some entry are identical to those of z^{**}.

(b) Show that $W^u_{loc}(z^{**})$ consists of those two-sided sequences all of whose entries to the left of some entry are identical to those of z^{**}.

(c) Show that there are infinitely many points homoclinic to z^{**}.

Exercise 11. Show that there are infinitely many numbers in the interval $(0, 1)$ that are homoclinic to 0 for the tent function T.

Exercise 12. Let $Q_\mu(x) = \mu x(1-x)$ for $0 \le x \le 1$.

(a) Show that there are no points homoclinic to the fixed point 0 whenever $0 < \mu < 4$.

(b) Show that there are infinitely many points homoclinic to 0 if $\mu = 4$.

(c) Show that there are infinitely many points homoclinic to the fixed point $p_\mu = 1 - 1/\mu$ if $\mu = 4$.

Exercise 13. Using the notation from Exercises 1 and 10, a point $z \in Z$ is **heteroclinic** if $z \in W_{loc}^s((0)) \cap W_{loc}^u((1))$. Describe the sequences that could exist for such a point, and prove they are dense in Z.

Chapter 3 has introduced dynamics of functions defined in $\mathbb{R}^2$. Besides the linear functions, which exhibit regularity, we have focused on three prominent functions: the baker's function, the Hénon map, and the Smale horseshoe map. For each of these three functions there is an attractor to which the iterates of all points in the domain converge, and which contains in a natural way a two-dimensional version of a Cantor set. In each, there is a stretching and some form of folding, which are characteristic of chaotic behavior. Finally, the baker's map and the Hénon map represent two-dimensional analogues of one-dimensional chaotic functions studied in Chapters 1 and 2.

4

Systems of Differential Equations

Objects in motion, like satellites in space, water molecules in a stream, and electrons in a diamond, have long intrigued scientists. During the past few centuries, scientists were under the impression that the motion of an object could, at least in theory, be completely understood if one could formulate accurately enough a set of equations that described certain attributes of the object such as its position, velocity and acceleration. That this premise was not universally true was observed around 1900 by the famous French mathematician Henri Poincaré (1854–1912) in his studies of the three-body problem.

However, in the early 1960's the American meteorologist Edward Lorenz (1917–2008) discovered that weather patterns modeled by a simple system of differential equations could be profoundly unpredictable. It was this discovery, along with the power of modern-day computers, that launched the new field of chaotic dynamics. One goal of this chapter is to describe and interpret the Lorenz system of differential equations.

> **Note:** A 2019 article in *Quanta* magazine, "The Hidden Heroines of Chaos" (Sokol) is highly recommended reading. It details the contributions of Ellen Fetter and Margaret Hamilton to the work of Lorentz. These women were given small acknowledgments by Lorentz in his papers, but the extent of their contributions were not fully investigated until recently. The article provides information about the often unsung women who were essential to the development of the theory we will cover in the remainder of the text.

Section 4.1 is a review of the important concepts relating to systems of differential equations, and includes an analysis of constant solutions of such systems. In Section 4.2 we apply the results of Section 4.1 to systems that are not linear but are what we will call almost linear systems. In the final sections of the chapter, we turn to two celebrated almost linear systems: the pendulum system that describes the motion of a pendulum, and the Lorenz system that has been applied to the study of weather.

DOI: 10.1201/9781032678757-4

4.1 Review of Systems of Differential Equations

Systems of differential equations are employed to describe a wide variety of physical phenomena, including the motion of a pendulum and particles in a weather system. Before we can analyze these applications, we need to understand the structure of solutions of such systems. That is the goal of the present section. If you do not have prior knowledge of differential equations, then you should probably consult a standard textbook on differential equations. The classic is Coddinton and Levinson (1984), though may find something like Judson's ODE project (freely available online)more accessible.

We begin with terminology.

Definition 4.1. Let $x_1, x_2, \ldots, x_n$ be differentiable functions of time t on an interval J of real numbers, and let $f_1, f_2, \ldots, f_n$ be functions of $x_1, x_2, \ldots,$ and x_n as well as t. The n differential equations

$$\frac{dx_1}{dt} = f_1(x_1, x_2, \ldots, x_n, t)$$

$$\frac{dx_2}{dt} = f_2(x_1, x_2, \ldots, x_n, t)$$

$$\cdots\cdots\cdots\cdots \tag{4.1}$$

$$\frac{dx_n}{dt} = f_1(x_1, x_2, \ldots, x_n, t)$$

form a **system of differential equations**. If $X : J \to \mathbb{R}^n$ is defined by

$$X(t) = \begin{pmatrix} x_1(t) \\ x_2(t) \\ \cdots \\ x_n(t) \end{pmatrix} \tag{4.2}$$

and if X satisfies (4.1), then X is a **solution** of the system. If t_0 is in J and X is a solution for all $t \geq t_0$, then the element $X(t_0)$ of $\mathbb{R}^n$ is an **initial condition** of a solution X. Usually t_0 will be 0.

Since $x_1, x_2, \ldots, x_n$ are differentiable functions of t, it follows that as t increases, $X(t)$ traces out a curve in $\mathbb{R}^n$, called the **trajectory**, or **orbit**, of X.

If $n = 1$ in (4.1), the system is simply $dx/dt = f(x, t)$, and we will write a solution as x. One well-known example of such an equation appears in calculus:

$$\frac{dx}{dt} = kx \tag{4.3}$$

where k is nonzero constant. Solutions of the equation in (3) are functions of the form $x(t) = ce^{kt}$, which represent exponential growth if $k > 0$ and exponential decay if $k < 0$.

If $n = 2$ in 4.36, we write the system as

$$\frac{dx}{dt} = f(x, y, t)$$

$$\frac{dy}{dt} = g(x, y, t)$$

For such a system, $x(t)$ might denote the position and $y(t)$ the velocity of an object at time t. Analogously, if $n = 3$, then we write

$$\frac{dx}{dt} = f(x, y, z, t)$$

$$\frac{dy}{dt} = g(x, y, z, t)$$

$$\frac{dz}{dt} = h(x, y, z, t)$$

In this case, $x(t), y(t)$, and $z(t)$ might represent the coordinates of an object at time t, or perhaps the position, velocity, and acceleration of the object at time t.

In the same way that fixed points play a significant role in the analysis of the dynamics of a function, special solutions assist in the analysis of solutions of systems of differential equations.

Definition 4.2. A **critical point** (or **equilibrium point**, or **stationary point**) of a system of differential equations is a constant solution, that is, a solution X such that $X(t) = X(t_0)$ for all t. If X is a critical point, then we identify the critical point with the vector $X(t_0)$.

It can be shown that if X is a solution and $dX/dt = \mathbf{0}$ for some $t = t_0$, then X is constant for all t, and hence is a critical point. For example, if

$$\frac{dx}{dt} = (x - 2)(x + 3)$$

then the critical points of the equation are the constant solutions $X(t) = 2$ and $X(t) = -3$, for all t.

4.1.1　Linear Differential Equations

A differential equation of the form

$$\frac{dx}{dt} = a(t)x + g(t) \tag{4.4}$$

where a and g are continuous functions of t on a given interval J, is called a **linear differential equation**, because the right-hand side of the equation is

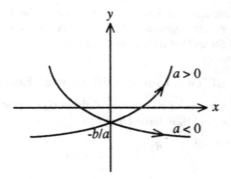

FIGURE 4.1
General solutions to a linear ODE

linear in x. Equation (4.4) can be solved completely, in the sense that we can
find a formula that describes every solution of (4.4). Such a formula is called
a **general solution**.

In the special case

$$\frac{dx}{dt} = ax + b \tag{4.5}$$

in which a and b are constant with $a \neq 0$, the general solution is particularly
easy to identify:

$$x(t) = -\frac{b}{a} + ce^{at}, \text{ for all } t \text{ in } J \tag{4.6}$$

where c is any real constant. To check that x given in (4.6) is indeed a solution
of (4.4), we just take the derivative of x in (4.6) and then see that (4.5) is
satisfied:

$$\frac{dx}{dt} = ace^{at} = a\left(-\frac{b}{a} + ce^{at}\right) + b = ax + b$$

That every solution to (4.5) has the form of (4.6) follows from the theory
of differential equations. We conclude that (4.6) gives the general solution of
(4.5). The graphs of two solutions are given in Figure 4.1; the graph curves
up if $a > 0$, and curves down if $a < 0$. From Figure 4.1 one might correctly
assume that the system in (4.5) has no critical point if $a \neq 0$ and $b \neq 0$.

The general solution of the general linear differential equation

$$\frac{dx}{dt} = a(t)x + g(t)$$

is given by

$$x(t) = e^{A(t)}\left(\int_{u_0}^t e^{-A(u)}g(u)\,du + c\right), \quad \text{where } A(t) = \int_{u_0}^t a(u)\,du \tag{4.7}$$

where c is any real number and u_0 is in J. You can check that (4.7) defines a solution to the differential equation. Only in rare cases is it possible to find a reasonable formula for $x(t)$ if $a(t)$ is not constant.

4.1.2   Systems of Two Linear Differential Equations

We turn to those systems of two linear differential equations with constant coefficients a, b, c, and d that have the form

$$\frac{dx}{dt} = ax + by$$

$$\frac{dy}{dt} = cx + dy$$

(4.8)

If $X(t) = \begin{pmatrix} x(t) \\ y(t) \end{pmatrix}$ is a solution, then the system in (8) can be rewritten as

$$\frac{dX}{dt} = \begin{pmatrix} a & b \\ c & d \end{pmatrix}\begin{pmatrix} x \\ y \end{pmatrix}$$

(4.9)

The matrix $\begin{pmatrix} a & b \\ c & d \end{pmatrix}$ is the **associated matrix** for the system.

Evidently if $X(t) = \mathbf{0}$ for all t, then X is a solution of the system in (4.8). The solution is called the **zero solution** and is denoted by $\mathbf{0}$, as is the corresponding critical point.

By (4.6), the general solution of the linear differential equation $dx/dt = ax$ is given by $x(t) = ce^{at}$. Thus we might hope that a nontrivial solution of the system in (4.8) would be related to functions of the form

$$X(t) = \begin{pmatrix} x(t) \\ y(t) \end{pmatrix} = \begin{pmatrix} re^{\lambda t} \\ se^{\lambda t} \end{pmatrix}$$

(4.10)

for appropriate constants λ, r, and s. For this to be the case, X must satisfy

$$\begin{pmatrix} \lambda re^{\lambda t} \\ \lambda se^{\lambda t} \end{pmatrix} = \begin{pmatrix} dx/dt \\ dy/dt \end{pmatrix} \overset{(8)}{=} \begin{pmatrix} ax + by \\ cx + dy \end{pmatrix} = \begin{pmatrix} are^{\lambda t} + bse^{\lambda t} \\ cre^{\lambda t} + dse^{\lambda t} \end{pmatrix}$$

$$= \begin{pmatrix} a & b \\ c & d \end{pmatrix}\begin{pmatrix} re^{\lambda t} \\ se^{\lambda t} \end{pmatrix}$$

so that

$$\lambda\begin{pmatrix} re^{\lambda t} \\ se^{\lambda t} \end{pmatrix} = \begin{pmatrix} a & b \\ c & d \end{pmatrix}\begin{pmatrix} re^{\lambda t} \\ se^{\lambda t} \end{pmatrix}, \text{ or equivalently, } \lambda\begin{pmatrix} r \\ s \end{pmatrix}$$

$$= \begin{pmatrix} a & b \\ c & d \end{pmatrix}\begin{pmatrix} r \\ s \end{pmatrix}$$

(4.11)

Consequently if $X(t) = \begin{pmatrix} re^{\lambda t} \\ se^{\lambda t} \end{pmatrix}$ and $X(t)$ is a solution of the system in (8), then λ is an eigenvalue and $X(t)$ an eigenvector of the associated matrix. Since λ is an eigenvalue,

$$0 = \det \begin{pmatrix} a - \lambda & b \\ c & d - \lambda \end{pmatrix} = (a - \lambda)(d - \lambda) - bc$$

Therefore λ satisfies the **characteristic equation**

$$\lambda^2 - (a + d)\lambda + ad - bc = 0 \tag{4.12}$$

If the characteristic equation has a real solution λ, then we see immediately that $X(t) = \begin{pmatrix} re^{\lambda t} \\ se^{\lambda t} \end{pmatrix}$ furnishes a solution of the system in (4.8). However, as we will see later in the section, the system in (4.8) has nonzero solutions even when the characteristic equation has no real solutions. In fact, the theory of differential equations tells us that there are always two solutions of the system in (4.8) that are not constant multiples of one another, that is, there are always two **independent solutions** of the system.

The system in (4.8) has a critical point, namely **0**. Are there any others? The answer is provided by Theorem 4.3.

Theorem 4.3. 0 is the only critical point of the system in (4.8) if and only if 0 is *not* an eigenvalue of the associated matrix.

Proof. Let $\lambda = 0$ be an eigenvalue of the associated matrix. Then there is a nonzero eigenvector $\begin{pmatrix} r \\ s \end{pmatrix}$. If X is defined by $X(t) = \begin{pmatrix} x(t) \\ y(t) \end{pmatrix} = \begin{pmatrix} r \\ s \end{pmatrix}$ for all t, then $dx/dt = 0 = dy/dt$, so that

$$\frac{dX}{dt} = \begin{pmatrix} dx/dt \\ dy/dt \end{pmatrix} = \begin{pmatrix} 0 \\ 0 \end{pmatrix} = \begin{pmatrix} a & b \\ c & d \end{pmatrix} \begin{pmatrix} r \\ s \end{pmatrix}$$

Thus X is a nonzero solution, so is a critical point of the system. Conversely, let $\begin{pmatrix} r \\ s \end{pmatrix}$ be a nonzero critical point. If we let $X(t) = \begin{pmatrix} x(t) \\ y(t) \end{pmatrix} = \begin{pmatrix} r \\ s \end{pmatrix}$, then by (9),

$$\begin{pmatrix} 0 \\ 0 \end{pmatrix} = \begin{pmatrix} dx/dt \\ dy/dt \end{pmatrix} = \frac{dX}{dt} = \begin{pmatrix} a & b \\ c & d \end{pmatrix} \begin{pmatrix} r \\ s \end{pmatrix}$$

It follows that 0 is an eigenvalue of the matrix. This completes the proof of the theorem. □

In our analysis of linear systems we will study those systems with only one critical point, namely **0**. This means, by Theorem 4.3, that the associated matrices we will encounter will not have an eigenvalue equal to zero.

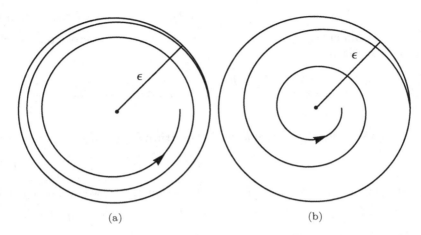

(a)   (b)

FIGURE 4.2
Stable and unstable solutions to ODEs

Fixed points of maps correspond to equilibrium solutions of systems of differential equations. The attributes of attracting and repelling fixed points of maps relate to the notion of stability for systems of differential equations.

Definition 4.4. Consider the linear system in (4.8), with critical point **0**.

a. **0** is **stable** if for each $\epsilon > 0$ there is a $\delta > 0$ such that if X is any solution satisfying $||X|| < \delta$, then $||X(t)|| < \epsilon$ for all $t > 0$.

b. **0** is **asymptotically stable** if **0** is stable, and if there is a $\delta > 0$ such that for each solution X, if $||X(0)|| < \delta$, then $\lim_{t \to \infty} ||X(t)|| = 0$.

c. **0** is **unstable** if it is not stable.

Intuitively, **0** is stable if any solution X that is close enough to **0** at time $t = 0$ remains close to **0** for all $t > 0$ (Figure 4.2(a)). In the same way, **0** is asymptotically stable if any solution X that is close enough to **0** at time $t = 0$ approaches **0** as t increases without bound (Figure 4.2(b)). By definition, if **0** is asymptotically stable, then it is stable. The converse, however, is not true, as we will note in Case 4 below.

We will classify the critical point **0** by the values of nonzero eigenvalues of the associated matrix. If there are two eigenvalues, we denote them by λ and μ, and the corresponding eigenvectors by $\begin{pmatrix} re^{\lambda t} \\ se^{\lambda t} \end{pmatrix}$ and $\begin{pmatrix} ve^{\mu t} \\ we^{\mu t} \end{pmatrix}$, respectively. It follows from the theory of differential equations that the general solution of (8) is given by linear combinations of the two eigenvectors, that is,

$$X(t) = p \begin{pmatrix} re^{\lambda t} \\ se^{\lambda t} \end{pmatrix} + q \begin{pmatrix} ve^{\mu t} \\ we^{\mu t} \end{pmatrix} = \begin{pmatrix} pre^{\lambda t} + qve^{\mu t} \\ pse^{\lambda t} + qwe^{\mu t} \end{pmatrix} \tag{4.13}$$

where p and q are arbitrary real numbers.

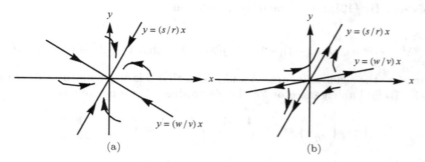

FIGURE 4.3
Portraits for $0 > \lambda > \mu$ and $0 < \mu < \lambda$

Case 1. λ and μ are real and distinct, with $\lambda\mu > 0$.

The general solution is given by (4.13). Since $\lambda \neq \mu$, the two eigenvector solutions are not multiples of one another (Exercise 13).

If $0 > \lambda > \mu$, then $e^{\lambda t}$ and $e^{\mu t}$ converge to 0 as t increases without bound, so $\lim_{t \to \infty} X(t) = \mathbf{0}$. Because all trajectories approach the origin as t increases, the critical point $\mathbf{0}$ is called an **asymptotically stable solution**, or an **asymptotically stable node**. An asymptotically stable node is analogous to an attracting fixed point of a map, since all solutions approach the node as t increases without bound. By hypothesis, $\lambda > \mu$, so that if t is large, then the terms containing $e^{\lambda t}$ dominate those containing $e^{\mu t}$. Thus if X is any nonzero solution that is not a multiple of the eigenvector $\begin{pmatrix} ve^{\mu t} \\ we^{\mu t} \end{pmatrix}$, then for large t,

$$X(t) \approx p \begin{pmatrix} re^{\lambda t} \\ se^{\lambda t} \end{pmatrix}$$

for an appropriate nonzero constant p. But this vector has slope s/r. A diagram that shows the direction of solutions as t increases is often called the **portrait** of the solutions. If $0 > \lambda > \mu$, the portrait of the solutions is as in Figure 4.3(a). If $0 < \mu < \lambda$, then $\lim_{t \to \infty} \|X(t)\| = \infty$ for every solution. Therefore all trajectories point away from the origin (Figure 4.3(b)). In order to portray the behavior of the trajectories, we need only reverse the arrows. In this case, $\mathbf{0}$ is an **unstable node**.

Example 4.1.1. Consider the system

$$\frac{dx}{dt} = -4x + y$$

$$\frac{dy}{dt} = 3x - 2y$$

Show that $\mathbf{0}$ is asymptotically stable, and sketch a portrait of the solutions.

Solution. By (12), the characteristic equation is

$$\lambda^2 - (-4-2)\lambda + (-4)(-2) - (1)(3) = 0, \quad \text{that is,} \quad \lambda^2 + 6\lambda + 5 = 0$$

Therefore the eigenvalues are -1 and -5, so that $\mathbf{0}$ is an asymptotically stable node. To find an eigenvector for the eigenvalue -1, we notice that

$$(-1)\begin{pmatrix} x \\ y \end{pmatrix} = \begin{pmatrix} -4 & 1 \\ 3 & -2 \end{pmatrix} \begin{pmatrix} x \\ y \end{pmatrix} = \begin{pmatrix} -4x + y \\ 3x - 2y \end{pmatrix}$$

Thus $-x = -4x + y$, so that $y = 3x$. Therefore $\begin{pmatrix} 1 \\ 3 \end{pmatrix}$ is an eigenvector for the eigenvalue -1. In a similar fashion, we find that $\begin{pmatrix} 1 \\ -1 \end{pmatrix}$ is an eigenvector for the eigenvalue -5. Figure 4.3(a) gives an idea of the portrait of the solutions of the system, with the solutions approaching the origin asymptotically along the line $y = 3x$, with the exception of those solutions along the line $y = -x$. $\square$

Case 2. $\mu < 0 < \lambda$.

The fact that $\mu < 0$ implies that trajectories that begin near the line $y = wx/v$ (but not on the line) tend to approach the origin before becoming slave to the term involving $e^{\lambda t}$ and moving away from the origin (Figure 4.4). Therefore, if a solution is the form of (4.13) with $p \neq 0$, then for large values of t,

$$X(t) \approx p \begin{pmatrix} re^{\lambda t} \\ se^{\lambda t} \end{pmatrix}$$

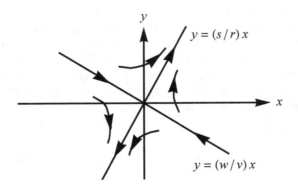

FIGURE 4.4
Portraits for $\mu < 0 < \lambda$

The critical point 0 is called a **saddle point** and is unstable. The line $y = wx/v$ is a stable manifold for the critical point, and the line $y = sx/r$ is an unstable manifold. (See Section 3.3 for the definitions of stable and unstable manifolds.)

Example 4.1.2. Consider the system

$$\frac{dx}{dt} = -x + y$$

$$\frac{dy}{dt} = 3x + y$$

Show that 0 is a saddle point, and sketch a portrait of the solutions of the system.

Solution. By (12) the characteristic equation is

$$\lambda^2 - (-1 + 1)\lambda + (-1)(1) - (1)(3) = 0, \quad \text{that is,} \quad \lambda^2 - 4 = 0$$

so that the eigenvalues are 2 and -2. By a straightforward calculation we find that an eigenvector for 2 is $\begin{pmatrix} 1 \\ 3 \end{pmatrix}$, and an eigenvector for -2 is $\begin{pmatrix} 1 \\ -1 \end{pmatrix}$. Figure 4.4 represents a portrait for the solutions of the system. $\square$

Case 3. $\lambda = \mu \neq 0$.

Here there is only one eigenvector, and as a result, the critical point is called a **degenerate node**. The discriminant of the characteristic equation $\lambda^2 - (a + d)\lambda + ad - bc = 0$ vanishes, which means that

$$(a + d)^2 - 4(ad - bc) = 0$$

and consequently $\lambda = (a + d)/2$. In the event that $b = 0$ and $c = 0$, the characteristic equation reduces to $\lambda^2 - (a + d)\lambda + ad = 0$. By assumption in Case 3, there is only one root for the characteristic equation, which implies that $a = d$. Consequently the system of differential equations reduces to

$$\frac{dx}{dt} = ax$$

$$\frac{dy}{dt} = ay$$

Evidently the two equations are independent of one another, and have solutions $x = pe^{at}$ and $y = qe^{at}$. It follows that the general solution to the system is given by

$$X(t) = \begin{pmatrix} pe^{at} \\ qe^{at} \end{pmatrix} = e^{at} \begin{pmatrix} p \\ q \end{pmatrix}$$

where p and q are arbitrary real constants. Thus all trajectories are linear, with slope q/p if $p \neq 0$. They are pointed toward the origin if $a < 0$ and

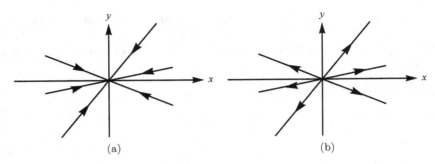

FIGURE 4.5
Portraits for $b = c = 0$ with $\mu = \lambda < 0$ and $\mu = \lambda > 0$

away from the origin if $a > 0$ (Figures 4.5(a) and (b)). This property of the solutions leads one to call **0** a **star solution**.

The other possibility is that $b \neq 0$ or $c \neq 0$. The general solution is more complicated to identify. One solution is given by

$$X_1(t) = \begin{pmatrix} re^{\lambda t} \\ se^{\lambda t} \end{pmatrix}$$

where $\begin{pmatrix} r \\ s \end{pmatrix}$ is an eigenvector corresponding to eigenvalue λ. To obtain a second solution, the theory tells us that we must first find values for v and w such that

$$\begin{pmatrix} a - \lambda & b \\ c & d - \lambda \end{pmatrix} \begin{pmatrix} v \\ w \end{pmatrix} = \begin{pmatrix} r \\ s \end{pmatrix}$$

Then a second solution is given by

$$X_2(t) = \begin{pmatrix} (v + rt)e^{\lambda t} \\ (w + st)e^{\lambda t} \end{pmatrix}$$

Since X_1 and X_2 are not multiples of one another, the general solution is given by

$$X(t) = pX_1(t) + qX_2(t) \tag{4.14}$$

where p and q are arbitrary real constants.

One can show that if $\lambda < 0$, then a given nonzero trajectory approaches the origin along the line $y = sx/r$. Therefore **0** is an asymptotically stable degenerate node (Figure 4.6(a)). Analogously, if $\lambda > 0$, then the trajectory recedes from the origin along the same line, and **0** is an unstable degenerate node (Figure 4.6(b)).

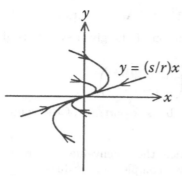

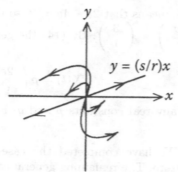

Portrait when $\mu = \lambda < 0$,
and $b \neq 0$ or $c \neq 0$

(a)

Portrait when $\mu = \lambda > 0$,
and $b \neq 0$ or $c \neq 0$

(b)

FIGURE 4.6
Degenerate node phase portraits

Example 4.1.3. Consider the system

$$\frac{dx}{dt} = 3x - 4y$$

$$\frac{dy}{dt} = x - y$$

Find the general solution of the system, and sketch a portrait for the system.

Solution. Since

$$\det \begin{pmatrix} 3 - \lambda & -4 \\ 1 & -1 - \lambda \end{pmatrix} = \lambda^2 - 2\lambda + 1 = (\lambda - 1)^2$$

it follows that the eigenvalue is 1. To find an eigenvector, we calculate that

$$1 \begin{pmatrix} r \\ s \end{pmatrix} = \begin{pmatrix} 3 & -4 \\ 1 & -1 \end{pmatrix} \begin{pmatrix} r \\ s \end{pmatrix} = \begin{pmatrix} 3r - 4s \\ r - s \end{pmatrix}$$

which yields $r = 2s$. Thus $\begin{pmatrix} 2 \\ 1 \end{pmatrix}$ is an eigenvector for 1. Next, we need to find v and w so that

$$\begin{pmatrix} 3 - \lambda & -4 \\ 1 & -1 - \lambda \end{pmatrix} \begin{pmatrix} v \\ w \end{pmatrix} = \begin{pmatrix} r \\ s \end{pmatrix}$$

Since $\lambda = 1, r = 2$, and $s = 1$, this reduces to

$$\begin{pmatrix} 2 & -4 \\ 1 & -2 \end{pmatrix} \begin{pmatrix} v \\ w \end{pmatrix} = \begin{pmatrix} 2 \\ 1 \end{pmatrix}$$

This means that $2v - 4w = 2$, so that $v = 1 + 2w$. As a result, we can choose $\begin{pmatrix} v \\ w \end{pmatrix} = \begin{pmatrix} 3 \\ 1 \end{pmatrix}$. By (14) the general solution of the given system is given by

$$X(t) = p \begin{pmatrix} 2e^t \\ e^t \end{pmatrix} + q \begin{pmatrix} (3 + 2t)e^t \\ (1 + t)e^t \end{pmatrix}$$

for any real constants p and q. Figure 4.6(b) is a portrait of the solutions. □

We have completed the cases for which the eigenvalues are real and nonzero. The remaining general case involves complex eigenvalues.

Case 4. The eigenvalues are not real.

If the solutions of the characteristic equation $\lambda^2 - (a + d)\lambda + ad - bc = 0$ are not real, then they have the form $\alpha + i\beta$ and $\alpha - i\beta$, where α and β are real. In this case the general solution takes a little more work to produce. The theory tells us to find an eigenvector for $\alpha + i\beta$ of the form $\begin{pmatrix} r + iv \\ s + iw \end{pmatrix} = \begin{pmatrix} r \\ s \end{pmatrix} + i \begin{pmatrix} v \\ w \end{pmatrix}$, where $r, s, v,$ and w are real numbers. If we let

$$X_1(t) = \left(\begin{pmatrix} r \\ s \end{pmatrix} \cos \beta t - \begin{pmatrix} v \\ w \end{pmatrix} \sin \beta t \right) e^{\alpha t} \text{ and}$$

$$X_2(t) = \left(\begin{pmatrix} r \\ s \end{pmatrix} \sin \beta t + \begin{pmatrix} v \\ w \end{pmatrix} \cos \beta t \right) e^{\alpha t}$$

then the general solution is given by

$$X(t) = pX_1(t) + qX_2(t) \tag{4.15}$$

where p and q are arbitrary real constants. All trajectories spiral toward the origin if $\alpha < 0$, and spiral away from the origin if $\alpha > 0$ (Figures 4.7(a)–(b)). For this reason, the critical point **0** is called a **spiral point**, or a **focus**, when $\alpha \neq 0$.

Example 4.1.4. Find the general solution of the system

$$\frac{dx}{dt} = x + 5y$$

$$\frac{dy}{dt} = -x - 3y$$

Solution. First, we find that

$$\det \begin{pmatrix} 1 - \lambda & 5 \\ -1 & -3 - \lambda \end{pmatrix} = \lambda^2 + 2\lambda + 2 = 0 \quad \text{if and only if} \quad \lambda = -1 \pm i$$

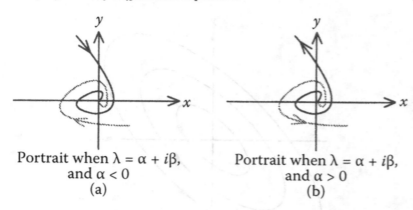

Portrait when $\lambda = \alpha + i\beta$, and $\alpha < 0$
(a)

Portrait when $\lambda = \alpha + i\beta$, and $\alpha > 0$
(b)

FIGURE 4.7
Spiral node phase portraits

Therefore the eigenvalues are $-1 + i$ and $-1 - i$. To find an eigenvector for $-1 + i$, we notice that if

$$(-1+i)\begin{pmatrix} v \\ w \end{pmatrix} = \begin{pmatrix} 1 & 5 \\ -1 & -3 \end{pmatrix}\begin{pmatrix} v \\ w \end{pmatrix} = \begin{pmatrix} v + 5w \\ -v - 3w \end{pmatrix}$$

then $-v + iv = v + 5w$, so $w = (-2 + i)v/5$. If $v = 5$, then $w = -2 + i$. Thus

$$\begin{pmatrix} 5 \\ -2+i \end{pmatrix} = \begin{pmatrix} 5 \\ -2 \end{pmatrix} + i\begin{pmatrix} 0 \\ 1 \end{pmatrix}$$

is an eigenvector for the eigenvalue $-1 + i$. By (15), the general solution is

$$X(t) = p\left(\begin{pmatrix} 5 \\ -2 \end{pmatrix}\cos t - \begin{pmatrix} 0 \\ 1 \end{pmatrix}\sin t\right)e^{-t} + q\left(\begin{pmatrix} 5 \\ -2 \end{pmatrix}\sin t + \begin{pmatrix} 0 \\ 1 \end{pmatrix}\cos t\right)e^{-t}$$

where as usual, p and q are real constants. $\qquad\square$

If the eigenvalues of the associated matrix are $\lambda = i\beta$ and $\lambda = -i\beta$ (that is, if $\alpha = 0$), then the solution in (4.15) is periodic, with period $2\pi/\beta$. Moreover, the trajectories can be shown to be elliptical. In view of these observations, one calls the solution **0** a **center** (Figure 4.8). Notice that a center critical point is an example of a stable but not asymptotically stable critical point.

We have now completed our analysis of systems of two linear differential equations with constant coefficients. In the following table we recapitulate our results concerning critical points.

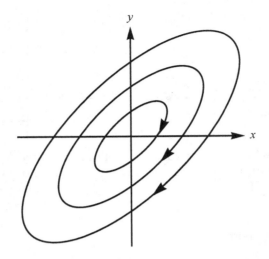

FIGURE 4.8
Portrait when $\lambda = \alpha + i\beta$ and $\alpha = 0$

4.1.3   Table of Results Illustrating the Relationship between Eigenvalues and Critical Points

EIGENVALUES	TYPE OF CRITICAL POINT
$0 < \mu < \lambda$	unstable node (trajectories recede from **0**)
$\mu < \lambda < 0$	asymptotically stable node (trajectories converge to **0**)
$\mu < 0 < \lambda$	saddle point (some trajectories approach **0**
before receding)	
$0 < \lambda = \mu$	unstable degenerate node (trajectories recede from **0**)
$\lambda = \mu < 0$	asymptotically stable degenerate node (trajectories converge to **0**)
$\lambda = \alpha + i\beta, \alpha > 0$	unstable spiral point (trajectories spiral away from **0**)
$\lambda = \alpha + i\beta, \alpha < 0$	asymptotically stable spiral point (trajectories spiral toward **0**)
$\lambda = \alpha + i\beta, \alpha = 0$	stable center (trajectories are elliptical, centered at **0**)

In the next section, we will study systems of differential equations that are not linear, but which are nearly linear.

4.1.4   Section 4.1 Exercises

In Exercises 1–12, find the eigenvalues of the matrix associated with the given system and then classify the critical point **0**. Then sketch a portrait of the solutions of the system.

Exercise 1.

$$\frac{dx}{dt} = -2x + 9y$$

$$\frac{dy}{dt} = x - 2y$$

Exercise 2.

$$\frac{dx}{dt} = 2x - y$$

$$\frac{dy}{dt} = 2x + 3y$$

Exercise 3.

$$\frac{dx}{dt} = -2x - y$$

$$\frac{dy}{dt} = -2y$$

Exercise 4.

$$\frac{dx}{dt} = x + 5y$$

$$\frac{dy}{dt} = -x - y$$

Exercise 5.

$$\frac{dx}{dt} = -3x + 2y$$

$$\frac{dy}{dt} = x - 4y$$

Exercise 6.

$$\frac{dx}{dt} = x + y$$

$$\frac{dy}{dt} = -9x + 3y$$

Exercise 7.

$$\frac{dx}{dt} = -2x - y$$

$$\frac{dy}{dt} = 4x - 2y$$

Exercise 8.

$$\frac{dx}{dt} = -3x$$

$$\frac{dy}{dt} = -3y$$

Exercise 9.

$$\frac{dx}{dt} = -2x - 3y$$

$$\frac{dy}{dt} = 3x + 4y$$

Exercise 10.

$$\frac{dx}{dt} = x + y$$

$$\frac{dy}{dt} = x + 2y$$

Exercise 11.

$$\frac{dx}{dt} = -x + 2y$$

$$\frac{dy}{dt} = x + 3y$$

Exercise 12.

$$\frac{dx}{dt} = -2x - y$$

$$\frac{dy}{dt} = x - 4y$$

Exercise 13. Let $X_1(t) = \begin{pmatrix} re^{\lambda t} \\ se^{\lambda t} \end{pmatrix}$ and $X_2(t) = \begin{pmatrix} ve^{\mu t} \\ we^{\mu t} \end{pmatrix}$. Show that X_2 is a constant multiple of X_1 (as a function) only if $\lambda = \mu$.

Exercise 14. Consider the system

$$\frac{dx}{dt} = ax + 10y$$

$$\frac{dy}{dt} = x + 3y$$

(a) Determine a value of a so that $\mathbf{0}$ is a center.

(b) Is the value of a that you found in part (a) unique? Explain your answer.

(c) Is there a value of a for which $\mathbf{0}$ is a spiral? Explain your answer.

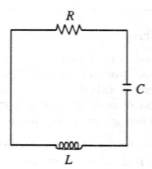

FIGURE 4.9
LRC circuit

Exercise 15. Consider the so-called LRC circuit shown in Figure 4.9. By Kirchhoff's Law, the current I in the circuit and the charge Q on the capacitor are related by the equation

$$L\frac{dI}{dt} + RI + \frac{Q}{C} = 0 \tag{4.16}$$

where $dQ/dt = I$, and where C, L, and R are positive constants representing the capacitance, inductance, and resistance, respectively.

(a) Rewrite (16) as a system involving dI/dt and dQ/dt.

(b) Suppose that R is practically 0. Then the solutions are essentially periodic. Determine the (approximate) period in terms of C and L.

(c) Classify the critical point **0** if it is not a center.

Exercise 16. Let A be the coefficient matrix of a system of differential equations. Note that the characteristic polynomial of a 2×2 matrix can be written as

$$\lambda^2 - T\lambda + D,$$

where T is the trace of A and D is the determinant.

(a) Prove that $D = \mu\lambda$ and that $T = \mu + \lambda$.

(b) Skech a two-dimensional graph, with the trace on the horizontal axis, and the determinant on the vertical axis. Add to this a graph of the line $T^2 - 4D = 0$.

(c) Including the axes and the line you just drew, you have divided the plane into 6 regions. Label each of these regions in terms of the type of node they will include. This should cover the cases in the table at the end of Section 4.1.

(d) You have drawn the Trace-Determinant plane for two-dimensions systems of ODEs.

4.2   Almost Linearity

In Section 4.1 we analyzed critical points for a general class of systems of
linear differential equations. Now we relax the linearity condition and study
a broader collection of systems called "almost linear" systems. As we will see
in Section 4.3, the motion of such objects as a swinging pendulum can be
described by an almost linear system of equations.

Definition 4.5. Let V be a subset of $\mathbb{R}^2$ that contains the origin in its interior.
Assume that F and G are real-valued functions on V that vanish at the origin
and whose partial derivatives are continuous and also vanish at the origin.
Then the system

$$\frac{dx}{dt} = ax + by + F(x, y)$$

$$\frac{dy}{dt} = cx + dy + G(x, y)$$

(4.17)

is **almost linear** at the origin, provided that a, b, c and d are real numbers
and $ad - bc \neq 0$.

The systems we will discuss in the present section have the form of (4.17),
in which F and G are continuous functions of x and y, but are independent
of t. Such systems are **autonomous.**

Suppose that the system in (1) is almost linear at the origin, and let

$$H\left(\begin{array}{c} x \\ y \end{array} \right) = \left(\begin{array}{c} ax + by + F(x, y) \\ cx + dy + G(x, y) \end{array} \right)$$

The fact that the partials of F and G vanish at the origin implies that

$$DH(\mathbf{0}) = \left(\begin{array}{cc} a & b \\ c & d \end{array} \right)$$

This is the **associated matrix** of the system in (4.17). We observe that the
associated matrix is the same as the associated matrix for the corresponding
linear system

$$\frac{dx}{dt} = ax + by$$

$$\frac{dy}{dt} = cx + dy$$

(4.18)

Acknowledging the relationship between the systems in (4.17) and (4.18), we
call the linear system in (4.18) the **auxiliary system** for the system in (4.17).

The notions of asymptotical stability, stability and nonstability that we
defined for linear systems in Section 4.1 carry forward without alteration for

almost linear systems. Thus **0** is **stable** if for every $\epsilon > 0$ there is a $\delta > 0$ such that if X is any solution satisfying $||X(0)|| < \delta$, then $||X(t)|| < \epsilon$ for all $t > 0$. Likewise, **0** is **asymptotically stable** if it is stable and any solution X such that $X(0)$ is suitably close to **0** has the property that $\lim_{t \to \infty} X(t) = \mathbf{0}$. Finally, **0** is **unstable** if it is not stable.

The relevance of these definitions is revealed in the following theorem, due to Lyapunov and proved in differential equations textbooks.

Theorem 4.6. Suppose that the system in (4.17) is almost linear at the origin. If **0** is asymptotically stable (respectively, unstable) for the auxiliary system, then **0** is asymptotically stable (respectively, unstable) for the almost linear system.

It can be shown in addition that if the critical point of the auxiliary system is a node, saddle point or spiral point, then the critical point of the almost linear system is of the same type. However, degenerate nodes and centers are not inherited by almost linear systems. More particularly, if a critical point of an auxiliary system is a center, then the corresponding critical point of the almost linear system may be asymptotically stable or unstable (Exercises 7–7). The reason that this can happen is that the eigenvalues of degenerate nodes and centers are very special numbers. Slight changes in the eigenvalues can alter the nature of the critical point.

Example 4.2.1. Consider the system

$$\frac{dx}{dt} = -2x + y + 2xy$$

$$\frac{dy}{dt} = x + y$$

Show that the system is almost linear at the origin, and that the critical point **0** is a saddle point.

Solution. On the one hand, the second equation of the system is linear. On the other hand, the first equation has the form

$$\frac{dx}{dt} = -2x + y + F(x,y), \quad \text{where} \quad F(x,y) = 2xy$$

Since F has continuous partial derivatives and $F_x(0,0) = 0 = F_y(0,0)$, the system is almost linear at the origin. The associated linear system is

$$\frac{dx}{dt} = -2x + y$$

$$\frac{dy}{dt} = x + y$$

and the associated matrix is

$$\begin{pmatrix} -2 & 1 \\ 1 & 1 \end{pmatrix}$$

Since the eigenvalues are $(-1 \pm \sqrt{13})/2$, it follows that $\mathbf{0}$ is an unstable saddle point for the linear system. We conclude from Theorem 4.6 and the comments following it that $\mathbf{0}$ is a saddle point for the given almost linear system. $\quad\square$

By Theorem 4.3, linear systems of differential equations can have critical points different from $\mathbf{0}$ only if 0 is an eigenvalue of the associated matrix. However, this is not true for other systems. For example, consider the system

$$\frac{dx}{dt} = 2x - y + 2xy$$

$$\frac{dy}{dt} = x + y + 1$$
(4.19)

Since $dx/dt = 0 = dy/dt$ if $x = 1$ and $y = -2$, it follows that $\begin{pmatrix} 1 \\ -2 \end{pmatrix}$ is a critical point of the system. You can check that the associated matrix has complex eigenvalues.

The simplest way to analyze the nature of critical points different from $\mathbf{0}$ is to perform a change of variables so that the translated critical point is the origin. We do this in the following way.

Suppose that $\mathbf{v_0} = \begin{pmatrix} x_0 \\ y_0 \end{pmatrix}$ is a critical point of the system

$$\frac{dx}{dt} = f(x, y)$$

$$\frac{dy}{dt} = g(x, y)$$

We say that the system is **almost linear** at $\mathbf{v_0}$ if by making the change of variables $x = x_0 + u$, $y = y_0 + v$, the resulting system

$$\frac{du}{dt} = f(u, v)$$

$$\frac{dv}{dt} = g(u, v)$$

is almost linear at the origin.

Example 4.2.2. Show that the system in (4.19) is almost linear at $\mathbf{v_0} = \begin{pmatrix} 1 \\ -2 \end{pmatrix}$.

Solution. Since $x_0 = 1$ and $y_0 = -2$, we make the change of variables

$$x = 1 + u \quad \text{and} \quad y = -2 + v$$

Then the system in (3) becomes

$$\frac{du}{dt} = 2(1 + u) - (-2 + v) + 2(1 + u)(-2 + v) = -2u + v + 2uv$$

$$\frac{dv}{dt} = (1 + u) + (-2 + v) + 1 = u + v$$

Since this system is almost linear at $\mathbf{0}$ by Example 4.2.1, the system in (3) is almost linear at $\mathbf{v_0}$. $\quad\square$

4.2.1 Limit Cycles

Until this point, the solutions of linear and almost linear systems we have encountered have been fixed points or ellipses, or have orbits that converge to fixed points or are unbounded as t increases without bound. However, for systems of two differential equations there is yet another possibility. Consider the system

$$\frac{dx}{dt} = x - y - x^3 - xy^2$$

$$\frac{dy}{dt} = x + y - x^2y - y^3 \qquad (4.20)$$

You can check that the system is almost linear at the critical point $\mathbf{0}$, which is unstable. What is the behavior of other solutions as t increases without bound? We will answer this question after converting the system in (4.20) into polar coordinates r and θ. In making such a conversion, we will need to find formulas for dr/dt and $d\theta/dt$ in terms of x and y and their derivatives.

Theorem 4.7. Let x and y be differentiable functions of t, and let

$$x = r \cos \theta \quad \text{and} \quad y = r \sin \theta$$

Then

$$\frac{dr}{dt} = \frac{x}{r} \frac{dx}{dt} + \frac{y}{r} \frac{dy}{dt} \qquad (4.21)$$

and

$$\frac{d\theta}{dt} = \frac{x}{r^2} \frac{dy}{dt} - \frac{y}{r^2} \frac{dx}{dt} \qquad (4.22)$$

Proof. With the help of the Chain Rule, we find that

$$\frac{dx}{dt} = \frac{dx}{dr} \frac{dr}{dt} + \frac{dx}{d\theta} \frac{d\theta}{dt} = (\cos \theta) \frac{dr}{dt} - (r \sin \theta) \frac{d\theta}{dt} = \frac{x}{r} \frac{dr}{dt} - y \frac{d\theta}{dt}$$

$$\frac{dy}{dt} = \frac{dy}{dr} \frac{dr}{dt} + \frac{dy}{d\theta} \frac{d\theta}{dt} = (\sin \theta) \frac{dr}{dt} + (r \cos \theta) \frac{d\theta}{dt} = \frac{y}{r} \frac{dr}{dt} + x \frac{d\theta}{dt} \qquad (4.23)$$

To solve for dr/dt we multiply the first equation in (4.23) by x and the second by y, and add. This yields

$$x\frac{dx}{dt} + y\frac{dy}{dt} = \frac{x^2}{r}\frac{dr}{dt} + \frac{y^2}{r}\frac{dr}{dt} = r\frac{dr}{dt}$$

from which (4.21) arises. To obtain (4.22) we multiply the first equation in (4.23) by y and the second by x and subtract the first from the second. We obtain

$$x\frac{dy}{dt} - y\frac{dx}{dt} = x^2\frac{d\theta}{dt} + y^2\frac{d\theta}{dt}$$

Dividing both sides by r^2 yields (4.22). Thus the proof is complete. $\qquad\square$

Now we are ready to convert the system in (4.20) into polar coordinates.

Theorem 4.8. In polar coordinates the system in (4.20) is

$$\frac{dr}{dt} = r(1 - r^2)$$

$$\frac{d\theta}{dt} = 1 \tag{4.24}$$

Proof. To find dr/dt, we use (4.21), substituting from the equations in (4.20) for dx/dt and dy/dt, and recalling that $x^2 + y^2 = r^2$. We obtain

$$\frac{dr}{dt} = \frac{x}{r}(x - y - x^3 - xy^2) + \frac{y}{r}(x + y - x^2y - y^3)$$

$$= \frac{1}{r}[(x^2 + y^2) - (x^4 + 2x^2y^2 + y^4)] = \frac{1}{r}[r^2 - (x^2 + y^2)^2]$$

$$= \frac{1}{r}(r^2 - r^4) = r(1 - r^2)$$

Similarly, to find $d\theta/dt$ we use (4.22) and substitute from (4.20):

$$\frac{d\theta}{dt} = \frac{x}{r^2}(x + y - x^2y - y^3) - \frac{y}{r^2}(x - y - x^3 - xy^2)$$

$$= \frac{1}{r^2}(x^2 + y^2) = 1$$

$\qquad\square$

What can we say about the behavior of solutions of the system in (4.24)? The answer to this question relies on two results about orbits of almost linear systems, which you can find in differential equations textbooks.

i. At most one orbit passes through any given point in the plane.

ii. The orbit of a solution that is not a critical point either approaches a critical point or a closed orbit as t increases without bound, or it recedes arbitrarily far from the origin as t increases.

The first result effectively is the Uniqueness Theorem of differential equations. The second result will give us significant information about the system in (4.24).

Now we are prepared to survey the solutions of (4.24) whose orbits are neither the origin nor the unit circle.

Case 1. $\|X(0)\| = 1$

By hypothesis, $r = 1$, so that from (4.24), $dr/dt = r(1 - r^2) = 0$. Then by the Uniqueness Theorem, $r = 1$ for all $t > 0$, so that the orbit of X lies on the unit circle centered at the origin. Since $d\theta/dt = 1$, the orbit travels around the unit circle with constant velocity, and travels in a counterclockwise fashion.

Case 2. $0 < \|X(0)\| < 1$

Since $0 < \|X(0)\| < 1$, the distance between $X(0)$ and the origin, which is r, satisfies $0 < r < 1$ for $t = 0$. Therefore $dr/dt = r(1 - r^2) > 0$ for all $t > 0$. Consequently $\|X\|$ is an increasing function of t. By (ii), we deduce that $\lim_{t \to \infty} \|X(t)\| = 1$.

Case 3. $\|X(0)\| > 1$

In this case, $dr/dt = r(1 - r^2) < 0$, so that $\|X\|$ is decreasing, and consequently by (11) we find that $\lim_{t \to \infty} \|X(t)\| = 1$.

Cases 2–3 tell us that the unit circle is a limit cycle for (4.24), in the sense that if $0 < \|X(t)\|$, then $X(t)$ approaches the unit circle as t increases without bound. More generally, a **limit cycle** in $\mathbb{R}^2$ is a closed curve C that is periodic and attracts the orbit of any solution X such that $X(0)$ is near C. In Figure 4.10 the dashed ellipse-like curve portrays a limit cycle.

An **attractor** of a system of differential equations is a closed bounded set toward which the orbit of each solution approaches as t increases without bound. The results of Section 4.1 show that the origin is the attractor for any linear system in which $\mathbf{0}$ is an asymptotically stable critical point. The other linear systems described in Section 4.1 have no attractor in the plane, since their orbits are either unbounded or concentric ellipses.

If an almost linear system has an attractor, it can be an asymptotically stable critical point, a limit cycle, or the union of critical points and limit cycles. The discussion of the almost linear system in (4.20) and its polar representation in (4.24) indicate that it has an attractor consisting of the unit circle.

Is it possible for an almost linear system to have a another type of attractor? The celebrated Poincaré-Bendixson Theorem, which appears in many differential equations books, provides the answer, for any autonomous system whose solutions lie in the xy plane.

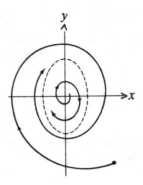

FIGURE 4.10
A limit cycle for an almost linear system of ODEs

Theorem 4.9. (Poincaré-Bendixson Theorem). Let $V \subseteq \mathbb{R}^2$, with V the union of a nonvoid open set and its boundary. Suppose that f and g have continuous derivatives throughout V, and let t_0 be a real number. Consider the system

$$\frac{dx}{dt} = f(x, y)$$

$$\frac{dy}{dt} = g(x, y)$$

and assume that the system has no critical points in V. Then any bounded trajectory that remains in V for all $t \geq t_0$ is a limit cycle or approaches a limit cycle.

The Poincaré-Bendixson Theorem implies that if an autonomous system in x and y has an attractor A, then A consists of a union of critical points and limit cycles. We will see in Section 4.4 that this result is no longer true if the solutions lie in three dimensions.

4.2.2   Section 4.2 Exercises

In Exercises 1–2, determine whether the system is almost linear at $\mathbf{0}$. If it is, then determine its type (node, saddle point, or spiral) and stability, where possible.

Exercise 1.

$$\frac{dx}{dt} = -x + \ln(1 + y^2)$$

$$\frac{dy}{dt} = 2x - 3y - x^2 y$$

Exercise 2.

$$\frac{dx}{dt} = -x + 2y + 1 - e^{xy}$$

$$\frac{dy}{dt} = x - y + x\sin y$$

In Exercises 3–6, determine whether the system is almost linear at each of its critical points. If it is, then determine its type and stability, where possible.

Exercise 3.

$$\frac{dx}{dt} = y$$

$$\frac{dy}{dt} = x - x^3$$

Exercise 4.

$$\frac{dx}{dt} = y$$

$$\frac{dy}{dt} = -x + x^3$$

Exercise 5.

$$\frac{dx}{dt} = 2x - x^2 - xy$$

$$\frac{dy}{dt} = -y + xy$$

Exercise 6.

$$\frac{dx}{dt} = y - x^3$$

$$\frac{dy}{dt} = 1 - xy$$

Exercise 7. Consider the system

$$\frac{dx}{dt} = y - x(x^2 + y^2)$$

$$\frac{dy}{dt} = -x - y(x^2 + y^2)$$

(a) Show that the system is almost linear at the critical point **0**.

(b) Show that **0** is a center point of the auxiliary system.

(c) Show that **0** is asymptotically stable. (*Hint*: Use polar coordinates and show that if $r(0) < 1$, then $\lim_{t \to \infty} r(t) = 0$.)

Exercise 8. Consider the system

$$\frac{dx}{dt} = y + x(x^2 + y^2)$$

$$\frac{dy}{dt} = -x + y(x^2 + y^2)$$

(a) Show that the system is almost linear at the critical point **0**.

(b) Show that **0** is a center point of the auxiliary system.

(c) Show that **0** is unstable.

(d) Show that the given almost linear system has no attractor.

Exercise 9. A nonlinear electric oscillator can be modeled by the van der Pol equation

$$\frac{d^2x}{dt^2} + \epsilon(x^2 - 1)\frac{dx}{dt} + x = 0$$

where x is related to the voltage in the circuit, and ϵ is a positive parameter related to the given circuit.

(a) Prove that the associated system of differential equations is almost linear at **0**. *Hint*: To render the equation as a system, let one equation be $dx/dt = y$; the other equation becomes

$$\frac{dy}{dt} + \epsilon(x^2 - 1)y + x = 0$$

(b) Show that **0** is unstable, whatever the value of ϵ is, and determine the values of ϵ for which **0** is a spiral point.

For the second-order differential equations appearing in Exercises 10–12, determine whether the associated system of differential equations is almost linear at the critical points. For any that are, classify the critical points where possible. Assume that all constants are positive.

Exercise 10. $\dfrac{d^2x}{dt^2} + a\dfrac{dx}{dt} - \dfrac{1}{2}x(1 - x^2) = 0$, which arises in the study of double-well potentials.

Exercise 11. $\dfrac{d^2x}{dt^2} + a\dfrac{dx}{dt} - b\sin x = 0$, which arises in the study of electric force fields.

Exercise 12. $\dfrac{d^2x}{dt^2} + a\dfrac{dx}{dt} - x^3 = 0$, which arises in nonlinear electric circuits.

Exercise 13. Consider the system

$$\frac{dx}{dt} = y$$

$$\frac{dy}{dt} = -x + \mu y(1 - 3x^2 - 2y^2)$$

Assume $\mu \in \mathbb{R}$, with $|\mu| < 2$.

(a) Show that the origin is the unique critical point.

(b) Determine how the qualitative nature of the critical point changes as μ changs from positive to negative.

(c) Use the Poincare-Bendixson theorem to show that, when $\mu > 0$, there is a periodic orbit in the set

$$V = \{(x,y) : 0 < x^2 + y^2 \leq \frac{1}{2}\}.$$

Keep the following in mind:

- Is the set in question the union of a nonempty open set and its boundary?
- Are there any critical points in V?
- Where are the trajectories bounded?
- Where must there be a limit cycle?

4.3 The Pendulum

A grandfather clock runs by the motion of a pendulum. In this section we will use the theory of almost linear systems to study the dynamics of a pendulum, not only when there is no external force applied on it, but also when there is a sinusoidal external force.

Suppose that a pendulum of length L has mass m, and its bob swings back and forth in a vertical plane. Let the angle of the pendulum with the vertical at time t be denoted by $\theta(t)$ (Figure 4.11). We will assume that g represents acceleration due to gravity, and for the present we will assume that there is no damping and no external force on the pendulum. This would be the case if the pendulum is located in a vacuum, under no influence save acceleration due to gravity. Under these conditions, the motion of the pendulum is governed by the second-order differential equation

$$mL^2 \frac{d^2\theta}{dt^2} + mgL \sin \theta = 0 \tag{4.25}$$

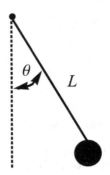

FIGURE 4.11
A pendulum

By letting $x = \theta$ and $y = d\theta/dt$, we transform (4.25) into the system

$$\frac{dx}{dt} = y$$

$$\frac{dy}{dt} = -\frac{g}{L}\sin x$$

(4.26)

We will call the system in (4.26) the **pendulum system**. Notice that the pendulum system is not linear because $\sin x$ does not have the form cx. Finding exact solutions for the pendulum system in (4.26) is no easier than finding exact solutions for the differential equation in (4.25). Indeed, one cannot solve either exactly!

Although the pendulum system is not linear, it is almost linear, as the following theorem shows.

Theorem 4.10. The pendulum system in (4.26) is almost linear at the origin.

Proof. We observe that the first equation in (4.26) is linear. Next, we rewrite the second equation as

$$\frac{dy}{dt} = -\frac{g}{L}x + G(x,y)$$

where

$$G(x,y) = \frac{g}{L}x - \frac{g}{L}\sin x$$

It is apparent that G has continuous partial derivatives, and that $G(0,0) = 0$. Since

$$G_x(x,y) = \frac{g}{L} - \frac{g}{L}\cos x \quad \text{and} \quad G_y(x,y) = 0$$

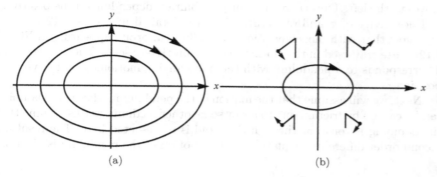

(a)   (b)

FIGURE 4.12

Portrait of the auxiliary system and directions of motion for an undamped pendulum

it follows that $G_x(0,0) = 0 = G_y(0,0)$. Consequently the pendulum system is almost linear at the origin.   □

Because the pendulum system is almost linear at the origin, Theorem 4.6 tells us that we can use the auxiliary system

$$\frac{dx}{dt} = y$$

$$\frac{dy}{dt} = -\frac{g}{L}x$$

(4.27)

to analyze the nature of the critical point **0** of (4.26). First, we observe that the associated matrix of (3) is

$$\begin{pmatrix} 0 & 1 \\ -g/L & 0 \end{pmatrix}$$

The eigenvalues of this matrix are $\lambda = \pm\sqrt{g/L}\, i$. It follows that the eigenvalues are pure imaginary, and thus **0** is a center for the auxiliary system in (4.27). We conclude that **0** is a stable critical point for the auxiliary system.

From the results in Case 4 of Section 4.1, we know that the orbits of solutions for the auxiliary system are ellipses, as shown in Figure 4.12(a). This figure, in which the vertical axis measures dx/dt, is called a **phase plane portrait**. (The expression "phase plane" comes from physics.) It is a reasonable representation of the motion of the undamped pendulum when the angular displacement and the angular velocity are small.

The angular displacement x from the downward vertical (denoted by θ in (4.25)) is assumed to be positive when the pendulum lies to the right of the vertical and negative when to the left. It follows that the angular velocity dx/dt is positive when the pendulum bob moves to the right and is negative when it moves to the left. Thus there are four possibilities, depending on the positivity and negativity of x and of dx/dt. They are detailed in Figure 4.12(b). One can show that with our conventions, the elliptical orbits appearing in Figure 4.12(a) are traversed in the clockwise direction (Exercise 1). The critical point **0** corresponds to a pendulum with the bob hanging motionless in the vertical position.

Next, we will assume that the motion of the pendulum is damped. Damping can be caused by friction due to air or some other medium. Let us assume that the damping is constant through time and is represented by c. The resulting second-order differential equation for the motion of the pendulum is given by

$$mL^2 \frac{d^2\theta}{dt^2} + cL \frac{d\theta}{dt} + mgL \sin\theta = 0$$

Again letting $x = \theta$ and $y = d\theta/dt$, we transform the differential equation into the **pendulum system**

$$
\begin{aligned}
\frac{dx}{dt} &= y \\[2mm]
\frac{dy}{dt} &= -\frac{g}{L} \sin x - \frac{c}{mL} y
\end{aligned}
\qquad (4.28)
$$

The pendulum system is not linear, but is almost linear at the origin (Exercise 2). The auxiliary system is

$$
\begin{aligned}
\frac{dx}{dt} &= y \\[2mm]
\frac{dy}{dt} &= -\frac{g}{L} x - \frac{c}{mL} y
\end{aligned}
\qquad (4.29)
$$

with associated matrix

$$\begin{pmatrix} 0 & 1 \\ -g/L & -c/(mL) \end{pmatrix}$$

The eigenvalues are

$$\frac{-c/mL \pm \sqrt{c^2/(mL)^2 - 4g/L}}{2} = \frac{-c \pm \sqrt{c^2 - 4gm^2 L}}{2mL}$$

Let the eigenvalues be denoted by λ and μ, where

$$\lambda = \frac{-c + \sqrt{c^2 - 4gm^2 L}}{2mL} \quad \text{and} \quad \mu = \frac{-c - \sqrt{c^2 - 4gm^2 L}}{2mL}$$

The nature of the critical point **0** depends on the values of λ and μ, which in turn depend on c, m, and L. For the analysis of the critical point when the

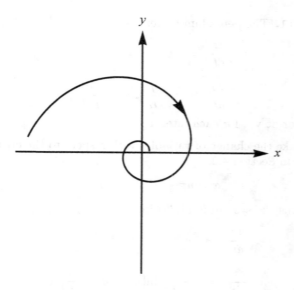

FIGURE 4.13
Solution curve that tends to the origin

damping parameter is allowed to vary, we will assume that m and L are fixed but positive numbers.

We already discussed the case $c = 0$, in which there is no damping, so we will assume that $c > 0$. Since the real parts of both λ and μ are negative numbers (Exercise 3), all solutions of the auxiliary system in (4.29) are asymptotically stable. By Theorem 4.6 the same is true for solutions of the pendulum system. In other words, all solutions tend to $\mathbf{0}$ with time (Figure 4.13). This is realistic, because if there is resistance due to air, then the pendulum winds down as time progresses. Anyone who regularly winds up a grandfather or cuckoo clock is aware of this phenomenon.

We remark that the tendency of the orbit of a solution to spiral while it approaches the zero solution diminishes as c increases from 0 toward $\sqrt{4gm^2L}$. As c increases further, the trajectory of a solution does not spiral noticeably as it converges to the zero solution.

The pendulum system in (4.28) has critical points different from the zero solution. In fact, let n be any integer, and let $\mathbf{v}_n = \begin{pmatrix} n\pi \\ 0 \end{pmatrix}$. Then using $x = n\pi$ and $y = 0$ in (4), we find that

$$\frac{dx}{dt} = 0 \quad \text{and} \quad \frac{dy}{dt} = -\frac{g}{L} \sin(n\pi) - \frac{c}{mL}(0) = 0$$

It follows that $\mathbf{v}_n$ is a critical point. In Theorem 4.11 we confirm the almost linearity of the system at $\mathbf{v}_n$.

Theorem 4.11. The pendulum system

$$\frac{dx}{dt} = y$$

$$\frac{dy}{dt} = -\frac{g}{L}\sin x - \frac{c}{mL}y \tag{4.30}$$

is almost linear at $\mathbf{v}_n$, for each integer n.

Proof. We make the change of variables with respect to the critical point $\mathbf{v}_n$, as prescribed in Section 4.2:

$$x = n\pi + u \quad \text{and} \quad y = v$$

This transforms the system in (4.30) into

$$\frac{du}{dt} = v$$

$$\frac{dv}{dt} = -\frac{g}{L}\sin(n\pi + u) - \frac{c}{mL}v \tag{4.31}$$

The first equation in (4.31) is linear. The second equation becomes

$$\frac{dv}{dt} = -\frac{g}{L}\sin u - \frac{c}{mL}v = -\frac{g}{L}u - \frac{c}{mL}v + \left(\frac{g}{L}u - \frac{g}{L}\sin u\right) \quad \text{if } n \text{ is even}$$

and

$$\frac{dv}{dt} = \frac{g}{L}\sin u - \frac{c}{mL}v = \frac{g}{L}u - \frac{c}{mL}v + \left(-\frac{g}{L}u + \frac{g}{L}\sin u\right) \quad \text{if } n \text{ is odd}$$

Therefore if n is even, then the system in (4.31) is almost linear at the origin because it is (4.28) with x and y replaced by u and v. If n is odd, then again the system in (4.31) is almost linear, by an analogous proof (Exercise 4). Consequently, the pendulum system is almost linear at $\mathbf{v}_n$ for each integer n. □

From Theorem 4.11 and the analysis of the critical point $\mathbf{0}$, we know that the critical point $\mathbf{v}_n$ for the pendulum system in (4.30) is asymptotically stable whenever n is an even integer. This is reasonable from a physical standpoint, since if n is even and if $x(0) = n\pi$ and $y(0) = 0$, then the pendulum is in the vertical position, pointed downward with the bob at rest.

The nature of the critical point $\mathbf{v}_n$ is far different if n is an odd integer. To be specific, let $n = 1$, so that the critical point is $\mathbf{v}_1 = \begin{pmatrix} \pi \\ 0 \end{pmatrix}$. Using the solution of Theorem 4.11, we find that the auxiliary system is

$$\frac{du}{dt} = v$$

$$\frac{dv}{dt} = \frac{g}{L}u - \frac{c}{mL}v \tag{4.32}$$

The associated matrix is

$$\begin{pmatrix} 0 & 1 \\ g/L & -c/(mL) \end{pmatrix}$$

Consequently the eigenvalues, which we denote by λ and μ, are given by

$$\lambda = \frac{-c + \sqrt{c^2 + 4gm^2 L}}{2mL} \quad \text{and} \quad \mu = \frac{-c - \sqrt{c^2 + 4gm^2 L}}{2mL}$$

Because the expression inside the square root is positive, λ and μ are real numbers. A moment's reflection shows that $\mu < 0 < \lambda$. Consequently the critical point **0** of (4.32) is an unstable saddle point, so that v_1 is also an unstable saddle point. The critical point v_1, in which the angle $x = \pi$, corresponds to the pendulum standing upright, with the bob above the anchor. Obviously this is an unstable position.

The results so far are predicated on the absence of an external applied force. Now suppose that there is an applied force F, such as a periodic push or impulse, thrust on the pendulum. Assume further that F is a function of t. Then the second-order equation becomes

$$mL^2 \frac{d^2\theta}{dt^2} + cL \frac{d\theta}{dt} + mgL \sin\theta = F(t) \tag{4.33}$$

The auxiliary system is

$$\frac{dx}{dt} = y$$

$$\frac{dy}{dt} = -\frac{g}{L} \sin x - \frac{c}{mL} y + \frac{1}{mL^2} F(t)$$

Assume that c, m, and L are held constant. If F is small, then the critical points may well have the same form as when there is no external force. However, if the force F is substantial, then solutions may exhibit very complicated behavior.

For example, consider the pendulum equation

$$\frac{d^2\theta}{dt^2} + \alpha \frac{d\theta}{dt} + \sin\theta = \mu \cos t \tag{4.34}$$

which can be obtained from (4.33) by taking suitable values for c, m, and L, and by letting F denote a sinusoidal force. If $\alpha = 1/5$ and $\mu = 2$, then (4.34) becomes

$$\frac{d^2\theta}{dt^2} + \frac{1}{5} \frac{d\theta}{dt} + \sin\theta = 2 \cos t$$

It turns out that the corresponding system has two critical points. The damping is diminished but the external force is augmented if we let $\alpha = 1/10$ and $\mu = 7/4$ in (4.34). In this case there are four critical points for the corresponding system.

Note: For more details about basins of attraction for the pendulum system, see the article by Grebogi, Ott, and Yorke (1986). This article lays the foundation for what would become the OGY method, used by physicists in the control theory of chaotic dynamical systems.

4.3.1   Section 4.3 Exercises

Exercise 1. Show that under our conventions, orbits of an undamped pendulum are traversed clockwise.

Exercise 2. Show that the system in (4.28) is almost linear at the origin.

Exercise 3. Show that the eigenvalues of the auxiliary system associated with (4.28) have negative real parts.

Exercise 4. Show that the system in (4.31) is almost linear when n is an odd integer.

In Exercises 5–8 we will study further the motion of the undamped pendulum, given by

$$mL^2 \frac{d^2\theta}{dt^2} + mgL \sin\theta = 0 \qquad (4.35)$$

Also let the angular velocity $d\theta/dt = \omega$ when $t = 0$.

Exercise 5.

(a) Multiply both sides of (11) by $2\dfrac{d\theta}{dt}$ and integrate to obtain

$$\left(\frac{d\theta}{dt}\right)^2 = c + \frac{2g}{L} \cos\theta$$

(b) Suppose that the angle $\theta = 0$. Show that

$$\frac{d\theta}{dt} = \omega \sqrt{1 - \frac{4g}{\omega^2 L} \sin^2 \frac{\theta}{2}}$$

Exercise 6. Assume that $\omega > 2\sqrt{g/L}$. Show that the pendulum rotates completely around its point of suspension, and has minimum angular velocity when $\theta = \pi$, that is, when the pendulum is in the upward vertical position.

Exercise 7. Assume that $\omega < 2\sqrt{g/L}$. Find the largest angle, θ_0, that the pendulum attains.

Exercise 8. Assume that $\omega = 2\sqrt{g/L}$. Show that the pendulum rises toward the upward vertical position, which it approaches as t increases without bound.

(a) Use the program PENDULUM to numerically solve the ODE in Exercises 5–8. Verify that this technique yields the same results as the last 4 exercises.

(b) This program also allows you to damp the pendulum. Explore this parameter, and see how the graph changes. Explain these changes qualitatively.

4.4 The Lorenz System

When a pot of water is heated on the stove, those particles near the bottom warm faster than the ones near the top. If the temperature difference is minimal, then the fluid near the bottom becomes lighter and rises in an orderly fashion (Figure 4.14(a)). However, if the temperature difference is larger, then the rising warmer water from the bottom and the falling cooler water from the top initiate what is called **convection**, which is a circulating flow (Figure 4.14(b)). This is the same phenomenon that occurs in our atmosphere. The air closer to the ground is heated by the sun's rays, and when the temperature differential is great enough, those air particles rise by convection. In this way warmer, polluted air escapes from the lower atmosphere to the upper atmosphere, where it is dispersed.

The almost linear system that we will study in this section models convection. For a specific fluid enclosed in a fixed region with constant height, the

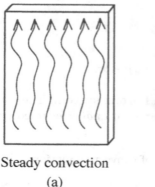
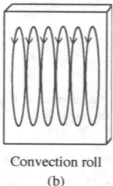

Steady convection   Convection roll

(a)   (b)

FIGURE 4.14
Illustration of types of convection

rate of convection depends on several constants relative to the region and the fluid:

> coefficient of thermal expansion, denoted by α
> kinematic viscosity, denoted by ν
> thermal conductivity, denoted by κ

These constants are automatically positive. In addition to the three listed above there is another constant: acceleration due to gravity, denoted by g. Finally, the rate of convection depends on the temperature difference between top and bottom of the region. We let

$$\Delta T = \text{(temperature at the bottom)} - \text{(temperature at the top)}$$

In order to have convection, we must have $\Delta T > 0$, which means that the bottom of the region must be warmer than the top.

Early in the 20th century Lord Rayleigh (1916) systematically studied convective currents in a region with constant depth H. He discovered that convective motion develops if a quantity R_a, now called the **Rayleigh number** and given by

$$R_a = g\alpha H^3 (\Delta T)\nu^{-1}\kappa^{-1}$$

exceeds a critical number $R_c = \pi^4 a^{-2}(1 + a^2)^3$. Here a is a number that is related to the region under consideration. The minimum possible value of R_c is $27\pi^4/4$, and occurs for $a = 1/\sqrt{2}$ (Exercise 1).

Over forty years later Barry Saltzman (1962) derived a system of differential equations that modeled convection. From that system a year later Edward Lorenz created the following famous stripped-down version:

$$\frac{dx}{dt} = \sigma y - \sigma x$$

$$\frac{dy}{dt} = rx - y - xz \qquad (4.36)$$

$$\frac{dz}{dt} = xy - bz$$

We will refer to this system as the **Lorenz system**. The variable t refers to time. However, the variables x, y, and z do *not* refer to coordinates in space. In fact,

x is proportional to the intensity of convective motion

y is proportional to the temperature difference between ascending and descending currents

z is proportional to the distortion (away from linearity) of the vertical temperature profile

The constants in (4.36) are given by

$\sigma = \nu/\kappa$, called the **Prandtl number**

$r = R_a/R_c$

$b = \dfrac{4}{1+a^2}$, where b is a constant related to the given space

For a given region and liquid, b and σ are positive constants. We will assume that $\sigma > 1$. The constant r depends, among other constants, on the temperature difference ΔT. We will assume that $r > 0$, which by the definition of r means that the bottom of the region is warmer than the top. By the comments accompanying the definitions of R_a and R_c, convection will occur if $R_a/R_c > 1$, that is, if $r > 1$. With this in mind, we observe that **0** is a critical point of the Lorenz system. The proof that the system is almost linear at **0** is straightforward, as we will learn presently.

Note: Lorenz (1963) made a detailed study of the system in (4.36), with the help of a computer and the research assistants mentioned earlier. As we study the Lorenz system, we will use the fact that the concepts pertaining to systems of two differential equations apply to systems of three differential equations, with only trivial modifications. For each term, review the similar definition for lower dimensional systems, and be sure you can see why they are not in conflict.

Theorem 4.12. The Lorenz system is almost linear at **0**.

Proof. We write the system in (1) as follows:

$$\frac{dx}{dt} = -\sigma x + \sigma y + F(x, y, z)$$

$$\frac{dy}{dt} = rx - y + G(x, y, z)$$

$$\frac{dz}{dt} = -bz + H(x, y, z)$$

where $F(x, y, z) = 0$, $G(x, y, z) = -xz$, and $H(x, y, z) = xy$. It follows that $F(0,0,0) = G(0,0,0) = H(0,0,0) = 0$, and all first partials of F, G, and H vanish at the origin. Thus the Lorenz system is almost linear at **0**. $\qquad\square$

The auxiliary system for the Lorenz system is

$$\frac{dx}{dt} = -\sigma x + \sigma y$$

$$\frac{dy}{dt} = rx - y \tag{4.37}$$

$$\frac{dz}{dt} = -bz$$

and the associated matrix A_0 is given by

$$A_0 = \begin{pmatrix} -\sigma & \sigma & 0 \\ r & -1 & 0 \\ 0 & 0 & -b \end{pmatrix}$$

To evaluate the eigenvalues of A_0, we need to know how to find the determinant of a 3×3 matrix. If B is a 3×3 matrix B given by

$$B = \begin{pmatrix} a & b & c \\ d & e & f \\ g & h & j \end{pmatrix}$$

then by definition,

$$\det B = aej + bfg + cdh - ceg - bdj - afh$$

Applying this formula to $A_0 - \lambda I$ (where I is the 3×3 identity matrix), we find that

$$\det(A_0 - \lambda I) = \det \begin{pmatrix} -\sigma - \lambda & \sigma & 0 \\ r & -1 - \lambda & 0 \\ 0 & 0 & -b - \lambda \end{pmatrix}$$

$$= (-\sigma - \lambda)(-1 - \lambda)(-b - \lambda) + r\sigma(b + \lambda)$$

$$= -(b + \lambda)[\lambda^2 + (\sigma + 1)\lambda + \sigma(1 - r)]$$

The eigenvalues of A_0 are the solutions of the equation $\det(A_0 - \lambda I) = 0$, of which there are (in theory) three:

$$\lambda_1 = \frac{-(\sigma + 1) + \sqrt{(\sigma - 1)^2 + 4r\sigma}}{2} \tag{4.38}$$

$$\lambda_2 = \frac{-(\sigma + 1) - \sqrt{(\sigma - 1)^2 + 4r\sigma}}{2} \tag{4.39}$$

$$\lambda_3 = -b \tag{4.40}$$

The eigenvalues of a 3×3 matrix indicate stability or instability of the critical point $\mathbf{0}$ in the same way they do for 2×2 matrices. In particular, if all the eigenvalues have negative real parts, then $\mathbf{0}$ is asymptotically stable. By contrast, if an eigenvalue has a positive real part, then $\mathbf{0}$ is unstable. We apply this information to the critical point $\mathbf{0}$ of the matrix associated with Lorenz system.

Case 1. $0 < r < 1$

Since $\sigma > 1$ and $0 < r < 1$, it follows that $(\sigma - 1)^2 + 4r\sigma > 0$, so that λ_1 and λ_2 are real numbers. The hypothesis that $r - 1 < 0$ implies that

$$\sqrt{(\sigma - 1)^2 + 4r\sigma} = \sqrt{(\sigma + 1)^2 + 4(r - 1)\sigma} < \sqrt{(\sigma + 1)^2} = \sigma + 1$$

so that all three eigenvalues are negative numbers. Thus if $0 < r < 1$, then $\mathbf{0}$ is an asymptotically stable critical point. Physically this means that as time passes, convection dies down and the system approaches a steady state of no convection. This conforms to the result of Rayleigh which states that convection begins as r rises above 1.

Case 2. $r = 1$

When $r = 1$, the eigenvalues simplify to $\lambda_1 = 0$, $\lambda_2 = -(\sigma + 1)$, and $\lambda_3 = -b$. Therefore two eigenvalues are negative, and the other is 0, which means that $\mathbf{0}$ is a **neutrally stable solution**. (This case is somewhat analogous to the case in which $\mu = 3$ for the quadratic family $\{Q_\mu\}$; indeed, if $\mu = 3$, then μ is a bifurcation point for Q_μ, at which the attracting fixed point becomes unstable.)

Case 3. $r > 1$

If $r > 1$, then by calculations similar to those in Case 1, we find that

$$\sqrt{(\sigma - 1)^2 + 4r\sigma} > \sigma + 1$$

It follows that $\lambda_1 > 0$, $\lambda_2 < 0$, and $\lambda_3 < 0$. Therefore $\mathbf{0}$ is an unstable solution. In the terminology of hydrodynamics, convection occurs.

When r increases and passes through 1, the asymptotic stability of $\mathbf{0}$ gives way to instability. Thus 1 is a point of bifurcation for the family of Lorenz systems (with parameter r). Might the incipient instability of $\mathbf{0}$ foreshadow the emergence of other, new critical points when $r > 1$? Theorem 4.13 gives the answer.

Theorem 4.13. If $r > 1$, then in addition to $\mathbf{0}$ there exist two critical points $\mathbf{p}$ and $\mathbf{q}$ for the system in (4.36), given by

$$\mathbf{p} = \begin{pmatrix} \sqrt{b(r-1)} \\ \sqrt{b(r-1)} \\ r-1 \end{pmatrix} \quad \text{and} \quad \mathbf{q} = \begin{pmatrix} -\sqrt{b(r-1)} \\ -\sqrt{b(r-1)} \\ r-1 \end{pmatrix}$$

Proof. Suppose that $\mathbf{v} = \begin{pmatrix} x \\ y \\ z \end{pmatrix}$ is a critical point. If $x = 0$, then by the first equation of (4.36) and by the hypothesis that $\sigma \neq 0$, we know that $y = 0$.

Since $b \neq 0$, it follows from the third equation in (4.36) that $z = 0$. Therefore if $x = 0$, then $\mathbf{v} = \mathbf{0}$. Now suppose that $\mathbf{v}$ is a critical point and $\mathbf{v} \neq \mathbf{0}$. By our preceding comments, $x \neq 0$. Therefore from the first equation in (4.36),

$$0 = \frac{dx}{dt} = \sigma y - \sigma x = \sigma(y - x)$$

so that $y = x$. Consequently we deduce from the second equation in (4.36) that

$$0 = \frac{dy}{dt} = rx - y - xz = rx - x - xz = x(r - 1 - z)$$

Since $x \neq 0$ by assumption, it follows that $z = r - 1$. Therefore the third equation in (4.36) yields

$$0 = \frac{dz}{dt} = xy - bz = x^2 - bz = x^2 - b(r - 1)$$

Consequently $x^2 = b(r - 1)$. As a result,

$$x = y = \pm\sqrt{b(r-1)} \quad \text{and} \quad z = r - 1$$

so that, by definition, $\mathbf{p}$ and $\mathbf{q}$ are critical points of the Lorenz system. □

One can show that the Lorenz system is almost linear at $\mathbf{p}$, with associated matrix

$$A_{\mathbf{p}} = \begin{pmatrix} -\sigma & \sigma & 0 \\ 1 & -1 & -\sqrt{b(r-1)} \\ \sqrt{b(r-1)} & \sqrt{b(r-1)} & -b \end{pmatrix} \tag{4.41}$$

and with eigenvalues that are the solutions of the characteristic equation

$$\lambda^3 + (\sigma + b + 1)\lambda^2 + b(\sigma + r)\lambda + 2b\sigma(r - 1) = 0 \tag{4.42}$$

(Exercise 7).

Notice that the Lorenz system is symmetric with respect to the z axis. This means that if

$$X(t) = \begin{pmatrix} x(t) \\ y(t) \\ z(t) \end{pmatrix} \quad \text{and} \quad Y(t) = \begin{pmatrix} -x(t) \\ -y(t) \\ z(t) \end{pmatrix}$$

and if X is a solution, then Y is a solution. Consequently the associated matrices $A_{\mathbf{p}}$ and $A_{\mathbf{q}}$ have identical characteristic equations, and hence the same eigenvalues for any given constants σ, b, and r (Exercise 5). Therefore both $\mathbf{p}$ and $\mathbf{q}$ are asymptotically stable (or unstable), or neither is. As a result, we will only analyze the dynamics of $\mathbf{p}$.

Henceforth we will assume that $\sigma > b + 1$, so that $\sigma - b - 1 > 0$, and let

$$r^* = \frac{\sigma(\sigma + b + 3)}{\sigma - b - 1}$$

Notice that $r^* > 1$ since $\sigma > 1$ and $\sigma > b + 1$. It is known that if $1 < r < r^*$, then all three roots of (4.42) have negative real parts, so that $\mathbf{p}$ is asymptotically stable. By contrast, if $r > r^*$, then two roots of (4.42) have positive real parts, so that $\mathbf{p}$ is unstable.

It follows from the preceding comment that if $r > r^*$, then the three critical points $\mathbf{0}$, $\mathbf{p}$, and $\mathbf{q}$ are unstable critical points. A natural question to ask is what the orbits of solutions look like when $r > r^*$. Following Lorenz, we will let $\sigma = 10$ and $b = 8/3$ (which means that $\sigma > b+1$), so that $r^* = 470/19 \approx 24.74$. The family of Lorenz systems in which $\sigma = 10$, $b = 8/3$, and $r > 1$ will be called the **Lorenz family**. Since $\mathbf{p}$ and $\mathbf{q}$ become unstable when r passes through r^*, it follows that the Lorenz family has a bifurcation at r^*.

To be able to visualize the behavior of solutions of the Lorenz system, let us follow the lead of Lorenz and fix $r = 28$, so that $r > r^*$. Since all three critical points are unstable for this value of r, it is difficult to conjecture what the orbit of an arbitrary solution might look like.

Assume that the initial point of a solution is near the unstable critical point $\mathbf{0}$. The orbit is illustrated in Figure 4.15. To describe the dynamics

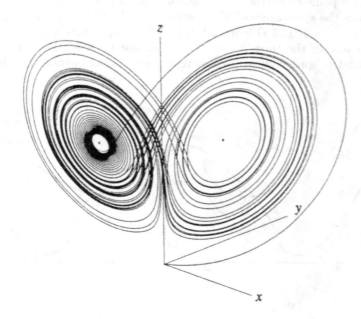

FIGURE 4.15
Lorenz attractor

of the orbit, we observe that at first the three coordinates x, y, and z grow rapidly. Similar signs of x and y indicate that warmer fluid is rising and colder fluid is sinking. When convection becomes strong enough, there is a reversal, in which warmer fluid rises above the colder fluid at the top; the signs of y and x change. After the orbit comes close enough to $\mathbf{p}$ (the dot with positive x and y coordinates, around which the orbit circles), it spirals outward until it is deflected toward $\mathbf{q}$ (the other dot), and once again spirals outward until recaptured near $\mathbf{p}$. This process continues indefinitely.

The surface on which the orbit resides is (essentially) the attractor A_L of the Lorenz system, called the **Lorenz attractor**. It may appear that A_L lies in a single plane, but that is not quite true. In fact, each time the orbit spirals around the critical point $\mathbf{p}$ (or $\mathbf{q}$), it travels on a different leaf, never retracing its steps. The fact that tiny differences in initial conditions can lead to far different values after a period of time means that the attractor is chaotic. This sensitive dependence led Lorenz to give the title "Predictability: Does the Flap of a Butterfly's Wings in Brazil Set Off a Tornado in Texas?" for an address to the American Association for the Advancement of Sciences in 1979. The title, along with the shape of the Lorenz attractor, has led to the nickname **butterfly effect** for the Lorenz system and its attractor.

The Lorenz system has been widely studied during the past half century, and various bifurcations of the system have been located for values of r much larger than 28. We will say that an orbit is of type ab^2 if it spirals around one critical point once (thus the single a) and around another critical point twice (thus the appearance of b^2). Franceschini (1980) identified a stable ab^2-orbit when $r \approx 100.75$ (Figure 4.16(a)), which represents a stable period-3 orbit because of the three loops. If r is lowered to 99.65, there is a stable a^2ba^2b-orbit (Figure 4.16(b)), which represents a stable period-6 orbit. Thus

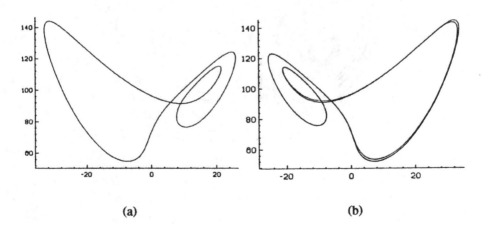

(a)   (b)

FIGURE 4.16
Stable periodic orbits in the Lorentz attractor

period-doubling has occurred. In fact, as r decreases further, there is a period-doubling cascade that terminates when r is approximately 99.24. Franceschini found that the ratios of successive bifurcation points approach the Feigenbaum constant, which is approximately 4.67.

4.4.1 Section 4.4 Exercises

Exercise 1. Let $f(a) = \pi^4 a^{-2}(1+a^2)^3$. Show that $f(1/\sqrt{2}) = 27\pi^4/4$ is the minimum value of f.

Exercise 2. Suppose that X is a solution of the Lorenz system and that $X(0)$ lies on the z axis. Show that the entire orbit lies on the z axis and that $\lim_{t \to \infty} X(t) = 0$.

Exercise 3. Show that when viewed from above, the orbits revolving around the z axis do so in a clockwise direction.

Exercise 4. Let $r > 1$. Show that the eigenvalues of A_0 satisfy $-\lambda_2 > \lambda_1 > -\lambda_3$ if and only if $r > 1 + b(\sigma + 1 + b)/\sigma$.

Exercise 5. Let σ, b, and r be fixed. Show that $A_\mathbf{p}$ and $A_\mathbf{q}$ have the same characteristic equation, and hence the same eigenvalues.

Exercise 6. Let $r > 1$. Show that the Lorenz system is almost linear at $\mathbf{p}$, with associated matrix given by (6).

Exercise 7. Show that the eigenvalues of $A_\mathbf{p}$ satisfy (7).

Exercise 8. Show that if (7) has complex roots, then the real root must be negative. (*Hint*: If $\alpha + i\beta$ is one complex root, then $\alpha - i\beta$ is a second complex root.)

Exercise 9.

(a) Show that the characteristic equation of $A_\mathbf{p}$ has two real roots only if

$$r < \frac{\sigma^2 - b\sigma + b^2 + 2\sigma + 2b + 1}{3b}$$

(b) Let $\sigma = 10$ and $b = 8/3$. Find a number r_0 such that if $r > r_0$, then two of the eigenvalues of $A_\mathbf{p}$ are complex numbers.

Exercise 10. Let $a > 0$ and $c > 0$. Show that the equation $\lambda^3 + a\lambda^2 + c\lambda + ac = 0$ has two pure imaginary roots.

Exercise 11. Find a value of r_0 such that $A_{\mathbf{p}}$ has a pure imaginary eigenvalue for $r = r_0$.

Exercise 12. Use the computer to determine a period-doubling window in the interval $145 < r < 166$. Can you say anything interesting about the periodic orbits?

Exercise 13. Find an open research problem related to the Lorentz attractor. Summarize the research into this problem, and evaluate why it has eluded a solution.

The system that Lorenz introduced to the world about 50 years ago has promoted a vast amount of research. In 1982 Colin Sparrow published the book *The Lorenz Equations: Bifurcations, Chaos, and Strange Attractors*, which contains virtually everything announced in this section, plus a great deal more. However, there are still open questions concerning the Lorenz system, as there are in chaotic dynamics.

One might say that the appearance of the Lorenz system in 1963 was the catalyst for modern research in chaotic dynamics, and for the emergence of a vast number of applications of chaotic dynamics in divergent fields. It will be interesting to see how the area of chaotic dynamics grows and is applied in the next half century.

5

Introduction to Fractals

In the study of chaotic dynamics one often encounters sets on the real line, in the plane, or in space that have very complicated and interesting structures. In Section 2.4 we discussed two such sets in the interval $[0, 1]$: the Cantor (ternary) set, and the set of numbers in the interval $[0, 1]$ whose iterates under Q_μ remain in $[0, 1]$ for a given $\mu > 4$. In Chapters 5–7 we will discuss several different kinds of complicated sets, many of which can be realized only with the help of a computer.

Chapter 5 is devoted to very complicated curves and shapes, generally in the plane, and strategies for understanding them and drawing them with computer assistance. Section 5.1 focuses on the notion of self-similarity, and relates it to the Cantor set and its two-dimensional compatriot, the Devil's staircase. Section 5.2 is devoted to planar sets like the Sierpiński gasket, the Koch curve and Menger sponge that were so complicated that a hundred years ago mathematicians called them "geometric monsters." Nowadays they are referred to as "fractals." In Section 5.3 we turn to other sets in the plane like the Peano curve and the Hilbert curve that appear to fill up space and are, in fact, called "space-filling" curves. Sections 5.4–5.6 introduce new kinds of dimensions, such as similarity dimension, capacity dimension, and Lyapunov dimension, that do not have to be integers. It is from these definitions that we will derive an operative definition of "fractal."

Note: If you mention to someone outside of the mathematics department that you are studying "fractals," then the person will likely have some questions! We recommend the PBS NOVA documentary "Hunting the Hidden Dimension," available free online. This can help bridge the abstract nature of fractals with their relationship to the geometry of nature.

5.1 Self-Similarity

One type of complicated figure or set that we want to study is composed of a collection of miniature look-alikes, each of which is itself composed of a collection of miniature look-alikes, etc.

DOI: 10.1201/9781032678757-5

FIGURE 5.1
Illustration of similarity function

To make the phrase "miniature look-alikes" more precise, consider a subset S of $\mathbb{R}^n$, where normally we will assume that $n = 1$ or $n = 2$. As in Section 3.2, we let the **distance** between two points $\mathbf{v}$ and $\mathbf{w}$ be denoted by $||\mathbf{v} - \mathbf{w}||$. Now we are ready for the definition of similarity.

Definition 5.1. Let S be a subset of $\mathbb{R}^n$. If a function $F : S \to S$ has the property that for some $s > 0$, we have

$$||F(\mathbf{v}) - F(\mathbf{w})|| = s||\mathbf{v} - \mathbf{w}|| \text{ for all } \mathbf{v} \text{ and } \mathbf{w} \text{ in } S$$

then F is a **similarity** of S, and s is the **similarity constant**.

The similarity constant may satisfy $0 < s < 1$ or $s \geq 1$. If $s > 1$, then F expands distances, and if $s = 1$, then F is distance-preserving. We will not be interested in s such that $s \geq 1$ because then distances between distinct points cannot converge through iteration.

By definition, if F is a similarity with $0 < s < 1$, then F shrinks all distances by the same factor, namely s. This means that the image under F of an ellipse or a triangle has the same shape, but is contracted by a factor of s (Figure 5.1). In other words, the image of a shape under F is an identical miniature of the shape.

Notice that in the first map, F merely translates the left-hand shape, whereas in the second map, F translates and rotates. In either case, the shape in the image is similar to the original shape.

Now we are ready for the definition of a self-similar set.

Definition 5.2. Let S be a subset of $\mathbb{R}^n$. If there are similarities $F_1, F_2, \ldots, F_m$ on S such that

$$S = F_1(S) \cup F_2(S) \cup \cdots \cup F_m(S)$$

and the images $F_1(S), F_2(S), \ldots, F_m(S)$ are nonoverlapping (except possibly for simple boundaries), then S is a **self-similar set**.

We mention that the similarity constants $s_1, s_2, \ldots, s_m$ of the m functions $F_1, F_2, \ldots, F_m$ need not be identical. Also, if a set S is self-similar, then S is composed of m miniatures (for some integer $m \geq 2$).

Let us show that the Cantor (ternary) set C is self-similar, with $m = 2$ and $s = 1/3$ (Figure 5.2). Indeed, the part C_1 of C that is located in the

-- --   -- --   -- --   -- --

FIGURE 5.2
Cantor set

FIGURE 5.3
Sierpiński gasket

FIGURE 5.4
Other self-similar sets

interval $[0, 1/3]$ is an exact miniature of C, with a reduction in size of $1/3$, that is, $s = 1/3$. Similarly, the part C_2 of C located in the interval $[2/3, 1]$ is a miniature of C with size reduction of $1/3$. Thus C is self-similar, with $m = 2$ and $s = 1/3$. You can also show that C is self-similar with $m = 4$ and $s = 1/9$, or $m = 8$ and $s = 1/27$, etc.

Another famous self-similar set appears in Figure 5.3. It is a **Sierpiński gasket**, after the Polish mathematician Waclaw Sierpiński (1882–1969). The Sierpiński gasket, which we will denote by S, is composed of three identical miniature triangles (left bottom, right bottom, and center top), each of which has side length equal to $1/2$ of the original triangle S. Thus S is self-similar, with $m = 3$ and $s = 1/2$.

There are many other self-similar sets, including ones with different values of s for the different similarity functions. Figure 5.4 shows two, each of which has two values of s. In the first, $m = 3$, and in the second, $m = 4$. Can you guess what the values of s are for the two sets?

There are more self-similar sets depicted in the exercises; also, we will study several famous self-similar sets in the upcoming sections. By contrast, in Figure 5.5 we exhibit two sets that are impossible to draw by hand, but

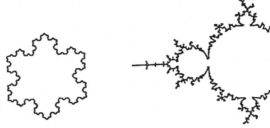

(a) The Koch snowflake   (b) Mandelbrot set boundary

FIGURE 5.5
Two non-self-similar sets

which are not self-similar. The left-hand set is the Koch snowflake; it has wonderful symmetry, but it is not self-similar. The reason it is not self-similar is that no proper subset of the Koch snowflake is a closed curve, so that there is no proper subset that looks exactly like the Koch snowflake itself. Although it is not quite so obvious, the same remark applies to the right-hand set, which depicts the boundary of the Mandelbrot set. We will return to the Koch snowflake in Section 5.2, and the Mandelbrot set in Chapter 7.

5.1.1   The Cantor Set Revisited

Recall from Section 2.4 that the Cantor (ternary) set C has several special properties:

   i. C is a closed subset of the interval $[0, 1]$.

  ii. C is totally disconnected (i.e., C contains no nonempty open intervals).

 iii. C is perfect (i.e., each point of C is the limit of a sequence of distinct points in C).

 iv. C has an uncountable number of elements.

From our discussion above, we have a fifth property:

  v. C is self-similar.

Next, we turn to the representation of any number in $[0, 1]$ in terms of a special series reminiscent of a geometric series. Label the three subintervals $[0, 1/3]$, $[1/3, 2/3]$, and $[2/3, 1]$ of $[0, 1]$ by I_0, I_1, and I_2. Then split each of the subintervals I_i into three equal subintervals, and label them I_{i0}, I_{i1}, and I_{i2}. Next, split each subinterval I_{ij} into 3 equal subintervals, and label them I_{ij0}, I_{ij1}, and I_{ij2}. Continue the process indefinitely. See Figure 5.6.

Now let x be any number in $[0, 1]$. Assign to x the numbers $x_1, x_2, x_3, \ldots$ as follows. If x is in I_i, let $x_1 = i$. If x is in I_{ij}, then let $x_2 = j$, etc. For

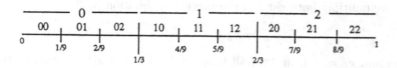

FIGURE 5.6

Triadic expansion setup

example, if $x = 2/5 = .4$, then x is in $I_1 = [1/3, 2/3]$, so $x_1 = 1$. Also x is in $I_{i0} = [1/3, 4/9]$, so $x_2 = 0$. And the process continues.

We comment that if x lies on the border of two subintervals, then there is a choice for the assignment, and that causes no problem. For example, if $x = 2/3$, then x_1 can be assigned either 1 or 2. No problem.

By the process we have described, it follows that for any x in $[0, 1]$, there is at least one representation of x:

$$x = \sum_{k=1}^{\infty} \frac{x_k}{3^k} = \frac{x_1}{3} + \frac{x_2}{9} + \frac{x_3}{27} + \cdots$$

Alternatively, we can associate x with the expression $x \approx .x_1 x_2 x_3 \ldots$, where the expansion is not the usual decimal expansion, but rather is a so-called **triadic expansion** for x.

How does this relate to the Cantor set C? Notice that if x is in C, then x is not in the middle open third of $[0, 1]$, so we can choose x_1 to *not* be 1. Similarly, we can choose $x_2, x_3, \ldots$ to *not* be 1. The result is that each number in the Cantor set C has a triadic expansion in which all of the numerators $x_1, x_2, \ldots$ are 0 or 2. In fact, the Cantor set consists of all numbers in $[0, 1]$ with a triadic expansion consisting only of 0's and 2's.

At this point we describe a method of determining an appropriate triadic expansion $.x_1 x_2 \ldots$ for any number in $[0, 1]$ (not just those numbers in C). Take, for example, the number $4/7$. We first write

$$\frac{4}{7} = \frac{x_1}{3} + \frac{x_2}{9} + \frac{x_3}{27} + \frac{x_4}{81} + \cdots$$

Then we multiply both sides by 3 to obtain

$$\frac{12}{7} = x_1 + \frac{x_2}{3} + \frac{x_3}{9} + \frac{x_4}{27} + \cdots$$

Since $x_k = 0$, 1, or 2 for all k, it follows that $x_1 = 1$. This means that

$$\frac{12}{7} = 1 + \frac{x_2}{3} + \frac{x_3}{9} + \frac{x_4}{27} + \cdots$$

so that

$$\frac{5}{7} = \frac{x_2}{3} + \frac{x_3}{9} + \frac{x_4}{27} + \cdots$$

Next, we multiply both sides of the preceding equation by 3:

$$\frac{15}{7} = x_2 + \frac{x_3}{3} + \frac{x_4}{9} + \cdots$$

Once again $x_k = 0$, 1, or 2 for all k, so we deduce that $x_2 = 2$. Subtracting 2 from both sides, and multiplying by 3, we obtain

$$\frac{3}{7} = x_3 + \frac{x_4}{3} + \cdots$$

so that $x_3 = 0$. Thus we know x_1, x_2, and x_3. Continuing the process, we could derive the numbers $x_4, x_5, \ldots$, in order to find the triadic expansion for 4/7:

$$\frac{4}{7} = .120102\overline{120102} \cdots$$

We comment that the triadic expansion for 4/7 is repeating, with period no longer than 6 digits. More generally, for any (reduced) rational number p/q in $[0, 1]$, the triadic expansion has no more than q digits. (See Exercise 8.)

5.1.2   The Length of the Cantor Set

By using another infinite process, we can define the length of C. Recall that C is obtained by taking the closed unit interval $[0, 1]$, deleting the middle open third, then deleting the 2 middle open thirds of what is left, then deleting the 4 middle open thirds of what remains, etc. Thus the length L of C is given by the following equation:

$$L = 1 - (\frac{1}{3} + \frac{2}{3^2} + \frac{2^2}{3^3} + \cdots)$$

Therefore

$$L = 1 - \sum_{k=1}^{\infty} \frac{2^{k-1}}{3^k} = 1 - \frac{1}{3} \sum_{k=0}^{\infty} \frac{2^k}{3^k} = 1 - \frac{1}{3} \frac{1}{1 - (2/3)} = 0$$

Thus the length of C is 0. As a result, C is a set with the most interesting property of having an uncountable number of points (as we showed in Section 2.4) but having zero length!

5.1.3   The Devil's Staircase

Finally, we turn to the "devil's staircase," which is the graph of a continuous function f on $[0, 1]$ created by means of the Cantor set. The function f is defined by letting its values be constant on each subinterval deleted in the construction of C. Specifically, we let

$$f(x) = \frac{1}{2} \text{ for } x \text{ in } [\frac{1}{3}, \frac{2}{3}]$$

$$f(x) = \frac{1}{4} \text{ for } x \text{ in } [\frac{1}{9}, \frac{2}{9}] \qquad f(x) = \frac{3}{4} \text{ for } x \text{ in } [\frac{7}{9}, \frac{8}{9}]$$

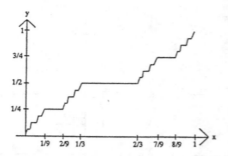

FIGURE 5.7
Devil's staircase

and continue the process with $f(x) = (2k - 1)/2^n$ for x in the kth deleted subinterval in the nth stage (at which time there are 2^n subintervals of length $1/3^n$ retained for C). Finally, $f(x)$ is defined for all other values of x in a way that makes f continuous and increasing on $[0, 1]$. (See Figure 5.7.)

The question we pose now is: What is the length L of the graph of f? Since $f(0) = 0$ and $f(1) = 1$, and the straight-line distance between $(0,0)$ and $(1,1)$ is $\sqrt{2}$, it would appear that

$$\sqrt{2} \leq L \leq 2$$

We will show not only that this is true, but also that, in fact, $L = 2$. First, we have a lemma.

Lemma 5.1. Suppose that g is an increasing, continuous function defined on $[a, b]$, and assume that $g(a) = c$ and $g(b) = d$. If the graph of g makes an angle of θ with the horizontal axis, then you can readily check that the length of the graph of g, denoted L_0, satisfies

$$L_0 = \sqrt{(b - a)^2 + (d - c)^2} = (d - c)\sqrt{\cot^2 \theta + 1} = (d - c)\csc \theta$$

Theorem 5.3. $L = 2$.

Proof. Using Lemma 5.1, we find that $L \geq L_1$, where L_1 is the length of the first graph in Figure 5.8 and is given by

$$L_1 = 2\left(\frac{1}{2} \csc \theta_1\right) + \frac{1}{3}$$

where the fraction $1/3$ above represents the length of the horizontal part. Similarly, $L \geq L_2$, where L_2 is the length of the second graph in Figure 5.8 and is given by

$$L_2 = 2^2\left(\frac{1}{2^2} \csc \theta_2\right) + \frac{1}{3} + \frac{2}{3^2}$$

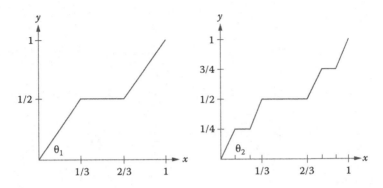

FIGURE 5.8
Stages 1 and 2 in finding the length of the Devil's staircase

Note again that the final two fractions represent the lengths of the horizontal parts.

For any positive integer $n \geq 1$, L_n is the length of the graph after n stages and is given by

$$L_n = 2^n \left(\frac{1}{2^n} \csc \theta_n \right) + \frac{1}{3} + \frac{2}{3^2} + \cdots + \frac{2^{n-1}}{3^n}$$

As n increases, θ_n approaches $\pi/2$. Therefore

$$L = \lim_{n \to \infty} L_n = \lim_{n \to \infty} (\csc \theta_n) + \frac{1}{3} \sum_{k=0}^{\infty} \left(\frac{2}{3} \right)^k = 1 + \left(\frac{1}{3} \right)(3) = 2$$

□

5.1.4   Section 5.1 Exercises

Exercise 1. Explain why shapes in Figure 5.5(a) and (b) are not self-similar.

Exercise 2. Each of the shapes in Figure 5.9 is self-similar. For each, determine the appropriate values of m and s.

Exercise 3. Find the rational number corresponding to the triadic expansion.

(a) $.2\overline{2} \cdots$

(b) $.012\overline{012} \cdots$

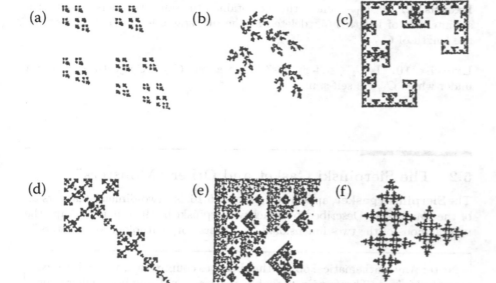

FIGURE 5.9
Illustrations for Exercise 2 of Section 5.1

Exercise 4. Find a triadic expansion for

(a) 8/27

(b) 4/5

(c) 7/12

Exercise 5. What is the form of the ternary expansion for those numbers in of the form $k/3^n$, where n is a positive integer and $k \leq 3^n$?

Exercise 6. Suppose that C^* is obtained by deleting *closed* middle third subintervals, instead of open middle thirds, and continuing the process indefinitely. Would C^* be empty, have a countable number of elements, or an uncountable number of elements? Support your answer.

Exercise 7. How would you define a two-dimensional Cantor set on the square $[0,1] \times [0,1]$? What would it look like graphically?

Exercise 8. Prove that if x is any (reduced) rational number of the form p/q, then the triadic expansion of x is repeating, with period no more than q digits.

Exercise 9. Let $C_{1/5}$ denote the set obtained the same way as is C, but with subintervals of length $1/5^n$ deleted at the nth stage, for each n. Determine the length of $C_{1/5}$.

Exercise 10. Is $C_{1/5}$ self-similar? What about $C_{1/4}$? Conjecture a criteria under which $C_{1/k}$ is self-similar.

5.2 The Sierpiński Gasket and Other "Monsters"

The **Sierpiński gasket**, appearing in Figure 5.10, is a two-dimensional version of the Cantor set. Described by Waclaw Sierpiński in 1915, it is perhaps the most famous of the two-dimensional "monsters" or "pathological monsters."

Note: Any mathematical object that behaves counter to intuition is called **pathological**. The first is probably the irrational numbers, where the Pythagoreans were stumped by the notion that a line could be drawn without rational length. Later examples include the Weierstrass function, the uncountability of the reals, the Peano space-filling curve, or, of course, the Cantor set. We'll investigate a few of these in the rest of this chapter. A common theme you'll see in the monsters we study is self-similarity. Not all such monsters exhibit this property, and it is an interesting subject to pursue. The books *Counter-Examples in Analysis* and *Counter-Examples in Topology* are recommended to the interested reader.

We use the process of creating the Cantor set by deleting middle thirds of subintervals as a model for creating the two-dimensional Sierpiński gasket. Indeed, for the Sierpiński gasket, we begin with a filled-in triangle S (which we often draw as an equilateral triangle!). Notice that S is composed of 4 congruent sub-triangles (left bottom, right bottom, center top, and middle), and

FIGURE 5.10
Sierpiński gasket

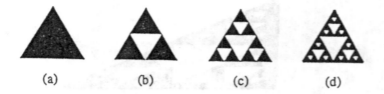

(a)   (b)   (c)   (d)

FIGURE 5.11

Constructing the Sierpiński gasket

delete the middle sub-triangle (Figure 5.11(a)) to obtain the shape in Figure 5.11(b). We continue the process (Figure 5.11(c)), and ultimately obtain the Sierpiński gasket (Figure 5.11(d)).

Recall that although the Cantor set is scattered through the interval $[0, 1]$ on the real line, it has 0 length. In an analogous way, we can show that the Sierpiński gasket has 0 area. This we do now.

Suppose that the initial triangle in Figure 5.11(a) has area 1. We will use the fact that at every stage in the construction of the Sierpiński gasket, each shaded triangle is composed of four congruent sub-triangles. Thus each of the sub-triangles has area equal to one-fourth of the area of its "parent" triangle.

Using this information, we find that at stage 1 (Figure 5.11(b)), the area of the shaded region is $1 - 1/4 = 3/4$. At stage 2 (Figure 5.11(c)), the area is smaller, due to the elimination of three smaller sub-triangles, each of which has area $1/4^2$. Therefore the area at stage 2 is

$$1 - \frac{1}{4} - 3(\frac{1}{4^2}) = \frac{9}{16}$$

Continuing the process indefinitely, we deduce that the area A of the Sierpiński gasket is given by

$$A = 1 - \frac{1}{4} - 3\frac{1}{4^2} - 3^2(\frac{1}{4^3}) - \cdots = \frac{3}{4} - \sum_{k=1}^{\infty} \frac{3^k}{4^{k+1}}$$

$$= \frac{3}{4} - \frac{1}{4}\sum_{k=1}^{\infty}(\frac{3}{4})^k = \frac{3}{4} - \frac{1}{4}\frac{3/4}{1 - (3/4)} = \frac{3}{4} - \frac{1}{4}(\frac{3/4}{1/4}) = 0$$

Consequently the area of the Sierpiński gasket is 0. This result may be surprising because it appears from Figure 5.11(d) that the gasket should have positive area. In fact, computer renditions are imperfect and cannot produce images that distinguish between, for example, the Sierpiński gasket and later stages in the construction. In Exercise 1 you are asked to determine the areas of the figures appearing at the 10th and 16th stages, and then you can tell why it is so hard to distinguish those stages from the final product.

We mention that the shape that is obtained from any given filled-in triangle by deleting successive sub-triangles in the manner described above for the Sierpiński gasket is normally called a Sierpiński gasket. Thus the Sierpiński

FIGURE 5.12
Another shape for the Sierpiński gasket

gasket illustrated in Figure 5.12 is a variant of the original Sierpiński gasket, and it also has 0 area.

5.2.1 The Chaos Game

The Sierpiński gasket appears in an infinite process called the **chaos game**, introduced by the modern mathematician Michael Barnsley. The process proceeds as follows:

Let B, C, and D be three noncollinear points arbitrarily placed in the plane, and let x_0 be any point whatsoever in the plane, as in Figure 5.13(a). Next, assign each side of a fair (six-sided) die one of the three letters B, C, or D, so that each letter appears twice on the die. Then roll the die. Let the face that shows up be denoted by A_1 (so that A_1 is B, C, or D, depending on the roll). Define x_1 to be the midpoint of the line segment joining x_0 and the point corresponding to A_1. Proceed inductively: if the nth roll of the die produces the face denoted A_n, then x_n is the midpoint of the line segment joining x_{n-1} and the point corresponding to A_n. Figure 5.13(b) shows the

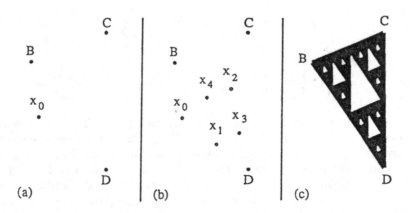

FIGURE 5.13
The chaos game

first four steps for particular x_0, B, C, and D. (Here the four first rolls of the die yielded D, C, D, and B, respectively.) If we eliminate the first dozen or so points in the process and then recognize the next 10,000 points in the process, then amazingly a Sierpiński gasket with outer vertices B, C and D arises, as we can see in Figure 5.13(c). In Section 6.5 we will be able to determine why the Sierpiński gasket arises from the chaos game.

5.2.2   The Sierpiński Carpet

A square version of the Sierpiński gasket is the **Sierpiński carpet**, introduced by Sierpiński in 1916. It is obtained by taking a square, eliminating the middle sub-square of nine congruent sub-squares, and continuing the process indefinitely. What results is the Sierpiński carpet (Figure 5.14).

One can show that the Sierpiński carpet has 0 area by using the same method as we used in showing that the Sierpiński gasket has 0 area (Exercise 4). Given the results with both the Sierpiński gasket and the Sierpiński carpet, you might wonder if the area would still be 0 if one took out smaller sub-squares. (See Exercise 5.)

The Sierpiński gasket and the Sierpiński carpet are somewhat alike, except that the carpet is triangular-shaped and the carpet is rectangular-shaped. The question we address now is whether there is any substantive difference between the gasket and the carpet. It turns out that there is at least one substantial difference, and we can determine this difference by studying what are called branching numbers for points in the sets. Let p be a point in a set S. Then the **branching number** $B(p)$ of p is the *minimum number* of points of S other than p that an arbitrary small circle of radius ϵ centered at p can include. It follows that $B(p)$ equals 0, 1, 2, ..., or ∞, depending on the point p and the set S.

Note that $B(p)$ represents the number of points *on* the circle, and p is interior to the circle. In particular, the branching number of a point is 0 if the point is isolated. Next, let S be the line segment in Figure 5.15(a). Every

FIGURE 5.14
Sierpiński carpet

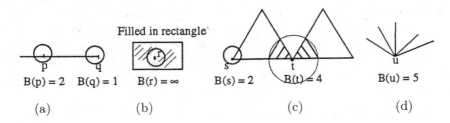

FIGURE 5.15
Different values for the branching number

small circle with interior point p cuts S twice, so $B(p) = 2$. By contrast, every small circle with interior point q cuts S just once, so $B(q) = 1$. In a similar vein, if S is a filled-in rectangle and r is a point interior to S, as in Figure 5.15(b), then $B(r) = \infty$.

Next, suppose that S is as in Figure 5.15(c). On the one hand, s is the vertex of an isolated angle, so $B(s) = 2$. On the other hand, if there is a series of nested triangles shrinking down toward the point t, then $B(t) = 4$. Finally, a 5-legged spider at u in Figure 5.15(d) has $B(u) = 5$. More generally, an n-legged spider emanating at u would have $B(u) = n$.

Since every point p in the Sierpiński gasket is either interior to a line or is a vertex, it follows that $2 \leq B(p)$. It is possible to prove that $2 \leq B(p) \leq 4$ for every point p in the Sierpiński gasket. In particular, each of the three outer vertices of the gasket has branching number equal to 2 (like s in Figure 5.15(c)), and the inner vertices have branching number 4 (like t in Figure 5.15(c)). What is less obvious but valid nevertheless is the fact that all other points have branching number 3. Thus in particular there is no point p in the Sierpiński gasket for which $p = 5$.

By contrast, the branching numbers for the Sierpiński carpet are very different. Although it is immediate that $2 \leq B(p)$ for every point p in the carpet, what is amazing is that if p is a corner, then $B(p) = \infty$! (See Exercise 6.) This means that, in particular, there are infinitely many nonintersecting paths lying in the carpet, all of which lead to a given corner point.

5.2.3   The Menger Sponge

There is a three-dimensional version of the Sierpiński carpet – the **Menger sponge**. It is obtained by starting with a cube, which is evidently composed of 27 congruent sub-cubes, each dimension of which is 1/3 of the dimension of the original cube. In step 1 we delete the sub-cube at the middle of each of the six faces and also the sub-cube in the middle of the cube; this means that seven sub-cubes are deleted. Then for each of the 20 sub-cubes that remain, the same procedure is applied. This process continues forever, yielding the

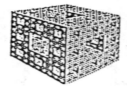

FIGURE 5.16
Menger sponge

Menger sponge (Figure 5.16). This shape, which looks rather like a cube of Swiss cheese, was introduced by the Austrian mathematician Karl Menger in 1926. The Menger sponge is evidently self-similar. (Question: What are the smallest m and s for the Menger sponge?) Each face of the sponge is a Sierpiński carpet. Finally, as you might suspect by now, the volume of the sponge is 0.

> **Note:** Modern computers have made visualizing objects like the Menger sponge much easier. You can even find an interactive version of this object in the game Minecraft!

5.2.4 The Koch Curve

The self-similar sets that we have discussed so far (the Cantor set, Sierpiński gasket, Sierpiński carpet, and the Menger sponge) have been constructed by successively deleting ever smaller subregions from the original shape. The last of the "geometric monsters" we study are curves in the plane created by "accretion," that is, by endowing the original figure with additional ornaments.

To create the **Koch curve** (or **von Koch curve**), which will be denoted by K, we begin with the unit line segment $[0, 1]$ in Figure 5.17(a). Step 1 involves erecting an equilateral triangle over the middle third of the interval, and then deleting the base. The resulting figure, which appears in Figure 5.17(b), consists of 4 line segments of length $1/3$. Step 2 involves replacing each of the 4 line segments from Step 1 with 16 line segments of length $1/9$ (Figure 5.17(c)).

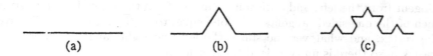

(a)   (b)   (c)

FIGURE 5.17
Constructing the Koch curve

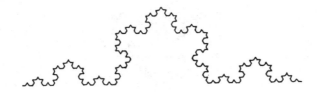

FIGURE 5.18
Koch curve

Continuing the process indefinitely, we obtain the Koch curve K, of which the curve in Figure 5.18 is a reasonable approximation after a large number of steps. This curve was described in 1904 by the Swedish mathematician Helge von Koch (1870–1924).

Once we notice that the Koch curve K is endowed with four miniatures (two horizontal and two diagonal), it is apparent that K is self-similar, with $m = 4$ and $s = 1/3$. What is not so obvious is that the Koch curve has infinite length. Indeed, the initial line segment in Figure 5.17(a) has length $L_0 = 1$. After the first step the length $L_1 = 4(1/3) = 4/3$, and after the second step the length $L_2 = (4/3)(4/3) = (4/3)^2$. After the nth step the length $L_n = (4/3)^n$. Taking the limit yields the length L_K of the Koch curve as ∞.

To the eye, it appears that there is nothing smooth about the Koch curve. In fact, the curve has no tangent at any of its points, that is, the curve is **nowhere differentiable**. It was the search for a "simple" curve that is nowhere differentiable that led von Koch to the curve K and the proof that K is nowhere differentiable.

Theorem 5.4. The Koch curve K is nowhere differentiable.

Partial proof. Von Koch claims that there are three kinds of points p in K:

i. p is a vertex, at some stage.

ii. p lies on one of the line segments but is not a vertex.

iii. The remaining points.

Now if p is a vertex at the nth stage of constructing K, like C, Q, or R in Figure 5.19, then at that stage the vertex has a tangent from the right that is distinct from the tangent from the left. For example, C has a horizontal tangent from the left, and a diagonal tangent from the right. For any vertex, each of the one-sided tangents has slope equal to $-\sqrt{3}$ or $\sqrt{3}$ or 0. Whatever the vertex, and whatever the stage, the left and right tangents differ by at least $\sqrt{3}$, so there is no two-sided tangent for p.

If p is like B in Figure 5.19, on a line segment appearing at some stage in the construction of K (e.g., B is on AC) but not a vertex, then we use the fact that the line segments AC and CQ are equal and angle $ACQ = 120°$,

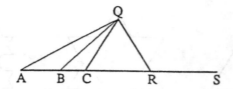

FIGURE 5.19
No tangent at B, C, Q, or R

so angle $QAC = 30°$. It follows that $30° <$ angle $QBC < 60°$. This always happens as diagonal line segments like CQ approach B from the right. Since the left-hand tangent has slope 0 and the right-hand tangent has slope at least $30°$, it follows that there is no two-sided tangent at such a point.

Finally, von Koch proved that for any other points of K, the slopes of left-hand and right-hand tangents would necessarily differ by at least $60°$. See p. 43 of *Classics on Fractals*, edited by Gerald Edgar (1993), for details. □

5.2.5 The Koch Snowflake

The **Koch snowflake** is created by employing three carbon copies of the Koch curve. The snowflake is displayed in Figure 5.20. The arrows point toward the extremities of the three copies of the Koch curve.

Since the Koch curve has infinite length, it follows directly that the boundary of the Koch snowflake also has infinite length. The Koch snowflake is a classical example of a planar shape whose boundary has infinite length while its interior has finite area.

One way to find the area A of the snowflake is to refer to the ideas used in creating the Koch curve. Indeed, the snowflake can be derived by starting with

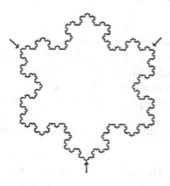

FIGURE 5.20
Koch snowflake

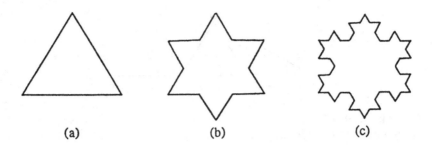

(a)   (b)   (c)

FIGURE 5.21
An alternative way to create the Koch snowflake

an equilateral triangle (Figure 5.21(a)), and then on each side performing the steps employed in creating the Koch curve. Figures 5.21(b) and 5.21(c) show the first two steps in the process. Alternatively, one can find the area of the region between the Koch curve and the line $[0, 1]$ on the x-axis, then multiply by 3 and add the area of the inscribed triangle. In Exercise 9 you are asked to determine the value of the area A of the snowflake.

5.2.6   Section 5.2 Exercises

Exercise 1. Determine the area of the nth stage of the Sierpiński gasket construction when

(a) $n = 10$

(b) $n = 16$

Exercise 2. For the Sierpinski carpet, compute the perimeter of the removed squares at each stage of the process. Express this as a series and show it does not converge.

Exercise 3. Choose a square at any stage of the construction of the Sierpinski carpet, and a random point on the boundary of the removed square. Show that any arbitrarily small neighborhood of the point you selected contains points of the carpet from the border of a different removed square. Conjecture an apparent paradox about the perimeter of the carpet compared to its area.

Exercise 4. Find the volume V of the Menger sponge whose edge length is 1.

Exercise 5. Suppose in the chaos game that instead of 3 given noncollinear points in the plane we started with 4 given points no three of which are collinear. If you apply the chaos game procedure (with a four-sided die!), what figure would result?

Exercise 6. For the Sierpiński carpet, find the self-similarity constants m and s, and then determine the area A.

Exercise 7. A square of side length 1 is composed of n^2 congruent sub-squares, for an arbitrary positive integer $n \geq 2$. At the first stage, delete the center sub-square. Proceed in the same way with the remaining sub-squares, and continue the process indefinitely. Find the area A of the resulting figure S.

Exercise 8. Show that $B(p) = \infty$ for each of the four outer vertices of the Sierpiński carpet. Would that be true for any other points in the carpet? Explain your answer.

Exercise 9. Suppose that S consists of the rational numbers in the interval $(0, 1)$, and let p be in S. Determine $B(p)$.

Exercise 10. Consider the Sierpiński carpet.

(a) Describe the form of pairs of ternary expansions for points in the carpet.

(b) Find the form of the pairs in (a) that correspond to vertices in the carpet.

(c) Show that not all points in the carpet are vertices.

Exercise 11. Find the area A of the region enclosed by the Koch snowflake that is derived from an initial triangle of area 1.

Exercise 12. Consider the region R bounded above by the Koch curve and below by the x-axis. Find the area A of R.

Exercise 13. Consider the following 5 sets of binomial expressions:

$$1$$
$$1a + 1b$$
$$1a^2 + 2ab + 1b^2$$
$$1a^3 + 3a^2b + 3ab^2 + 1b^3$$
$$1a^4 + 4a^3b + 6a^2b^2 + 4ab^3 + 1b^4$$
$$1a^5 + 5a^4b + 10a^3b^2 + 10a^2b^3 + 5ab^4 + 1b^5$$

If we delete all the a's, b's, and $+$'s, then what remains is the top 5 lines of the **Pascal's triangle**, after the 17th century French mathematician Blaise

Pascal (1623–1662):

$$1$$

$$1 \quad 2 \quad 1$$

$$1 \quad 3 \quad 3 \quad 1$$

$$1 \quad 4 \quad 6 \quad 4 \quad 1$$

$$1 \quad 5 \quad 10 \quad 10 \quad 5 \quad 1$$

Now you are asked to create the top 12 lines of Pascal's triangle, then delete all even coefficients and retain all remaining numbers. What shape seems to emerge?

Exercise 14. Rewrite Pascal's triangle modulo 2 (even numbers go to 0, odd numbers go to 1). Prove that the first 2^{n+1} rows are exactly 3 copies of the first 2^n rows. Explain why this relationship gives us the Sierpinski triangle from Exercise 12.

5.3 Space-Filling Curves

One of the really amazing results in mathematics is the fact that there is a continuous function that maps the closed unit interval $[0, 1]$ onto the closed unit square $[0, 1] \times [0, 1]$ in the plane. That is, there exists a continuous function on the one-dimensional closed interval whose range is a two-dimensional square. Since the continuous image of a continuous function on a real interval is called a **curve**, it follows that the image of such a function is also a curve, called a **space-filling curve** because it fills out the entire square. The existence of such a function was proved over a century ago, in 1890, by the Italian mathematician Giuseppe Peano (1858–1932).

It is impossible to actually draw the Peano curve. In fact, we cannot even draw a really good approximation to the curve. Nevertheless, by using a sequence of continuous functions and a limit for the sequence of functions, we will show the curve exists and has the properties we have mentioned. Let us indicate how the curve is derived.

Initially we will construct the images of three functions, f_1, f_2, and f_3, each of which can be considered as having domain $[0, 1]$ and range in the unit square $[0, 1] \times [0, 1]$. The image of f_1 is depicted in Figure 5.22(a). Note that $f_1(x) = (x, x)$ for $0 \leq x \leq 1$, and that in Figure 5.22(a) the arrow points northeast, which tells us the direction of the image of $[0, 1]$ for increasing values of x in $[0, 1]$.

For the image of f_2, we break up the unit square into 9 congruent subsquares, and assign the image of $[0, 1]$ to be the polygonal curve appearing in Figure 5.22(b). The arrows show the direction the curve is traversed, and the

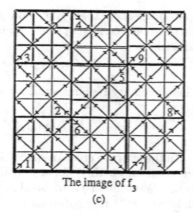

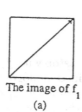

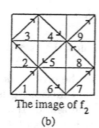

The image of f_1

(a)

The image of f_2

(b)

The image of f_3

(c)

FIGURE 5.22
Constructing the Peano space-filling curve

numbers in the sub-squares indicate the order in which the sub-squares are entered by the curve. Notice that as in Figure 5.22(a), the initial arrow is at the southwest corner (0, 0) and the final arrow is at the northeast corner (1, 1) The multiple-part rule for f_2 has the following form:

$$f_2(x) = \begin{cases} (3x, 3x) & \text{if } 0 \le x \le 1/9 \\ (2/3 - 3x, 3x) & \text{if } 1/9 < x \le 2/9 \\ (3x - 2/3, 3x) & \text{if } 2/9 < x \le 1/3 \\ \text{etc.} \end{cases}$$

We remark that by construction f_2 is a continuous function on $[0, 1]$, as you can see by tracing out the curve in Figure 5.22(b) according to the sequence of arrows. Moreover, the function's range bisects each of the sub-squares in Figure 5.22(b).

The image of the third function, f_3, appears in Figure 5.22(c), where the unit square is now divided into $81 = 9^2$ subintervals. As before, the arrows show the direction that the curve is traversed. We have placed a double arrow in the initial miniature sub-square and in the final miniature sub-square of each of the 9 sub-squares in Figure 5.22(b), and we have also identified those 9 sub-squares by number. This should help you to be able to trace out the curve from beginning to ending – through all 81 little squares!

Not only is f_3 a continuous function on $[0, 1]$, but its range intersects every one of the 81 miniature sub-squares. An additional important quality of f_3 is that f_3 and f_2 are close together for each x in $[0, 1]$. To be more specific, let us consider an x in the subinterval $[4/9, 5/9]$. For such an x, both $f_3(x)$ and $f_2(x)$ lie in the same sub-square 4 appearing in Figure 5.22(b); the maximum distance between any two points in that sub-square is the length of the diagonal, which is $\sqrt{2}/9$. This means that if we parametrize f_2 and f_3

appropriately so that $f_2(x)$ and $f_3(x)$ lie in the same sub-square, then the distance between $f_3(x)$ and $f_2(x)$, which we will denote by $D(f_3(x), f_2(x))$, satisfies

$$D(f_3(x), f_2(x)) \leq \frac{1}{9}\sqrt{2} \text{ for } 4/9 \leq x \leq 5/9$$

This inequality remains valid for all x in $[0, 1]$:

$$D(f_3(x), f_2(x)) \leq \frac{1}{9}\sqrt{2} \text{ for } 0 \leq x \leq 1$$

In the effort to create a space-filling curve the process does not stop with f_3, but continues with f_n for $n = 4, 5, 6, \ldots$. The function f_n is defined analogous to the definition of f_3, by breaking up the domain $[0, 1]$ and the range $[0, 1] \times [0, 1]$ into 9^{n-1} tiny sub-squares. The major features of f_n are the following:

i. f_n is continuous on $[0, 1]$.

ii. The image of f_n intersects each of the 9^{n-1} sub-squares of $[0, 1] \times [0, 1]$.

iii. For any m greater than n, the function values $f_n(x)$ and $f_m(x)$ are uniformly close together for x in $[0, 1]$. More precisely,

$$D(f_n(x), f_m(x)) \leq \frac{1}{3^{n-1}}\sqrt{2} \qquad \text{for } 0 \leq x \leq 1, \text{ for each } m > n \qquad (5.1)$$

It turns out that if a sequence of continuous functions $\{f_n\}_{n=1}^{\infty}$ has the property that $f_n(x)$ and $f_m(x)$ are close together for all large n and each $m > n$ and each x in $[0, 1]$, then the sequence of continuous functions converges to a unique function f that has the following properties:

1. f is continuous on $[0, 1]$.

2. For any n, the function values $f(x)$ and $f_n(x)$ are uniformly close together for x in $[0,1]$. In our case,

$$D(f(x), f_n(x)) \leq \frac{1}{3^{n-1}}\sqrt{2} \qquad \text{for } 0 \leq x \leq 1$$

This function f is the desired function that is space-filling. Indeed, its values are as close as we please to those of f_n when n is large enough, and f_n intersects each ever-smaller miniature sub-square as n increases. Our function f is the **Peano space-filling curve** on $[0, 1]$, or more simply, the **Peano curve**. In the final subsection of Section 5.3 we will discuss the technical details involved in showing that f exists and is continuous.

We should point out that f is *not* one-to-one, that is, there are distinct numbers x and z in $[0, 1]$ such that $f(x) = f(z)$. In fact, every vertex of each sub-square is the image of more than one number in $[0, 1]$. (See Exercise 2.)

5.3.1 Hilbert's Space-Filling Curve

A year after Peano produced his space-filling curve, the great German mathematician David Hilbert (1862–1943) produced a somewhat different space-filling curve. The first three stages appear in Figure 5.23.

The process in creating the Hilbert space-filling curve is quite analogous to that in constructing the Peano curve, except in the case of the Hilbert curve the nth stage involves 4^n sub-squares. (See Figure 5.23(a)–(c).) However, the important features of the Peano curve are retained with the Hilbert curve:

i. The nth function h_n is continuous on $[0, 1]$.

ii. The image of h_n intersects each of the 4^n miniature sub-squares of $[0, 1] \times [0, 1]$.

iii. For any m greater than n, the function values $h_n(x)$ and $h_m(x)$ are uniformly close together for x in $[0, 1]$. More precisely,

$$D(h_n(x), h_m(x)) \leq \frac{1}{2^n}\sqrt{2} \qquad \text{for } 0 \leq x \leq 1, \text{ for each } m > n$$

Reminiscent of the Peano curve process, the sequence $\{h_n\}_{n=1}^{\infty}$ converges to a function h that is continuous on $[0,1]$ and whose range is the complete (filled-in) square $[0, 1] \times [0, 1]$. The function h is called the **Hilbert space-filling curve**, or more succinctly, the **Hilbert curve**.

5.3.2 Cauchy Sequences

We defined the Peano curve and the Hilbert curve in terms of sequences of functions defined on $[0, 1]$ with range in $[0, 1] \times [0, 1]$. In order to make more

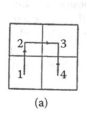

(a)

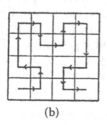

(b)

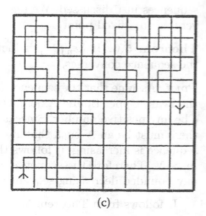

(c)

FIGURE 5.23
Constructing the Hilbert space-filling curve

precise the mathematical basis for defining these limiting functions, let us first recall from calculus the definition of a convergent sequence.

Definition 5.5. A sequence $\{x_n\}_{n=1}^{\infty}$ of real numbers **converges** to x if for each $\epsilon > 0$ there is an integer N such that whenever $n \geq N$, it follows that $|x_n - x| < \epsilon$. The number x is the **limit** of the sequence. If the sequence does not converge, then it **diverges**.

In calculus we encountered many sequences that converge. For example, it is easy to see that the sequence $\{2 - 1/n\}_{n=1}^{\infty}$, whose terms are 1, 3/2, 5/3, 7/4, ..., converges to 2. By contrast, the sequences $\{(-1)^n\}_{n=1}^{\infty}$ and $\{\sqrt{n}\}_{n=1}^{\infty}$ diverge.

Next, a sequence $\{x_n\}_{n=1}^{\infty}$ is **increasing** if $x_n \leq x_{n+1}$ for each n. Thus the sequence $\{2 - 1/n\}_{n=1}^{\infty}$ is increasing. A similar definition holds for decreasing sequences.

In the same vein, the sequence $\{x_n\}_{n=1}^{\infty}$ is **bounded** if there is a number M such that $|x_n| \leq M$ for all n. The number M is an **upper bound** for the sequence.

Sequences with upper bounds and lower bounds are prevalent, and have a very important property that is usually presented as an axiom. **LEAST**

UPPER BOUND AXIOM: Suppose that a set S of real numbers has an upper bound. Then it has a *least* upper bound. That is, there is an x^* such that $x \leq x^*$ for all x in S, and x^* is the smallest such number. Similarly a set of real numbers that has a lower bound has a *greatest* lower bound x_*.

An immediate consequence of the Least Upper Bound Axiom is the fact that a bounded sequence has both a least upper bound and a greatest lower bound.

Now we are ready to prove the first theorem relating the properties of sequences just discussed. We referred to it after Theorem 1.3 on iterates converging to a fixed point.

Theorem 5.6. If $\{x_n\}_{n=1}^{\infty}$ is increasing (or decreasing) and bounded, then the sequence converges.

Proof. Assume that the sequence $\{x_n\}_{n=1}^{\infty}$ is increasing and bounded. Then by the Least Upper Bound Axiom, the sequence has a least upper bound x^*. Let ϵ be any positive number. Since x^* is the *least* upper bound for the sequence, there must be an N such that $x_N \leq x^*$ and $|x_N - x^*| < \epsilon$. However, since the sequence is increasing, it follows that $x_N \leq x_m \leq x^*$ and $|x_m - x^*| < \epsilon$ for all $m > N$. Therefore the sequence converges to x^*. A similar proof shows that any bounded decreasing sequence converges. $\square$

It follows from Theorem 5.6 that a sequence such as

$$\left(\sum_{k=0}^{n} \frac{1}{k!} \right)_{n=1}^{\infty}$$

which corresponds to the sequence $\{x_n\}_{n=1}^{\infty}$ defined by

$$x_n = \frac{1}{0!} + \frac{1}{1!} + \frac{1}{2!} + \frac{1}{3!} + \cdots + \frac{1}{n!}$$

converges because it is increasing and is bounded by 3 (see Exercise 7). In fact, it converges to e.

Unfortunately it is not always possible to determine whether a sequence is increasing or decreasing, or even to determine exactly what the limit might be. For example, consider the function defined by $f(x) = x^3 + 3x - 1$. Because f has both negative and positive values and is continuous, we can conclude from the Intermediate Value Theorem that f has a zero z, which is unique because f is an increasing function. Suppose we wish to approximate z by using the bisection method. Since $f(0) < 0$ and $f(1) > 0$, we know that z lies in $[0, 1]$. Taking the midpoint $.5$ of $[0, 1]$, we find that $f(.5) > 0$, so that z lies in $[0, .5]$. Continuing with the bisections, we learn that z lies, successively, in $[0, .5]$, $[.25, .5]$, $[.25, .375]$, $[.3125, .375]$, etc. If we let $y_n =$ the midpoint of the nth subinterval in the process, then $\{y_n\}$ converges, since the nth subinterval has length $1/2^n$ and the subintervals are nested. However, $\{y_n\}$ is neither increasing nor decreasing, and it is not obvious what the limit of $\{y_n\}$ might be.

The following observation concerning this process and successive nested subintervals is relevant: If we take any sequence $\{x_n\}_{n=1}^{\infty}$ of numbers such that x_n lies in the nth subinterval for each n, then we notice that although the entries of $\{x_n\}_{n=1}^{\infty}$ likely do not seem to follow any pattern (i.e., increasing or decreasing, or alternating), the numbers differ from each other less and less as n increases. This kind of behavior of the sequence gives rise to the following definition, named for the great French mathematician Augustin–Louis Cauchy (1789–1857), who played a pivotal role in putting the calculus on a firm logical foundation.

Definition 5.7. A sequence $\{x_n\}_{n=1}^{\infty}$ is said to be **Cauchy**, or is a **Cauchy sequence**, if for every $\epsilon > 0$ there is an N such that whenever $n \geq N$ and $m \geq N$, it follows that $|x_n - x_m| < \epsilon$.

Next, we show that a Cauchy sequence is necessarily bounded. In the proof we will use the fact that a finite set is automatically bounded.

Theorem 5.8. A Cauchy sequence is bounded.

Proof. The idea is straightforward: If eventually all entries are close together, then that portion of the sequence is bounded, so the entire sequence is bounded. For the proof itself, suppose that $\{x_n\}_{n=1}^{\infty}$ is Cauchy. Then there is an N such that whenever $n \geq N$ and $m \geq N$, then $|x_n - x_m| < 1$, and in particular, $|x_N - x_m| < 1$. This means that for $m \geq N$,

$$x_N - 1 < x_m < x_N + 1$$

Consequently the maximum and minimum values of the finite set x_1, x_2, ..., x_N, $x_N - 1$, and $x_N + 1$ are upper bound and lower bound, respectively, for the given sequence.   □

Now we are ready for the main theorem of this subsection.

Theorem 5.9. A sequence of real numbers is a Cauchy sequence if and only if it is convergent.

Proof. On the one hand, suppose that $\{x_n\}_{n=1}^{\infty}$ converges, say to x. Let $\epsilon > 0$. Since the sequence converges to x, there is an N such that if $n \geq N$, then $|x_n - x| < \epsilon/2$. Thus for any $m \geq N$ and any $n \geq N$, we find that

$$|x_m - x_n| \leq |x_m - x| + |x - x_n| < \epsilon/2 + \epsilon/2 = \epsilon$$

Consequently the sequence is Cauchy.

Conversely, suppose that the sequence $\{x_n\}_{n=1}^{\infty}$ is Cauchy. From Theorem 5.8 we know that the sequence is bounded. Therefore any subsequence is also bounded, so by the Least Upper Bound Axiom has both a least upper bound and a greatest lower bound. Let $\{y_k\}_{k=1}^{\infty}$ and $\{z_k\}_{k=1}^{\infty}$ be defined as follows:

$$y_k = \text{least upper bound of } \{x_n\}_{n=k}^{\infty}$$

and

$$z_k = \text{greatest lower bound of } \{x_n\}_{n=k}^{\infty}$$

Then $\{y_k\}_{k=1}^{\infty}$ is bounded and is decreasing (see Exercise 6), so converges to a number y^*. Analogously, $\{z_k\}_{k=1}^{\infty}$ is bounded and is increasing, so converges to a number z^*. Since the original sequence $\{x_n\}_{n=1}^{\infty}$ is Cauchy, we can conclude that $y^* = z^*$, so that, in fact, $\{x_n\}_{n=1}^{\infty}$ converges to $y^*(= z^*)$.   □

One consequence of Theorem 5.9 relates to $\{x_n\}_{n=1}^{\infty}$, where x_n is in the nth subinterval obained in the employment of the bisection method to approximate a zero of a function f. Then $\{x_n\}_{n=1}^{\infty}$ is Cauchy because the successive nested subintervals shrink by a factor of $1/2$. By Theorem 5.9, the sequence converges to the corresponding zero of f.

A second consequence applies to sequences of points in the plane. In particular, a sequence $\{(x_n, y_n)\}_{n=1}^{\infty}$ **converges** to (x^*, y^*) if the distance

$$D((x_n, y_n), (x^*, y^*)) = \sqrt{(x_n - x^*)^2 + (y_n - y^*)^2} \qquad \text{converges to 0}$$

as n increases without bound. This occurs precisely when $\{x_n\}_{n=1}^{\infty}$ converges to x^* *and* $\{y_n\}_{n=1}^{\infty}$ converges to y^*.

The notion of a Cauchy sequence in the plane is analogous to that for real number sequences: A sequence $\{(x_n, y_n)\}_{n=1}^{\infty}$ in the plane is said to be **Cauchy**, or is a **Cauchy sequence**, if for every $\epsilon > 0$ there is an N such that whenever $n \geq N$ *and* $m \geq N$, it follows that

$$D((x_n, y_n) - (x_m, y_m)) < \epsilon$$

Using these definitions and results, one can show that Theorem 5.16 remains valid for sequences in the plane.

Theorem 5.10. A sequence of points in the plane is a Cauchy sequence if and only if it is convergent.

Proof. See Exercise 9. □

The previous discussion relates to the Peano space-filling curve. Recall that in the construction of the curve we created a sequence of functions $\{f_n\}_{n=1}^{\infty}$. We found that

$$D(f_n(x), f_m(x)) \leq \frac{1}{3^{n-1}}\sqrt{2} \qquad \text{for } 0 \leq x \leq 1, \text{ for each } m > n$$

(See () in this section.) Consequently for each x in $[0, 1]$, $\{f_n(x)\}_{n=1}^{\infty}$ is a Cauchy sequence. By Theorem 5.10, this sequence converges to a number depending on x. We will call that number $f(x)$. This result holds true for all x in $[0, 1]$, and the function f has range equal to the square $[0, 1] \times [0, 1]$ by our discussion earlier in this section. The only unfinished business is the proof that f is continuous.

Theorem 5.11. The function f, whose range is the Peano space-filling curve, is continuous.

Proof. Let $\epsilon > 0$, and let z be any number in $[0, 1]$. We must find an open interval I centered at z such that if y is in I and in $[0, 1]$, then $|f(y) - f(z)| < \epsilon$. By the inequality in the paragraph just before the theorem, if N is large enough so that $\sqrt{2}/3^{N-1} < \epsilon/3$, then

$$D(f_m(x), f_N(x)) < \epsilon/3 \text{ for all } x \text{ in } [0, 1]$$

and for all $m \geq N$. Consequently since $f(x) = \lim_{m\to\infty} f_m(x)$,

$$D(f(x), f_N(x)) \leq \epsilon/3 \text{ for all } x \text{ in } [0, 1]$$

By virtue of the fact that the image of f_N is a polygonal line, f_N is continuous. Thus there is an open interval I centered at z such that if y is in I and in $[0, 1]$, then $|f_N(z) - f_N(y)| < \epsilon/3$.

We are now ready to put this information together. For y in I and in $[0, 1]$, we have

$$D(f(z), f(y)) \leq D(f(z), f_N(z)) + D(f_N(z), f_N(y)) + D(f_N(y), f(y))$$
$$< \epsilon/3 + \epsilon/3 + \epsilon/3 = \epsilon$$

That completes the proof that f is continuous at the arbitrary number z in $[0, 1]$, so f is a continuous function. □

5.3.3   Section 5.3 Exercises

Exercise 1. Referring to the process of creating the Peano curve, use Figure 5.22(b) to help determine formulas for the values $f_2(x)$ for the following intervals:

(a) $1/3 < x \leq 4/9$

(b) $7/9 < x \leq 8/9$

Exercise 2. For the Peano curve, find a value of x in $[0, 1]$ such that $x \neq 1/9$ and $f(1/9) = f(x)$.

Exercise 3. Complete Figure 5.23(c), which gives the image of h_3 in the construction of the Hilbert curve. (*Hint*: Note that the curve must traverse each of the squares represented in Figure 5.23(b) in the order they are traversed by the graph of h_2.)

Exercise 4. Tell why

$$D(h_n(x), h_m(x)) \leq \frac{1}{2^{n-1}}\sqrt{2} \qquad \text{for } 0 \leq x \leq 1$$

for each m and n with $m > n$.

Exercise 5. Discuss a reasonable method for describing a three-dimensional space-filling curve that is the range of a function g whose domain is $[0, 1]$ and whose range is the unit cube $[0, 1] \times [0, 1] \times [0, 1]$.

Exercise 6. Show that the sequence $\{y_k\}_{k=1}^{\infty}$ appearing in the proof of Theorem 5.9 is decreasing.

Exercise 7. Show that the sequence $\{\sum_{k=0}^{n} 1/k!\}_{n=1}^{\infty}$ is bounded above by 3.

Exercise 8. Consider the sequence $\{y_n\}_{n=1}^{\infty}$ of partial sums of the harmonic series:

$$y_n = \sum_{k=1}^{n} \frac{1}{k} = 1 + \frac{1}{2} + \cdots \frac{1}{n}$$

We know that the harmonic series, and hence the sequence $\{y_n\}_{n=1}^{\infty}$, diverges.

(a) Show that $|y_n - y_{n+1}|$ approaches 0 as n increases.

(b) Does (a) contradict Theorem 5.9? Explain your answer.

Exercise 9. Prove Theorem 5.10.

Exercise 10. Consider the sequence $\{x_n\}_{n=1}^{\infty}$ defined by $x_1 = \sqrt{2}$, $x_2 = \sqrt{\sqrt{2} + 2}$, and in general,

$$x_{n+1} = \sqrt{x_n + 2}$$

(a) Show that if $1 < x_n < 2$, then $1 < x_{n+1} < 2$. Since $x_1 = \sqrt{2}$ and by the Law of Induction, this means that the sequence is bounded.

(b) Show that the sequence is increasing.

(c) By (a) and (b), the sequence has a limit x. Find the value of x. (*Hint:* Use the fact that $x_{n+1}^2 = x_n + 2$ for all n, and take limits.)

Exercise 11. Must a Cauchy sequence on a bounded set converge to a value in that set?

Exercise 12. If a sequence $\{x_n\}$ is Cauchy, must the sequence $\{x_n^2\}$ also be Cauchy? What about the converse?

5.4   Similarity and Capacity Dimensions

The word "dimension" is common to everyday life. Indeed, we say that a wire has dimension one, paper has two dimensions, and a watermelon has three dimensions. However, it might not be so clear what dimension we should assign to shapes like the Cantor set, Sierpiński gasket, or the Menger sponge. On the one hand, a glance at Figure 5.24 with the Sierpiński gasket tells us that the dimension of the gasket is surely at least 1 and no greater than 2, since the gasket contains line segments and lies in the plane. On the other hand, the gasket is full of holes and in no way appears to have the same fullness of a filled-in triangle with the same outer boundary. It might be tempting to determine the area of the gasket and compare it with the area of a triangle of the same outer boundary. However, the Sierpiński gasket has area 0, as we proved in Section 5.2.

In this section we will discuss two notions of dimension: similarity dimension and capacity dimension. The former applies, as the name indicates, to self-similar sets, whereas the latter applies to more general kinds of shapes. We

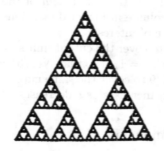

FIGURE 5.24
Sierpiński gasket

will see that both kinds of dimension can have values that are not necessarily integers.

Note: The colloquial notion of dimension has a wide variety of explanations. Most people on the street will tell you that the 4th dimension is time, which doesn't tend to match up with the intuition we develop as mathematicians. Manning (1914) delves into this interesting history. Manning notes that Aristotle and Pappus both speculated about the number of potential dimensions, and whether dimension beyond 3 could exist. Later, philosophers such as Moore and Kant would also speculate about dimension. A lovely resource that illustrates the thinking of its time, while being an amusing story, is Abbott's *Flatland* (1884). For the more visual learner, there have been animated films made to illustrate this, most notably *Flatland 2* or the episode "2-D Blacktop" of the television program *Futurama*.

The fundamental ingredient we will use in defining the dimension of a given shape involves covering the shape with regular kinds of objects like line segments, squares or cubes. By an **n-dimensional box** we mean a closed interval if $n = 1$, a filled-in square if $n = 2$, and a cube if $n = 3$. Normally we will refer to such a set simply as a **box** if n is understood.

Next, let S be a set in $\mathbb{R}^n$, where $n = 1$, 2, or 3. For convenience, we may assume that S has diameter 1 (so that S could lie in the unit interval, unit square, or unit cube if properly placed). For any constant $c > 0$, we define $N(c)$ by:

$N(c) =$ the smallest number of n-dimensional boxes of side length c required

to completely cover S

Of course $N(c)$ depends on the set S, so to be precise we really should write $N_S(c)$. However, to simplify notation we will suppress the S if the set is clear from context. Notice that c is the *side-length* of the box, in contrast to the diameter, or the area or volume in higher dimensions. In our discussion, side-length will be the measure of interest.

For example, in order to cover the cutout unit square S in Figure 5.25(a) it takes 8 boxes of side length $c = 1/3$, so that $N(1/3) = 8$. Similarly, it takes 72 boxes of side length $c = 1/9$ to cover the same cutout square (Figure 5.25(b)). As you can imagine, $N(c)$ increases as c decreases.

5.4.1   Similarity Dimension

In order to define the notion of similarity dimension, we first focus on a self-similar set S with a single self-similar constant s, such that S is the union of m essentially nonoverlapping miniatures (that is, nonoverlapping except

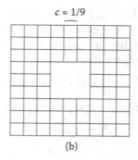

FIGURE 5.25
Covers for a cut-out square

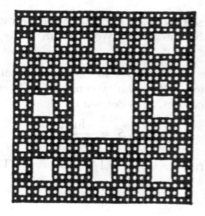

FIGURE 5.26
The Sierpiński carpet

possibly for simple boundaries), each one of which is a factor of s as large as S. Then $N(s) = m$. For example, in Figure 5.26 we have the Sierpiński carpet, which can be considered as the union of 8 miniatures of side length $1/3$, or the union of $64 = (8^2)$ miniatures of side length $1/9 = 1/3^2$, etc.

For such a self-similar set S, we deduce that $N(s^k) = m^k$ for each positive integer k. In other words, each time we reduce the side length of a box by a factor of s, we increase the number N by the multiple m. Of course, both $N(s^k)$ and $1/s^k$ increase without bound as k increases without bound.

How does $N(s^k)$ relate to s^k if the set S is a unit interval or a unit square? If S is the unit interval, then letting $s = 1/n$, we find that $N(1/n^k) = n^k$. Similarly, if S is the unit square, then letting $s = 1/n$ again, we find that $N(1/n^k) = n^{2k}$, as you can check.

In case of the unit interval and the unit square, the number of boxes needed to cover the set increases as a power of $1/s^k$, for appropriate k. For a more general self-similar S with m miniatures each with similarity constant s, we search for a nonnegative number D such that

$$N(s^k) = (1/s^k)^D, \text{ or equivalently, } \ln N(s^k) = D \ln(1/s^k)$$

Using the fact that $N(s^k) = m^k$, we find that D must satisfy

$$D = \frac{\ln N(s^k)}{\ln(1/s^k)} = \frac{\ln m^k}{\ln(1/s^k)} = \frac{\ln m}{\ln(1/s)}$$

Consequently for any $k = 1, 2, \ldots$,

$$N(s^k) = (1/s^k)^D \quad \text{if} \quad D = \frac{\ln m}{\ln(1/s)}$$

We say that $N(s^k)$ **scales** as $(1/s^k)^D$, because the ratio $N(s^k)/(1/s^k)^D$ is constant for all positive integers k. We are thus ready for the definition of similarity dimension.

Definition 5.12. Let S be self-similar, with m essentially nonoverlapping miniatures and a unique self-similarity constant s. Then the **similarity dimension** of S, denoted $\dim_s S$, is defined by

$$\dim_s S = \frac{\ln m}{\ln(1/s)}$$

For the Cantor ternary set C, $s = 1/3$ and $m = 2$. Therefore

$$\dim_s C = \frac{\ln 2}{\ln 3} \approx 0.63$$

In the same vein, the similarity dimension for the Sierpiński gasket G, with $s = 1/2$ and $m = 3$, is given by

$$\dim_s G = \frac{\ln 3}{\ln 2} \approx 1.58$$

Also, if $I =$ an interval $[a, b]$, and if I is covered by m subintervals of length $(b - a)/m$, then

$$\dim_s I = \frac{\ln m}{\ln(1/(1/m))} = \frac{\ln m}{\ln m} = 1$$

as you would expect.

The same formula in the definition allows us to calculate the similarity dimension for other self-similar "monsters," like the Sierpiński carpet, the Koch curve, and the Menger sponge.

The formula given above for the similarity dimension applies to a self-similar set S provided that S has a unique self-similarity constant. However, we can extend the definition of similarity dimension to a set S with more than one self-similarity constant, but the formula is more complicated, and will be given after we have defined the capacity dimension.

5.4.2   Capacity Dimension

Recall that if S is a self-similar set with single similarity constant s, then

$$N(s^k) = (1/s^k)^D$$

for an appropriate nonnegative number D. For other kinds of sets this formula may not be valid or even make sense. However, in the same spirit it does make sense to inquire whether there are numbers $C > 0$ and $D \geq 0$ such that $N(\epsilon) \approx C(1/\epsilon)^D$ as ϵ approaches 0. Of course such an approach would be consistent with the procedure for self-similar sets, where we substitute s^k for ϵ.

Continuing, we consider a given set S in $\mathbb{R}^n$, where $n = 1$, 2, or 3. Assume that $N(\epsilon) \approx C(1/\epsilon)^D$ as ϵ approaches 0. Taking logarithms of both sides, we find that

$$\ln N(\epsilon) \approx D \ln \frac{1}{\epsilon} + \ln C$$

Rearranging and solving for D, we obtain

$$D \approx \frac{\ln N(\epsilon)}{\ln(1/\epsilon)} - \frac{\ln C}{\ln(1/\epsilon)} \approx \frac{\ln N(\epsilon)}{\ln(1/\epsilon)}$$

provided that ϵ is very small (which implies that $\ln(1/\epsilon)$ is very large). If the limit of the final fraction above exists as ϵ approaches 0, we define D to be the capacity dimension of S.

Definition 5.13. Let S be a subset of $\mathbb{R}^n$, where $n = 1$, 2, or 3. The **capacity dimension** (or **box dimension**) of S is given by

$$\dim_c S = \lim_{\epsilon \to 0} \frac{\ln N(\epsilon)}{\ln(1/\epsilon)}$$

if the limit exists. If the capacity dimension of S exists and is *not* an integer, then S is said to have **fractal dimension**.

The c in $\dim_c S$ refers to the fact that the dimension is the "capacity" dimension. The term "fractal" was introduced by the mathematician Benoit Mandelbrot in the late 1970s in order to refer to a set with fractional dimension, that is, a noninteger dimension. Also, we should mention that "capacity dimension" is often referred to as **fractal dimension** in the mathematical literature.

If $D = \dim_c S$ exists, then whether or not the dimension is an integer, we say that the number $N(\epsilon)$ **scales** as $(1/\epsilon)^D$.

Now that we have defined capacity dimension it is time to relate it to the similarity dimension of a self-similar set with a single self-similarity constant. In the corollary to the following theorem we will show that in this case the capacity dimension and similarity dimension are identical.

Theorem 5.14. Consider a subset S of $\mathbb{R}^n$, where $n = 1$, 2, or 3, and let $0 < r < 1$. Then $\lim_{k\to\infty}\left[\ln N(r^k)/\ln(1/r^k)\right]$ exists if and only if $\dim_c S$ exists. In that case,

$$\dim_c S = \lim_{k\to\infty} \frac{\ln N(r^k)}{\ln(1/r^k)}$$

Proof. Let $0 < \epsilon < r$, and recall that $r < 1$. Let k be a positive integer so large that $r^{k+1} < \epsilon \le r^k$. Then $N(r^k) \le N(\epsilon) \le N(r^{k+1})$. Because the natural logarithm is an increasing function and $0 < r < 1$, we find that

$$\ln(1/r^k) \le \ln(1/\epsilon) < \ln(1/r^{k+1}) \text{ and } \ln N(r^k) \le \ln N(\epsilon) \le \ln N(r^{k+1})$$

Therefore

$$\frac{\ln N(r^k)}{\ln(1/r^k) + \ln(1/r)} = \frac{\ln N(r^k)}{\ln(1/r^{k+1})} \le \frac{\ln N(\epsilon)}{\ln(1/\epsilon)} \le \frac{\ln N(r^{k+1})}{\ln(1/r^k)}$$

$$= \frac{\ln N(r^{k+1})}{\ln(1/r^{k+1}) - \ln(1/r)}$$

As k increases without bound, the limit of the left side exists if and only if the limit of the right side does. Thus it follows that

$$\lim_{k\to\infty} \frac{\ln N(r^k)}{\ln(1/r^k)} \text{ exists if and only if } \lim_{\epsilon\to 0} \frac{\ln N(\epsilon)}{\ln(1/\epsilon)} \text{ exists}$$

and when they do exist, they are equal, and also equal $\dim_c S$. $\square$

Corollary 5.15. Let S be self-similar, with a single similarity constant s. Then $\dim_c S = \dim_s S$.

Proof. We only need to let $r = s$ in Theorem 5.14. $\square$

Suppose more generally that S is self-similar with miniatures having varying similarity constants. Then there is a natural extension to the notion of a similarity dimension, provided that the capacity dimension exists.

Theorem 5.16. Assume that the capacity dimension D of S exists, and that S is self-similar, with similarity constants $s_1, s_2, ..., s_m$. Then D satisfies the equation

$$s_1^D + s_2^D + \ldots s_m^D = 1$$

Proof. For the present, assume that $m = 2$. Let T_1 and T_2 be the constituent contractions of S, with similarity constants s_1 and s_2, respectively, so that $S = T_1(S) \cup T_2(S)$. For any $\epsilon > 0$,

$$N_S(\epsilon) = N_{T_k(S)}(s_k\epsilon), \text{ for } k = 1, 2$$

because $T_k(S)$ is a miniature of S whose size is a factor s_k of the original set S. Equivalently,

$$N_{T_k(S)}(\epsilon) = N_S(\epsilon/s_k), \text{ for } k = 1, 2$$

Thus far we have not needed to use the fact that the capacity dimension for S exists. Now we do need to apply that assumption, namely that for small ϵ, there are constants $C > 0$ and $D \geq 0$ such that $N_S(\epsilon) \approx C(1/\epsilon)^D$. Using the fact that $S = T_1(S) \cup T_2(S)$, we deduce that

$$N_S(\epsilon) = N_{T_1(S)}(\epsilon) + N_{T_2(S)}(\epsilon) = N_S(\epsilon/s_1) + N_S(\epsilon/s_2)$$
$$\approx C(1/(\epsilon/s_1))^D + C(1/(\epsilon/s_2))^D$$

Consequently for small $\epsilon > 0$,

$$C(1/\epsilon)^D \approx N_S(\epsilon) \approx C(1/\epsilon)^D (s_1^D + s_2^D)$$

so that by cancellation we conclude that $s_1^D + s_2^D = 1$. Thus we have proved the theorem for $m = 2$. The result for an arbitrary positive integer $m > 2$ follows by an application of the Law of Induction. □

From Theorem 5.16 we see that if S is self-similar with m similarity constants $s_1, s_2, ..., s_m$, then if it exists, the capacity dimension D of S has the property that

$$\dim_c S = D, \text{ where } D \text{ is the unique solution to the equation } \sum_{k=1}^{m} s_k^D = 1$$

The formula yields the same dimension as the similarity dimension when the self-similar set has but one similarity constant (see Exercise 7). When there are multiple distinct similarity constants, then solving for D requires a little more effort.

Before we turn to examples in which we determine (approximate) capacity dimensions, we should note that Theorem 5.16 provides a formula for the capacity dimension of any self-similar set, provided that the capacity dimension exists. Proving that the capacity dimension exists is not normally trivial; nevertheless, it appears to exist (at least) for most self-similar sets. In the sequel we will assume that the capacity dimension of a given self-similar set does exist, and we will then determine its value.

Example 5.4.1. Let S be the set pictured in Figure 5.27, with similarity constants $1/2$, $1/4$, $1/4$, and $1/4$, respectively. Find an approximate value of $\dim_c S$.

Solution. The formula for D becomes

$$3(\frac{1}{4})^D + (\frac{1}{2})^D = 1$$

We can solve this equation by letting f be defined by

$$f(D) = 3(\frac{1}{4})^D + (\frac{1}{2})^D - 1$$

and approximating the (unique) zero of f with the help of the Newton-Raphson method. We find that $D \approx 1.20$, and conclude that $\dim_c S \approx 1.20$. □

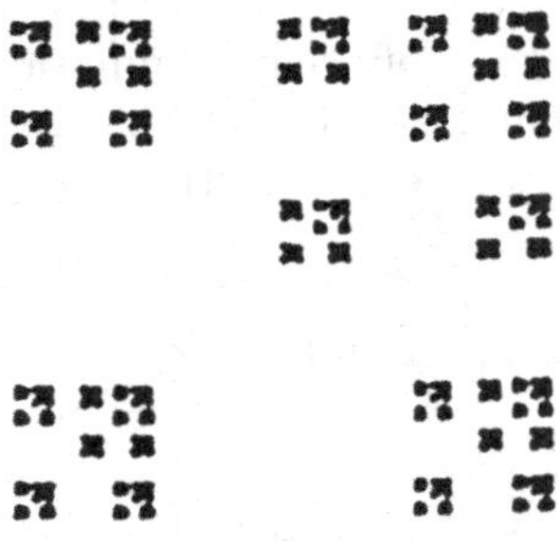

FIGURE 5.27
Self-similar set with similarity constants $1/2$, $1/4$, $1/4$, and $1/4$

Now we will find the capacity dimension of a nonself-similar set.

Example 5.4.2. Let S be the sequence $\{1, 1/2, 1/3, 1/4, \ldots\}$. Show that $\dim_c S = 1/2$.

Solution. Let n be an arbitrary integer with $n > 1$, and let $\epsilon = 1/[n(n-1)]$. Because of the size of ϵ, it takes no fewer than $n - 1$ intervals of length ϵ to cover the $n - 1$ numbers $1, 1/2, 1/3, \ldots, 1/(n-1)$. In addition, it takes $n - 1$ intervals of length ϵ to completely cover the interval $[0, 1/n]$. Thus

$$n - 1 \leq N(1/[n(n-1)]) \leq (n-1) + (n-1) = 2(n-1)$$

Since $\ln 1/\epsilon = \ln n(n-1) = \ln n + \ln(n-1)$, it follows that

$$\frac{\ln(n-1)}{\ln n + \ln(n-1)} \leq \frac{\ln N(1/[n(n-1)])}{\ln n(n-1)} \leq \frac{\ln 2 + \ln(n-1)}{\ln n + \ln(n-1)}$$

The left and right quotients approach $1/2$ as n increases without bound. Consequently $\dim_c S = 1/2$. □

Curves in the plane come in all varieties. At the one extreme, the curve may have capacity dimension 1, as occurs if the curve is a straight line. At the other extreme, a space-filling curve like the Peano curve or Hilbert curve has dimension 2, since it fills out a square in the plane. Next, we find the capacity dimension for a curve that represents the graph of a continuously differentiable function.

Theorem 5.17. Let f be a real-valued function that is continuously differentiable on $[0, 1]$, and let S be the graph of f. Then $\dim_c S = 1$.

Proof. Since f is assumed to have a continuous derivative on $[0, 1]$, by the Maximum-Minimum Theorem from calculus there is a positive constant M such that $|f'(x)| \leq M$ for all x in $[0, 1]$. Let $\epsilon > 0$ and let $P = \{0 =$

$x_0, x_1, x_2, \ldots, x_n = 1\}$ be a partition of $[0, 1]$. For each $k = 1, 2, \ldots, n$, we know from the Mean Value Theorem and the definition of M that the length L_k of the graph that lies between the points $(x_{k-1}, f(x_{k-1}))$ and $(x_k, f(x_k))$ satisfies the inequality

$$L_k \leq (x_k - x_{k-1})\sqrt{1 + M^2}$$

The reason is that the portion of the graph of f can be no longer than the length of the straight line segment with slope M and consisting of points (x, y) with $x_{k-1} \leq x \leq x_k$, which is the value on the right-hand side of the inequality (see Exercise 14).

If we let $x_k - x_{k-1} \approx \epsilon/\sqrt{1 + M^2}$ for each k, then it takes approximately $\sqrt{1 + M^2}/\epsilon$ intervals $[0, x_1], [x_1, x_2], \ldots, [x_{n-1}, x_n]$ to cover the domain $[0, 1]$ of f. This implies that

$$N(\epsilon) \leq \frac{\sqrt{1 + M^2}}{\epsilon}$$

Combined with the fact that $N(\epsilon) \geq 1/\epsilon$ (with equality only if the graph is a horizontal line), we conclude that

$$1 = \frac{\ln(1/\epsilon)}{\ln(1/\epsilon)} \leq \frac{\ln N(\epsilon)}{\ln(1/\epsilon)} \leq \frac{\ln\sqrt{1 + M^2} - \ln \epsilon}{-\ln \epsilon}$$

The limit on the right side is 1 as ϵ approaches 0. Therefore

$$\dim_c S = \lim_{\epsilon \to 0} \frac{\ln N(\epsilon)}{\ln(1/\epsilon)} = 1$$

which was what we were to prove. $\qquad\qquad\square$

We hasten to mention that graphs of continuously differentiable functions defined on finite closed intervals have capacity dimension 1, whether or not the interval is $[0, 1]$.

If $\dim_c S = D$, then $N(\epsilon)$ scales as $(1/\epsilon)^D$, so that for small values of ϵ,

$$\ln N(\epsilon) \approx D\ln(1/\epsilon)$$

If we plot $\ln N(\epsilon)$ against $\ln(1/\epsilon)$ for various small values of ϵ, letting $\ln N(\epsilon)$ run along the vertical axis and $\ln(1/\epsilon)$ along the horizontal axis, then we find that the points $(\ln(1/\epsilon), \ln(N(\epsilon)))$ lie approximately on a line with slope D. Indeed, let α and β be two distinct such small values of ϵ. Then since $N(\alpha) \approx C(1/\alpha)^D$ and $N(\beta) \approx C(1/\beta)^D$, it follows that

$$\frac{\ln N(\beta) - \ln N(\alpha)}{\ln 1/\beta - \ln 1/\alpha} \approx \frac{(D\ln 1/\beta + \ln C) - (D\ln 1/\alpha + \ln C)}{\ln 1/\beta - \ln 1/\alpha}$$

$$= \frac{D\ln 1/\beta - D\ln 1/\alpha}{\ln 1/\beta - \ln 1/\alpha} = D$$

The fact that we can find an approximate value for D by plotting $\ln N(\epsilon)$ against $\ln(1/\epsilon)$ (or equivalently, plotting $N(\epsilon)$ against $1/\epsilon$ on log-log paper) has been effective for determining reasonable values of dimensions in various sciences.

Note: The 1994 book *A Random Walk Through Fractal Dimension* by Brian Kaye contains a discussion of fractal dimension in a wide variety of applications. They include the following:

1. Ferrography: Experimental study of procedures for separating wear debris from lubrication oil is utilized in characterizing the shape and size of the fine particles.

2. Electrolysis: Fractal dimension of a crystal grown in an electric cell can be changed by altering the voltage applied to the cathode.

3. Thin films: Study of fractal dimensions of various crystals deposited by the sputtering process can assist in understanding how crystals form and grow, so that one can be more efficient in the commercial growth of thin films related to the manufacture of computer chips.

4. Colloid science: Here there is study of fine particles which are so tiny that their behavior is totally controlled by surface forces. Many colloidal fine particles are themselves made up of much smaller units, agglomerated into larger units which display fractal structure.

5. Pharmacology: The fractal nature of grains can be used in the study of "bioavailability," that is, how a drug in tablet form disintegrates in the body, and thus delivers the drug to the body.

6. Rock drilling: Fractal structure of a body breaking apart under impact stress produced by a high-velocity projectile is likely to be useful in studying how rock breaks apart when one drills into it.

7. Pigment loading: The fractal dimension of the surface of a pigment probably determines the strength of the composite material, and contributes to the "chunkiness" of the material.

5.4.3   Section 5.4 Exercises

Exercise 1. Find the similarity dimension of the Sierpiński carpet.

Exercise 2. Find the similarity dimension of the Koch curve.

Exercise 3. Find the similarity dimension of the Menger sponge.

Exercise 4. Describe a self-similar set whose similarity dimension is 1.5.

Exercise 5. Find the capacity dimension of the Barnsley castle in Figure 5.28.

FIGURE 5.28
Barnsley's castle

Exercise 6. Suppose that a self-similar set S has three similarity constants: $1/2$, $1/3$, and $1/4$. Assuming that the capacity dimension exists, approximate it.

Exercise 7. Prove that the formula $\sum_{k=1}^{m} s_k^D = 1$ yields the similarity dimension for any self-similar set S with a single similarity constant.

Exercise 8. Let $S = \{1, 1/2^2, 1/3^2, \ldots\}$. Show that $\dim_c S = 1/3$.

Exercise 9. Let $S = \{1, 1/2, 1/2^2, 1/2^3, \ldots\}$. Find $\dim_c S$.

Exercise 10. Find the capacity dimension of the devil's staircase.

Exercise 11. Find a bounded, infinite set S of real numbers for which $\dim_c S = 0$.

Exercise 12. Find a set S in the plane for which $\dim_c S$ does *not* exist.

Exercise 13. Suppose that S is in R^2 and that $\dim_c S$ exists. Prove that

$$\dim_c S = \lim_{\epsilon \to 0} \frac{N^*(\epsilon)}{\ln(1/\epsilon)}$$

where $N^*(\epsilon)$ = the smallest number of disks of radius ϵ that are needed to cover S. (*Hint:* If B_ϵ = the square of side length ϵ, and if D_δ = the disk of radius δ, find positive constants a and b such that $B_{a\epsilon} \supseteq D_\delta \supseteq B_{b\epsilon}$.)

Exercise 14. Show that the length of the line segment with positive slope M and consisting of all x such that $x_{k-1} \le x \le x_k$ is $(x_k - x_{k-1})\sqrt{1 + M^2}$.

Exercise 15. Complete the proof of Theorem 5.16 by applying the Law of Induction.

5.5   Lyapunov Dimension

The sets with which we illustrated the notion of capacity dimension in Section 5.4 have properties of self similarity, so that we were able to ascertain the capacity dimension algebraically.

Suppose we try to approximate the capacity dimension of the Hénon attractor A_H, say, with the grid shown in Figure 5.29. Notice that the number of rectangles that intersect A_H is approximately 100. If we let ϵ equal the height of each rectangle, then $\epsilon = 1/20$, so that

$$\frac{\ln N(\epsilon)}{\ln (1/\epsilon)} = \frac{\ln 100}{\ln 20} \approx 1.58$$

However, it is known that $\dim_c A_H \approx 1.26$, which can be confirmed by letting ϵ be microscopic and using self-similarity of the attractor near the saddle fixed point **p**.

A reason that one needs extremely fine grids in order to approximate the capacity dimension of A_H is that many of the squares in the grid meet

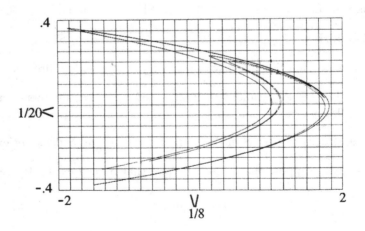

FIGURE 5.29
A grid to estimate capacity of the Hénon attractor

portions of A_H that are clearly one-dimensional (such as those curves on the left portion of the attractor), and relatively few squares meet the portion of the attractor close to the fixed point p which, however, is where the dimension rises markedly above 1. The capacity dimension of a given set is a "geometric dimension" because its value depends solely on the geometry of the set.

By contrast, we can study the frequency with which iterates of a given point for a given function enter a certain region in the plane. That region is called an "attractor of the function."

Definition 5.18. Let V be a subset of $\mathbb{R}^n$ and $F : V \to \mathbb{R}^n$, where $n = 1, 2$, or 3. Also let A be a subset of V. Then A is an **attractor** of F provided that the following conditions hold:

i. A is a closed subset of V such that $F(A) \subseteq A$, that is, A is an **invariant** subset of V.

ii. There is an open set U containing A such that whenever $\mathbf{v}$ is in U, then $F^{[n]}(\mathbf{v}) \to A$ as n increases without bound, in the sense that for each $\epsilon > 0$, there is a positive integer N such that if $k \geq N$, there exists a $\mathbf{w}_k$ in A such that $||F^{[k]}(\mathbf{v}) - \mathbf{w}_k|| < \epsilon$.

The invariance of A in V means that the iterates of any point in A are also in A. By the definition of attractor, attracting cycles are attractors, as is A_H.

In preparation to define the Lyapunov dimension, let us recall from (6) in Section 2.1 that the Lyapunov exponent of a one-dimensional function f at x is given by

$$\lambda(x) = \lim_{n \to \infty} \frac{1}{n} \ln |(f^{[n]})'(x)| \tag{5.2}$$

provided that the limit exists. For two dimensions, we let V be a subset of R^2, and suppose that $F : V \to R^2$ has continuous partial derivatives. Assume also that $\mathbf{v}_0$ is in V, with orbit $\{\mathbf{v}_n\}_{n=0}^{\infty}$. For each $n = 1, 2, \ldots$, we define $D_n F(\mathbf{v}_0)$ by the formula

$$D_n F(\mathbf{v}_0) = [DF(\mathbf{v}_{n-1})] [DF(\mathbf{v}_{n-2})] \cdots [DF(\mathbf{v}_0)] \tag{5.3}$$

where $DF(\mathbf{v}_k)$ denotes the 2×2 Jacobian matrix identified with the differential of F at $\mathbf{v}_k$ (see Definition 3.16). Then $D_n F(\mathbf{v}_0)$ is also a 2×2 matrix (depending on n). If $D_n F(\mathbf{v}_0)$ has nonzero real eigenvalues, we denote their absolute values by $d_{n1}(\mathbf{v}_0)$ and $d_{n2}(\mathbf{v}_0)$. For convenience, we will assume that $d_{n1}(\mathbf{v}_0) \geq d_{n2}(\mathbf{v}_0)$. Now we define the **Lyapunov numbers** $\lambda_1(\mathbf{v}_0)$ and $\lambda_1(\mathbf{v}_0)$ of F at $\mathbf{v}_0$:

$$\lambda_1(\mathbf{v}_0) = \lim_{n \to \infty} [d_{n1}(\mathbf{v}_0)]^{1/n} \quad \text{and} \quad \lambda_2(\mathbf{v}_0) = \lim_{n \to \infty} [d_{n2}(\mathbf{v}_0)]^{1/n} \tag{5.4}$$

provided that the limits exist. The Lyapunov numbers were originally defined by the Russian mathematician V. I. Oseledec in the 1960s. They measure

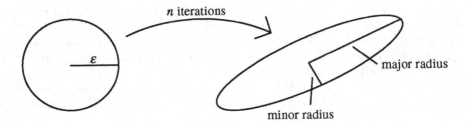

FIGURE 5.30
Change in circle radius under iteration

the rate at which a circle of small radius centered on the attractor is deformed through iteration. In particular, after n iterations a circle of radius ϵ centered at $\mathbf{v}_0$ is transformed into an ellipse whose major and minor radii are approximately $[\lambda_1(\mathbf{v}_0)]^n \epsilon$ and $[\lambda_2(\mathbf{v}_0)]^n \epsilon$, respectively (Figure 5.30). Since $d_{n1}(\mathbf{v}_0) \geq d_{n2}(\mathbf{v}_0)$ by prescription, it follows that $\lambda_1(\mathbf{v}_0) \geq \lambda_2(\mathbf{v}_0)$.

Now we are ready to define the Lyapunov dimension.

Definition 5.19. Let V be a subset of R^2, and let function $F : V \to R^2$ have coordinate functions with continuous partial derivatives. Also assume that F has an attractor A_F, and that $\mathbf{v}_0$ is in A_F. Finally, assume that $\lambda_1(\mathbf{v}_0) > 1 > \lambda_2(\mathbf{v}_0)$. Then the **Lyapunov dimension** of A_F at $\mathbf{v}_0$, denoted $\dim_L A_F(\mathbf{v}_0)$, is given by

$$\dim_L A_F(\mathbf{v}_0) = 1 - \frac{\ln \lambda_1(\mathbf{v}_0)}{\ln \lambda_2(\mathbf{v}_0)} \tag{5.5}$$

In the event that $\lambda_1(\mathbf{v}_0)$ and $\lambda_2(\mathbf{v}_0)$ are independent of $\mathbf{v}_0$ (except possibly for isolated points $\mathbf{v}_0$), we write λ_1 and λ_2 for $\lambda_1(\mathbf{v}_0)$ and $\lambda_2(\mathbf{v}_0)$, respectively. In that case we define the **Lyapunov dimension** of A_F by the formula

$$\dim_L A_F = 1 - \frac{\ln \lambda_1}{\ln \lambda_2} \tag{5.6}$$

The Lyapunov dimension was defined by Kaplan and Yorke (1978). By the definition of Lyapunov dimension and with the help of a computer, one can show that the Lyapunov dimension of the Hénon attractor satisfies $\dim_L A_H \approx 1.26$. Thus $\dim_L A_H \approx \dim_c A_H$.

Figure 5.31 depicts the attractor for the map F given by

$$F \begin{pmatrix} x \\ y \end{pmatrix} = \begin{pmatrix} x^2 - y^2 + 0.9x - 0.6013y \\ 2xy + 2x + 0.5y \end{pmatrix}$$

which has been studied by James Yorke and named the **Tinkerbell attractor**. By computer we find that $\dim_L A_F \approx 1.40$. Notice that the Tinkerbell attractor appears to be more complex than the Hénon attractor. This is borne out in its larger Lyapunov dimension.

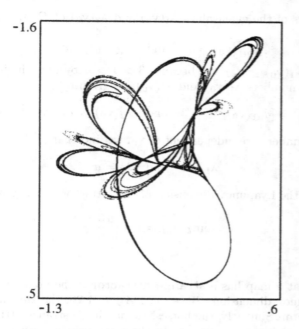

FIGURE 5.31
Tinkerbell attractor

Theoretical reasons why $\dim_L A_F$ should be approximately equal to $\dim_c A_F$ are discussed in Farmer, Ott and Yorke (1983). However, the Lyapunov dimension factors in the effect of iteration, as we suggested at the outset of the discussion.

Next, we will find the Lyapunov dimension of a map without the aid of a computer. It's worth doing this as an exercise at this point, though in practicality, you will eventually want to learn to write a program that will compute it for you.

Example 5.5.1. Let M be the horseshoe map discussed in Section 3.5. Find the Lyapunov dimension of the attractor A_M.

Solution. If $\mathbf{v}$ and $\mathbf{w}$ are in the attractor A_M and are very near to each other, then M shrinks the distance between $\mathbf{v}$ and $\mathbf{w}$ vertically by a factor $a < 1/3$, and expands the distance horizontally by a factor $b = 3$. Therefore

$$DM(\mathbf{v}) = \begin{pmatrix} a & 0 \\ 0 & 3 \end{pmatrix}$$

For each $\mathbf{v}$ in A_M, the eigenvalues of $DM(\mathbf{v})$ are a and 3. Now fix $\mathbf{v}_0$ in A_M. Then the iterates of $\mathbf{v}_0$ are also in A_M, so by () and results of Section 3.1, the absolute values of the eigenvalues $\lambda_1(\mathbf{v}_0)$ and $\lambda_2(\mathbf{v}_0)$ of $D_n M(\mathbf{v}_0)$ are given by

$$d_{n1}(\mathbf{v}_0) = 3^n \quad \text{and} \quad d_{n2}(\mathbf{v}_0) = a^n$$

Notice that $d_{n1}(\mathbf{v}_0) \geq d_{n2}(\mathbf{v}_0)$ because $3 > 1 > a$ by hypothesis. Therefore the Lyapunov numbers $\lambda_1(\mathbf{v}_0)$ and $\lambda_2(\mathbf{v}_0)$ are given by

$$\lambda_1(\mathbf{v}_0) = [d_{n1}(\mathbf{v}_0)]^{1/n} = 3 \quad \text{and} \quad \lambda_2(\mathbf{v}_0) = [d_{n2}(\mathbf{v}_0)]^{1/n} = a$$

Since these numbers are independent of $\mathbf{v}_0$, we find that

$$\lambda_1 = 3 \quad \text{and} \quad \lambda_2 = a$$

Finally, by () the Lyapunov dimension of the attractor A_M is given by

$$\dim_L A_M = 1 - \frac{\ln 3}{\ln a}$$

$\square$

We say that a map has a **strange attractor** if the attractor has a non-integer Lyapunov dimension. Since $\dim_L A_H \approx 1.26$, the Hénon map has a strange attractor. Similarly, the horseshoe map has a strange attractor whenever $(\ln 3)/(\ln a)$ is not an integer. In the same vein, a map has a **chaotic attractor** if the attractor has sensitive dependence on initial conditions or a Lyapunov number larger than 1. The Hénon map and the horseshoe map (with virtually any admissible value of a) have chaotic attractors. Although most attractors that are strange are also chaotic, Grebogi, Ott, Pelikan and Yorke (1984) exhibit attractors that are strange but not chaotic, and vice versa.

We end the section with three observations. The first is that except in special cases, we can only find a good approximation of the Lyapunov dimension by means of a computer.

The second observation is that the object of the Lyapunov dimension is to gather information as the iterates of a point run around the attractor. Thus there is no reason to expect that if $\mathbf{p}$ is a fixed point of the map F, then $\dim_L A_F(\mathbf{p})$ would equal $\dim_L A_F(\mathbf{v})$ for nonperiodic points $\mathbf{v}$ in A_F. For example, consider the Hénon map H, with $a = 1.4$ and $b = 0.3$. Then $\dim_L A_H(\mathbf{v}) \approx 1.26$ for practically every $\mathbf{v}$ in the attractor A_H, yet if $\mathbf{v}$ is the fixed point $\mathbf{p}$ on the attractor, then $\dim_L A_H(\mathbf{p})$ is *not* approximately 1.26 (Exercise 2). Thus when determining the Lyapunov dimension of an attractor, one should use a nonperiodic point whose orbit is expected to be dense in the attractor.

A third, more discomforting observation is that when one encounters an attractor that comes from observed data, then it may be very difficult to obtain a good measurement of the dimension for the attractor. This is especially true if the data is biological in nature.

Alan Garfinkel and several colleagues (1991) have studied electrocardiograms of cats under the influence of cocaine. Figure 5.32(a) displays a time

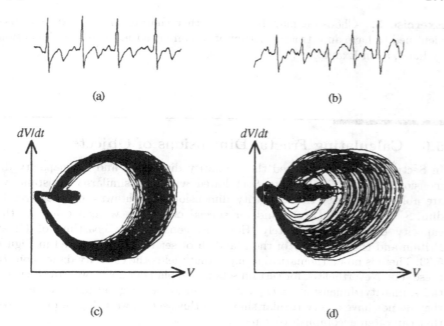

FIGURE 5.32
More cat brains on cocaine

series for the EKG before the cat ingested cocaine, and Figure 5.32(b) shows
the time series after ingestion. In Figures 5.32 (c) and (d) the EKG time series
is transformed into a "phase-plane plot," with the derivative of the voltage (as
a function of time) plotted against the voltage. The results are very interesting. The attractor on the left, which corresponds to the EKG before ingestion
of cocaine, appears much less dispersed than the one on the right, which corresponds to the EKG after ingestion. The fractal dimension of the attractor
on the left appears to be smaller than that for the attractor on the right. However, Garfinkel cautions that attempts to measure the Lyapunov dimension,
among other fractal dimensions, have given very imprecise information.

5.5.1   Section 5.5 Exercises

Exercise 1.   Let ϵ = the length of the horizontal sides of the rectangles
in Figure 5.29. Find an approximation to $\dim_c A_H$. How does your answer
compare with the one obtained at the outset of the section?

Exercise 2.   Let $\mathbf{p}$ be the fixed point of the Hénon map $H = H_{ab}$ given in
() and () of Section 3.4, respectively. Assume that $a = 1.4$ and $b = 0.3$. Show
that $\dim_L A_H(\mathbf{p})$ is not approximately 1.26.

Exercise 3. Choose a map from an earlier chapter of the text that was identified to have an attractor. Attempt to compute the Lyapunov dimension by hand, as in Example 1.

5.6   Calculating Fractal Dimensions of Objects

In Section 5.4 we introduced the similarity dimension and the capacity dimension. For self-similar sets S endowed with one similarity constant we are able to calculate the similarity dimension straightaway by the formula $\dim_s S = (\ln m)/(\ln(1/s))$, and for special other sets we can calculate the capacity dimension accurately. However, consider the coastline of Great Britain and the boundary of the Mandelbrot set that are depicted in Figure 5.33. There is no mathematical formula that yields the capacity dimensions of these sets. Nevertheless, for certain sets in the plane or space we can calculate the similarity dimension or capacity dimension with good accuracy even if they do not have a very regular shape. In this section we will give procedures that can often accomplish that result.

Recall that if $\dim_c S = D$, then $N(\epsilon)$ scales as $(1/\epsilon)^D$, so that for small values of ϵ,

$$\ln N(\epsilon) \approx D \ln(1/\epsilon)$$

As we showed in Section 5.4, if we plot $\ln N(\epsilon)$ against $\ln(1/\epsilon)$ for various small values of ϵ, where $\ln(1/\epsilon)$ runs along the horizontal axis and $\ln N(\epsilon)$ runs along the vertical axis, then the points $(\ln(1/\epsilon), \ln(N(\epsilon)))$ lie very near to a line with slope D. Thus if α and β are two distinct such small values of

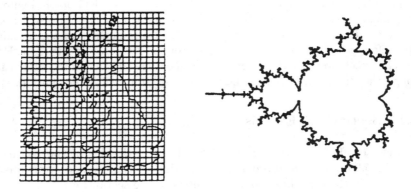

FIGURE 5.33
Coastline of Great Britain on a grid and the boundary of the Mandelbrot set

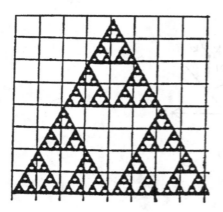
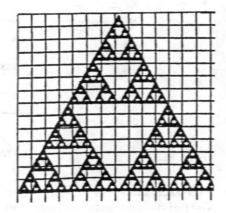

FIGURE 5.34
Estimating the dimension with meshes of 1/4 inch and 1/7 inch

ϵ, then

$$\frac{\ln N(\beta) - \ln N(\alpha)}{\ln 1/\beta - \ln 1/\alpha} \approx D$$

This means that if we plot the two points $(\ln(1/\alpha), \ln N(\alpha))$ and $(\ln(1/\beta), \ln N(\beta))$, then those points lie on a straight line whose slope should be essentially D if ϵ is very small and positive.

Let us test this method with the Sierpiński gasket, whose fractal dimension we know to be approximately 1.58. We place a grid of squares of side length 1/4 inch over the Sierpiński gasket and determine that 24 such squares appear to intersect the gasket. Then we perform the same operation with a grid of squares of side length 1/7 inch and find that 59 squares appear to intersect the gasket. (See Figure 5.34.) Using the formula above, we obtain

$$\frac{\ln N(1/4) - \ln N(1/7)}{\ln 1/(1/4) - \ln 1/(1/7)} = \frac{\ln 24 - \ln 59}{\ln 4 - \ln 7} \approx 1.61$$

Thus our estimate for the dimension of the gasket is 1.61, which is a very decent approximation to the real value of 1.58... (accurate to 2 places)!

In our estimates above, we counted each square that seemed to intersect the gasket nontrivially. However, sometimes a square may seem to touch an edge or a vertex of the given set, or perhaps it might not. Moreover, the number of squares needed to cover a set could depend on the exact placement of the squares. The upshot is that the calculations are predicated on judgment calls. You might count the squares that cover the gasket in the two parts of Figure 5.34 in order to see if you obtain the same results that we did.

Because two points determine a line, we can obtain an estimate for the capacity dimension of a set by using boxes of two different side lengths, as we did above, and determining the slope of the line that results. For more

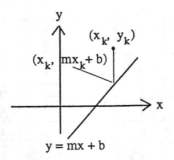

FIGURE 5.35
Vertical distance between point and line

definitive results we should make several sets of calculations, one each for pairs of different box sizes. In that way we could at least in theory be able to obtain more accurate information by comparing the results. For example, using a grid of squares of side length 1/3 inch overlaid on the gasket, we find the covering number to be 16. Comparing with the 1/7 inch and covering number of 59, we find that

$$\frac{\ln N(1/3) - \ln N(1/7)}{\ln[1/(1/3)] - \ln[1/(1/7)]} = \frac{\ln N(1/3) - \ln N(1/7)}{\ln 3 - \ln 7} \approx 1.54$$

This time our estimate 1.54 is not quite as close to the known dimension of 1.58 (accurate to 2 places).

How can we reconcile the difference in the results? In particular, if we produce n points in the plane from the box-counting method, how can we determine what is the most reasonable estimate for the dimension? The answer lies in what is called the "line of best fit," to be defined presently.

Suppose that $(x_1, y_1), (x_2, y_2), \ldots, (x_n, y_n)$ are distinct points in the plane, and let $y = mx + b$ be a line. One way of deciding which line best conforms to the n points is to take the sum of the squares of the vertical distances between the points and a given line. More precisely, let (x_k, y_k) be one of the points. Then its vertical distance to the line is $y_k - (mx_k + b)$ (Figure 5.35).

The measure of best fit for a line relative to the n points is given by

$$f(m, b) = \sum_{k=1}^{n} [y_k - (mx_k + b)]^2$$

The smaller $f(m, b)$ is for the set of points $(x_1, y_1), \ldots, (x_n, y_n)$, the better the approximation is. You can well imagine that there would be values m_0 and b_0 that minimize the value of $f(m, b)$. The line $y = m_0 x + b_0$ that results is called the **line of best fit**, or in statistics parlance, the **line of regression**. Figure 5.36 gives an example of the line of best fit. The important issue is that $m_0 = D$, the capacity dimension we seek.

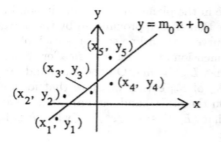

FIGURE 5.36
The line of best fit

One can determine the values of m_0 and b_0 by taking the partial derivative f_m of f (with respect to m) and the partial derivative of f_b of f (with respect to b), and setting them simultaneously equal to 0. The method we have just described is called the **method of least squares**.

Example 5.6.1. Find the line of best fit for the 3 points that have the form $(\ln(1/x_k), \ln N(x_k))$ when $x_1 = 1/7, x_2 = 1/4$ and $x_3 = 1/3$ for the Sierpiński gasket. Use it to estimate the capacity dimension of the Sierpiński gasket.

Solution. First, we calculate that the points are

$$(\ln 7, \ln 59) \approx (1.9459, 4.0775), \quad (\ln 4, \ln 24) \approx (1.3863, 3.1781), \text{ and}$$
$$(\ln 3, \ln 16) \approx (1.0986, 2.7726)$$

Then

$$f(m, b) \approx [4.0775 - (m(1.9459) + b)]^2 + [3.1781 - (m(1.3863) + b)]^2$$
$$+ [2.7726 - (m(1.0986) + b)]^2$$

Using either the partial derivatives or a linear regression statistical package, we find that the line of best fit is approximately $y = 1.549x + 1.055$. Thus the slope of the line, which is approximately 1.549, is the estimate of the capacity dimension of the Sierpiński gasket. $\square$

5.6.1 The Compass Dimension

Another measure of dimension that is particularly applicable to curves is the so-called compass dimension. Basically the idea is that we take a curve, which could be the graph of a function of a real variable, or the boundary of a shape, and fit a regular polygon that approximates the curve. We can use the fact that the lengths of the sides of a regular polygon are identical in order to find a different way of determining the dimension of the curve. Now we turn to the details.

Let C be a curve in the plane. From the discussion in the preceding paragraph, let us assume that C is approximated by the polygonal curve S_m, that is, by a polygonal curve with m line segments each of length s. We will analyze the capacity dimension of S_m for ever smaller values of s. Let the length of S_m be L_m, so that $L_m = ms$. On the one hand, assuming that the capacity dimension D_m of S_m exists, we know that for any positive integer k, $m^k = (1/s^k)^{D_m}$. On the other hand, suppose that there are constants $c > 0$ and $d_m \geq 0$ such that $(L_m)^k \approx c(1/s^k)^{d_m}$ when k is large. Together these features tell us that

$$c(\frac{1}{s^k})^{d_m} \approx (L_m)^k = m^k s^k = (\frac{1}{s^k})^{D_m} s^k = (\frac{1}{s^k})^{D_m-1}$$

Taking logarithms and rearranging, we deduce that

$$(D_m - 1)\ln\frac{1}{s} \approx \frac{\ln c}{k} + d_m \ln\frac{1}{s}$$

If k is large, then $(D_m - 1)\ln(1/s) \approx d_m \ln(1/s)$, which implies that $d_m \approx D_m - 1$.

It turns out that under normal conditions, as the length s of the sides of the polygons approximating the given curve C approach 0, the numbers d_m and D_m approach limits d and D, respectively. In that case, $d = D - 1$, or equivalently, $D = d + 1$.

Using the fact that $d_m = D_m - 1$ and $L_m = ms$, we find that

$$(L_m)^k \approx c(\frac{1}{s^k})^{D_m-1}$$

so that

$$\ln L_m \approx \frac{\ln c}{k} + d_m \ln\frac{1}{s}$$

If k increases without bound, the result is that

$$d_m \approx \frac{\ln L_m}{\ln(1/s)}$$

This leads us to make the following definition.

Definition 5.20. Let C be a curve in the plane, and let S_m be a regular polygon of m equal sides of length s and total length L_m that approximates C. Then the **compass dimension** $\dim_C S_m$ is given by

$$\dim_C S_m = 1 + d_m \approx 1 + \frac{\ln L_m}{\ln(1/s)}$$

One of the famous examples for the compass dimension involves rugged coastlines, like those of Great Britain or Norway or Maine. Peitgen et al (1992) have determined the following data for the coastline of Great Britain, where the "compass setting" refers to the length of the sides of the polygon that approximates the coastline:

compass setting	length
500 km	2600 km
100 km	3800 km
54 km	5770 km
17 km	8640 km

Using the line of best fit, or linear regression, one can calculate that the compass dimension is approximately 1.36.

5.6.2   Section 5.6 Exercises

Exercise 1. Approximate the capacity dimension of the Hénon attractor by counting boxes.

Exercise 2. Find the compass dimension of the Koch curve.

Exercise 3. Use the data for the coastline of Great Britain appearing in this section in order to approximate the compass dimension. Does your answer agree with that of Peitgen et al. (1986)?

Exercise 4. Approximate the compass dimension of the coastline of Maine.

6

Creating Fractal Sets

In Chapter 5 we introduced a number of exotic sets in $\mathbb{R}$, $\mathbb{R}^2$, and $\mathbb{R}^3$, including the self-similar Cantor set, Sierpiński set, and the Koch curve, as well as the space-filling Peano curve and Hilbert curve. A pertinent question is how to draw, or have the computer draw, approximations to those and other very complicated sets. In Chapter 6 we address this issue. In order to be able to give and understand algorithms for creating those sets, we study the notion of metric space in Section 6.1. Section 6.2 is devoted to the Hausdorff metric that is defined on closed and bounded sets of R and $\mathbb{R}^2$. In Section 6.3 we turn to special functions called contractions and affine functions that will be critical in order for the computer to produce fractal images. Section 6.4 presents a discussion of combinations of contractions called iterated function systems that are the major ingredients in the algorithms for creating fractal shapes that occupy Section 6.5. The main goal of the chapter is to understand the technique of creating fractal sets on the computer. We mention that many of the ideas in this chapter appeared in Barnsley's (1993) book *Fractals Everywhere*.

6.1 Metric Spaces

In this section we begin preparing the analysis that will tell us how mathematical functions can be used in order to generate fractals on the computer.

The first notion is that of distance, which we discussed in Section 2.4 with respect to sequences of 0's and 1's, and in Section 3.2 for vectors in $\mathbb{R}^2$. For the set of real numbers R, recall that the absolute value serves as the notion of distance:

$$|x - y| = \text{the distance between } x \text{ and } y$$

The absolute value has four fundamental properties:

i. $|x - y| \geq 0$

ii. $|x - y| = 0$ if and only if $x = y$

DOI: 10.1201/9781032678757-6

iii. $|x - y| = |y - x|$

iv. $|x - z| \leq |x - y| + |y - z|$

Properties (i)–(iii) indicate that the distance between two real numbers is always nonnegative, that the distance is 0 if and only if the two numbers x and y are identical, and that the distance between x and y is the same as the distance between y and x. Thus the first three properties follow immediately from the definition of absolute value. The fourth property, which is called a triangle property, can be proved in a straightforward manner (see Exercise 1).

Properties (i)–(iv) are the very features that we wish to generalize in order to be able to consider the "distance" between two sets, whether they be sets of real numbers, vectors, sequences, or other kinds of sets.

Definition 6.1. Let S be a set. Let $d : S \times S \to [0, \infty)$ be a function that assigns nonnegative numbers to pairs (x, y) of elements in S in such a way that

    i. $d(x, y) \geq 0$, for all x and y in S

    ii. $d(x, y) = 0$ if and only if $x = y$, for all x and y in S

    iii. $d(x, y) = d(y, x)$, for all x and y in S (**symmetry property**)

    iv. $d(x, z) \leq d(x, y) + d(y, z)$, for all x, y, and z in S (**triangle inequality**)

Then d is called a **metric** on the set S, and we say that the pair (S, d) is a **metric space**.

By the definition of metric and the properties of absolute value given in (i)–(iv), $(R, |\cdot|)$ is a metric space. The usual distance, or metric, d between points in $\mathbb{R}^2$ is given by

$$d((x_1, x_2), (y_1, y_2)) = \sqrt{(y_1 - x_1)^2 + (y_2 - x_2)^2}, \quad \text{for all } (x_1, x_2) \text{ and } (y_1, y_2) \text{ in } \mathbb{R}^2$$

The metrics just given for R and $\mathbb{R}^2$ are often referred to as the **Euclidean metric** on R and $\mathbb{R}^2$, respectively.

For the Euclidean metric on $\mathbb{R}^2$, properties (i)–(iii) are apparent. The fact that the triangle inequality holds follows directly from the relationship between the sides of a triangle (see the left triangle in Figure 6.1). In fact, the expression "triangle inequality" derives from the comparative lengths of the sides of a triangle. One can define the Euclidean metric analogously for $\mathbb{R}^3$, and in fact, for $\mathbb{R}^n$, for $n \geq 1$.

Because the Euclidean metric is a two-dimensional version of the absolute value metric on R, we normally represent it in the following way:

$$d(\mathbf{x}, \mathbf{y}) = ||\mathbf{x} - \mathbf{y}|| \text{ for all } \mathbf{x} \text{ and } \mathbf{y} \text{ in } \mathbb{R}^2$$

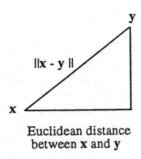

Euclidean distance
between **x** and **y**

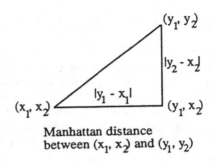

Manhattan distance
between (x_1, x_2) and (y_1, y_2)

FIGURE 6.1
Illustration of different metrics

Thus if $\mathbf{x} = (x_1, x_2)$ and $\mathbf{y} = (y_1, y_2)$, then

$$||\mathbf{x} - \mathbf{y}|| = \sqrt{(y_1 - x_1)^2 + (y_2 - x_2)^2}$$

The nonnegative number $||\mathbf{x} - \mathbf{y}||$ is referred to as the **distance** between **x** and **y**.

Another metric in $\mathbb{R}^2$ is the so-called **Manhattan metric**, which is defined to be $||\cdot||_1$, given by

$$||\mathbf{x} - \mathbf{y}||_1 = |y_1 - x_1| + |y_2 - x_2|, \text{ for } \mathbf{x} = (x_1, x_2) \text{ and } \mathbf{y} = (y_1, y_2) \text{ in } \mathbb{R}^2$$

(See the right triangle in Figure 6.1.) One reason why this is referred to as the Manhattan metric is that the vast majority of the streets in the main part of Manhattan, New York, run north-south or east-west, so that if one wishes to get from point A to point B in that part of Manhattan, then one can travel only in East-West or North-South directions, so the "distance" the person must travel is effectively the Manhattan distance between A and B.

From the properties of distance in $\mathbb{R}^2$, it is clear that

$$||\mathbf{x} - \mathbf{y}|| \le ||\mathbf{x} - \mathbf{y}||_1 \text{ for all } \mathbf{x} \text{ and } \mathbf{y} \text{ in } \mathbb{R}^2$$

However, we can say even more:

$$||\mathbf{x} - \mathbf{y}|| \le ||\mathbf{x} - \mathbf{y}||_1 \le \sqrt{2} \, ||\mathbf{x} - \mathbf{y}|| \text{ for all } \mathbf{x} \text{ and } \mathbf{y} \text{ in } \mathbb{R}^2$$

(see Exercise 2). This means that although the Euclidean distance between two points in $\mathbb{R}^2$ is no greater than the Manhattan distance, the Manhattan distance is no greater than a constant multiple (namely $\sqrt{2}$) of the Euclidean metric. Thus in some sense the Euclidean distance and the Manhattan distance are compatible with one another. That leads us to the notion of equivalent metrics.

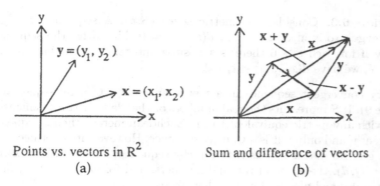

Points vs. vectors in R²   Sum and difference of vectors
(a)   (b)

FIGURE 6.2
Arithmetic with vectors

Definition 6.2. Let d and D be two metrics on a space S. Then d and D are **equivalent** if there are positive constants k and K such that

$$k\, d(x,y) \leq D(x,y) \leq K\, d(x,y) \text{ for all } x \text{ and } y \text{ in } S$$

A consequence of the definition and the remarks preceding it is that the Euclidean metric and the Manhattan metric on $\mathbb{R}^2$ are equivalent to one another. However, there are plenty of metrics on $\mathbb{R}^2$ that are *not* equivalent to the Euclidean metric. (See Exercises 6–7.)

We have represented points $\mathbf{x}$ and $\mathbf{y}$ in the plane as pairs of real numbers, that is, $\mathbf{x} = (x_1, x_2)$ and $\mathbf{y} = (y_1, y_2)$. Another way of representing $\mathbf{x}$ and $\mathbf{y}$ is as directed line segments, or vectors. In this way, we can consider $\mathbf{x} = (x_1, x_2)$ to be the directed line segment from the origin $(0,0)$ to the point (x_1, x_2), and similarly for $\mathbf{y} = (y_1, y_2)$ (see Figure 6.2(a)). There is a one-to-one correspondence between pairs of real numbers and vectors starting at the origin. Normally we associate all vectors that have the same length and direction. Thus the two vectors denoted by $\mathbf{x}$ in Figure 6.2(b) represent the same vector, as do the two vectors denoted by $\mathbf{y}$. This representation allows us to graphically display the sum $\mathbf{x} + \mathbf{y}$ and difference $\mathbf{x} - \mathbf{y}$ of two (non-zero) vectors (see Figure 6.2(b)).

We comment that if (S, d) is a metric space and if $T \subseteq S$, then (T, d) is also a metric space. For example, the set Q of all rational numbers is a metric space under the absolute value metric.

6.1.1   Complete Metric Spaces

The notions of convergent sequences and Cauchy sequences, defined for real numbers and points in the plane in Section 5.3, have their counterparts in general metric spaces.

Definition 6.3. Consider the metric space (S, d). A sequence $(x_n)_{n=1}^{\infty}$ in S **converges** to x in S if $\lim_{n \to \infty} d(x_n, x) = 0$. Moreover, the sequence is **Cauchy** if for each $\epsilon > 0$ there is a positive integer N such that whenever $n, m \geq N$, we have $d(x_n, x_m) < \epsilon$.

Every convergent sequence in a metric space is a Cauchy sequence (see Exercise 9). If S represents the real numbers or the plane or three-dimensional space, with any metric equivalent to the Euclidean metric, then a sequence is convergent if and only if it is a Cauchy sequence. However, it is not necessarily true for general metric spaces that Cauchy sequences converge. For example, consider (Q, d), the space of rational numbers, and for each n, let x_n denote the first n decimal places of π. In other words,

$$x_1 = .1, \quad x_2 = .14, \quad x_3 = .141, \text{ etc.}$$

Then $\{x_n\}_{n=1}^{\infty}$ is evidently a Cauchy sequence in Q, and converges to π, which is *not* in Q. It follows that there are sequences in Q that are Cauchy but are *not* convergent. A metric space S that has the property that each Cauchy sequence converges to a member of S qualifies for the name "complete."

Definition 6.4. The metric space (S, d) is **complete** if every Cauchy sequence in S converges to a member of S.

In different terminology, Theorems 5.9 and 5.10 show that both the real numbers and the plane with the usual metrics are complete. Similarly one can show that R^3 under the Euclidean metric is complete. A subset T of a metric space (S, d) is **closed** if whenever x is the limit of a sequence $(x_n)_{n=1}^{\infty}$ of members of T, then x actually is in T. In other words, a subset T of S is closed if it contains all of its limit points. For example, the interval $[0, 1]$ is closed in R, whereas the interval $(0, 1)$ is not closed because the sequence $(1/n)_{n=1}^{\infty}$ lies in $(0, 1)$ but converges to 0, which is not in $(0, 1)$. Likewise, the set Q is not closed in R (see Exercise 11).

For future reference we prove the following theorem that relates complete subspaces of a metric space with closed subspaces.

Theorem 6.5. Let S be a complete metric space, and let $T \subseteq S$.

    a. If T is closed, then T is complete.

    b. If T is complete, then T is closed.

Proof. To prove (a), let $(x_n)_{n=1}^{\infty}$ be a Cauchy sequence in T. Since S is complete by hypothesis, it follows that the sequence converges to an element x in S. Since T is closed, T contains all its limit points, so x is in T. Therefore T is complete.

To prove (b), let $(x_n)_{n=1}^{\infty}$ be a sequence in T that converges to an element x in S. Convergent sequences in S are Cauchy, so the sequence is a Cauchy sequence in S, and hence is a Cauchy sequence in T. Since T is assumed complete, this means that the limit x lies in T. Consequently T is closed. $\square$

A nonempty subset S of R^2 is **bounded** provided that there is a positive number M such that $\|\mathbf{x}\| \le M$ for all $\mathbf{x}$ in S. Sets in R or $\mathbb{R}^2$ that are *both* closed *and* bounded are called **compact**, and will be important to us in this chapter. For example, closed intervals and convergent sequences with their limits are compact in R, whereas closed disks and ellipses and line segments and finite sets are among the compact sets in $\mathbb{R}^2$. By contrast, the horizontal axis and open disks are not compact in $\mathbb{R}^2$. The notions of boundedness and compactness are analogous in R^n for any positive integer $n \ge 3$, when these spaces are endowed with their own Euclidean metrics.

The following theorem relates compactness in R with convergence of certain sequences in R.

Theorem 6.6. Let S be a subset of R. Then S is compact if and only if each sequence in S has a subsequence that converges to an element of S.

Proof. On the one hand, suppose that S is compact (i.e., closed and bounded), and let $(x_n)_{n=1}^{\infty}$ be a sequence in S. We will define a subsequence $(z_n)_{n=1}^{\infty}$ that is guaranteed to be Cauchy and converges to a number in S. We do it by using the bisection method algorithm:

Since S is bounded, it is contained in an interval $I_0 = [a, b]$. There are infinitely many members of the sequence $(x_n)_{n=1}^{\infty}$ in either the left closed half or the right closed half of I_0. Call such a half I_1, and pick z_1 to be an element x_{n_1} of the sequence that lies in I_1. Next, there are infinitely many members of the subsequence $(x_n)_{n=n_1+1}^{\infty}$ in either the left closed half or the right closed half of I_1. Call such a half I_2, and pick z_2 to be an element of the subsequence that lies in I_1. Continue the process indefinitely to obtain a subsequence $(z_n)_{n=1}^{\infty}$ of the original sequence. Notice that $(z_n)_{n=1}^{\infty}$ is Cauchy because the nth member is in I_n, which has length $(b - a)/2^n$. Since S is compact, it is closed in R and hence complete by Theorem 6.5, so $(z_n)_{n=1}^{\infty}$ converges to an element of S. Thus half of the proof is complete.

On the other hand, we assume that each sequence in S has a subsequence that converges to an element of S. We will show that this implies that S is closed and bounded. To show that S is closed, suppose that $(x_n)_{n=1}^{\infty}$ is a sequence in S that converges to a number x in R. By hypothesis there is a subsequence that converges to a number z in S. Evidently $x = z$, so x is in S. Thus S is closed. If S were *not* bounded, then there would be a sequence $(x_n)_{n=1}^{\infty}$ in S such that $|x_n| > n$ for all n. But then no subsequence could converge, which contradicts the assumption. Therefore S is bounded. This completes the proof of the theorem. $\square$

Note: Theorem 6.6 is sometimes referred to as the **Bolzano-Weierstrass Theorem**, after the Bohemian mathematician-priest Bernhard Bolzano (1781–1848) and the great German mathematician Karl Weierstrass (1815–1897), who started out as a secondary high school teacher and later as a famous college teacher who brought precision and rigor to the area of mathematical analysis. Weierstrass was mentioned earlier, for creating a pathological monster. His Weierstrass function is often mentioned in first-year calculus courses to illustrate that not all continuous functions are differentiable.

The two-dimensional version of Theorem 6.6 is also valid:

Corollary 6.7. Let S be a subset of $\mathbb{R}^2$. Then S is compact if and only if each sequence in S has a subsequence that converges to an element of S.

Proof. See Exercise 12.   □

Finally, we present two features of compact sets in $\mathbb{R}^2$ that will be needed in Section 6.4.

Theorem 6.8.

   (a) A finite union of compact sets in $\mathbb{R}^2$ is also compact.

   (b) Let S be compact in $\mathbb{R}^2$, and let $T : \mathbb{R}^2 \to \mathbb{R}^2$ be continuous. Then $T(S)$ is compact.

Proof. To prove (a), let A and B be compact in $\mathbb{R}^2$, and let $S = A \cup B$. If $\{x_n\}_{n=1}^{\infty}$ is any sequence in S, then there is a subsequence $\{x_{n_k}\}_{k=1}^{\infty}$ that is totally in A, or is totally in B. Without loss of generality, assume that the subsequence is in A. Since A is assumed to be compact, Corollary 6.7 asserts that there is a sub-subsequence converging to an element of A. This means that there is a subsequence of the original sequence that converges to an element of A. It follows that S is compact. With only minor modifications, one can prove that any finite union of compact sets in $\mathbb{R}^2$ is compact.

To prove (b), let S be compact, and $\{y_n\}_{n=1}^{\infty}$ a sequence in $T(S)$. Then there is a sequence $\{x_n\}_{n=1}^{\infty}$ in S such that $T(x_n) = y_n$ for each n. Since S is compact, there is a subsequence $\{x_{n_k}\}_{k=1}^{\infty}$ that converges to an element z in S. But then, because T is assumed to be continuous, $\{T(x_{n_k})\}_{k=1}^{\infty} = \{y_{n_k}\}_{k=1}^{\infty}$ converges to $T(z)$ in $T(S)$. Thus $T(S)$ is compact.   □

6.1.2   Section 6.1 Exercises

Exercise 1. Prove that the absolute value on R satisfies the triangle inequality.

Exercise 2.   Prove that on $\mathbb{R}^2$, the Euclidean metric and the Manhattan metric are equivalent.

Exercise 3. Find a value of y such that the Manhattan and Euclidean distances from the origin to $(1, y)$ satisfy

$$||(0,0) - (1,y)||_1 = \frac{5}{4} ||(0,0) - (1,y)||$$

Exercise 4. Let $\mathbf{x}$ and $\mathbf{y}$ be arbitrary points in R^2.

(a) Show that

$$||\mathbf{x} + \mathbf{y}||^2 + ||\mathbf{x} - \mathbf{y}||^2 = 2||\mathbf{x}||^2 + 2||\mathbf{y}||^2$$

(b) What property of parallelograms does the equation in (a) tell us?

Exercise 5. Let $\mathbf{x} = (x_1, x_2)$ and $\mathbf{y} = (y_1, y_2)$. Define $||\mathbf{x} - \mathbf{y}||_\infty$ on R^2 by

$$||\mathbf{x} - \mathbf{y}||_\infty = \text{the maximum of } |y_1 - x_1| \text{ and } |y_2 - x_2|$$

(a) Show that the formula defines a metric on R^2.

(b) Describe $||\mathbf{x} - \mathbf{y}||_\infty$ in geometric terms.

(c) Is the metric $||\mathbf{x} - \mathbf{y}||_\infty$ equivalent to the Euclidean metric on R^2? Explain why it is, or why it is not.

Exercise 6. Let $d(x, y) = |x^3 - y^3|$ for all real numbers x and y. Show that d represents a metric on R, and that it is *not* equivalent to the absolute value metric on R.

Exercise 7. Let $d(x, y) = |\sqrt{x} - \sqrt{y}|$ for all real numbers x and y. Show that d represents a metric on R, and that it is *not* equivalent to the absolute value metric on R.

Exercise 8. Let $d(x, y) = |xy|$ for all real numbers x and y. Show that (R, d) is *not* a metric space with this particular d.

Exercise 9. Prove that every convergent sequence in a metric space is a Cauchy sequence.

Exercise 10. Prove that every Cauchy sequence is bounded.

Exercise 11. Prove that a sequence is Cauchy if and only if it is convergent.

Exercise 12. Let $\{x_n\}$ be a sequence, and a a positive real number. Suppose for some $0 < r < 1$, and some chosen metric, we have

$$d(x_n, x_{n+1}) < ar^n$$

for all $n > 0$. Prove that this sequence is Cauchy.

Exercise 13. Show that $(\mathbb{R}^2, ||\cdot||)$ is a complete metric space.

Exercise 14. Show that the set Q of rationals is not closed in R.

Exercise 15. Prove Corollary 6.7. (*Hint:* For the first part, notice that S is contained in a suitably large square in the plane. Split the square into four congruent sub-squares, and use an argument analogous to that employed in the first part of the proof of Theorem 6.6. The second part of the proof is quite like its counterpart in the proof of Theorem 6.6.)

Exercise 16.

(a) Show that the finite union of compact subsets of $\mathbb{R}^2$ is compact.

(b) Find an example of a sequence of compact sets of $\mathbb{R}^2$ such that its union is *not* compact.

Exercise 17. Show that the countable intersection of compact subsets of $\mathbb{R}^2$ is compact.

6.2   The Hausdorff Metric

In Section 6.1 we discussed several metrics, including the absolute value on R, and the Euclidean and Manhattan metrics on R^2. In the present section we define the distance between pairs of nonempty compact (i.e., closed and bounded) subsets of $\mathbb{R}^2$. Informally, the distance between two compact sets will be considered small if the sets are nearly identical, that is, if every point in one compact set is near to a point of the other compact set, and vice versa.

The process of defining the Hausdorff metric has several stages. To begin the process, let $\mathcal{K}$ denote the collection of all compact subsets of $\mathbb{R}^2$. Thus $\mathcal{K}$ contains each set with only a finite number of elements, all bounded closed intervals, all bounded, closed disks and rectangles, and many other sets in the plane.

First, we will define the distance from a point to a compact set in $\mathbb{R}^2$.

Definition 6.9. Let B be a nonempty member of $\mathcal{K}$, and let $\mathbf{v}$ be any point in $\mathbb{R}^2$. We define the **distance** from $\mathbf{v}$ to B by the formula

$$d(\mathbf{v}, B) = \text{the minimum value of the distance } ||\mathbf{v} - \mathbf{b}||, \text{ for all } \mathbf{b} \text{ in } B$$

Unfortunately there is nothing that appears to automatically guarantee that the minimum value of distances in Definition 6.9 exists. Our first result rectifies that situation, thanks to the fact that by assumption B is compact.

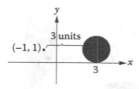

FIGURE 6.3
Distance from $(-1, 1)$ to the disk

Theorem 6.10. $d(\mathbf{v}, B)$ exists for every $\mathbf{v}$ in R^2 and every member B of $\mathcal{K}$.

Proof. Let $\mathbf{v}$ be in R^2 and B in $\mathcal{K}$. Define p by the formula

$$p = \text{the largest number} \leq \|\mathbf{v} - \mathbf{b}\| \text{ for all } \mathbf{b} \text{ in } B$$

Then p exists and is a nonnegative number. By the definition of p, there is a sequence

$$(\mathbf{b}_n)_{n=1}^{\infty} \subseteq B$$

such that

$$p \leq \|\mathbf{v} - \mathbf{b}_n\| < p + \frac{1}{n} \text{ for all } n$$

Since B is compact, Corollary 6.7 tells us that there is a subsequence $(\mathbf{b}_{n_k})_{k=1}^{\infty}$ that converges to an element $\mathbf{b}_\infty$ in B. Consequently the sequence $\|\mathbf{v} - \mathbf{b}_{n_k}\|$ converges to $\|\mathbf{v} - \mathbf{b}_\infty\|$ and also converges to p. It follows that $d(\mathbf{v}, B)$ exists, and

$$d(\mathbf{v}, B) = p = \|\mathbf{v} - \mathbf{b}_\infty\|$$

That completes the proof.

□

Let us compute the distance between a point and a compact set by choosing B to be the closed, filled-in disk of radius 1 centered at the point $(3, 1)$ in R^2 (Figure 6.3). If $\mathbf{v} = (-1, 1)$, then it follows that $d(\mathbf{v}, B) = 3$ because the point in B closest to $(-1, 1)$ is $(2, 1)$, which is at a distance of 3 from $(-1, 1)$ (Figure 6.3).

By the definition, $d(\mathbf{v}, B)$ is larger or smaller, depending on the distance from $\mathbf{v}$ to the nearest point of B. In particular, since B is compact, $d(\mathbf{v}, B) = 0$ if and only if $\mathbf{v}$ is in B (Exercise 1).

Next, we define the distance from one compact set to another compact set.

Definition 6.11. Let A and B be members of $\mathcal{K}$. Then the **distance from A to B**, denoted $d(A, B)$, is defined by

$$d(A, B) = \text{the maximum value of the set of numbers } d(\mathbf{a}, B), \text{ for } \mathbf{a} \text{ in } A$$

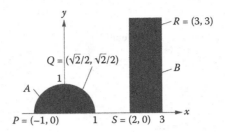

FIGURE 6.4
Distance between compact sets

The fact that the maximum of the numbers does, in fact, exist can be proved in a manner similar to the proof of Theorem 6.10. (See Exercise 2.)

For any two compact sets A and B, in order to find the distance $d(A, B)$ *from A to B*, we need to locate a point in A that has maximal distance from *any* point in B, and then compute the distance from that point to B. For example, let A be the half unit disk centered at the origin, and B the filled-in rectangle of base length 1 and height 3, as in Figure 6.4. To find $d(A, B)$, we first notice that the point on the disk A that is farthest away from B is $P = (-1, 0)$. Now the distance between P and B is evidently the distance from P to S, which is 3. Thus $d(A, B) = 3$.

Next, we will determine the distance $d(B, A)$ from B to A, for A and B in the preceding paragraph. Observe that the point of the rectangle B that is farthest away from the half disk A is the point $R = (3, 3)$. Its distance to the half disk is the distance between $R = (3, 3)$ and the point $Q = (\sqrt{2}/2, \sqrt{2}/2)$. By using the distance formula in the plane, we find that distance to be $\sqrt{19 - 6\sqrt{2}} \approx 3.24$. Thus $d(A, B) = 3 \neq d(B, A) \approx 3.24$. The fact that $d(A, B) \neq d(B, A)$ is not unusual. In fact, the distance from a compact set A to another compact set B is normally different from the distance from B to A. As a result, $d(A, B)$ does not represent a metric on K because it does not satisfy the symmetry relation (iii), which is $d(A, B) = d(B, A)$. (There are very special conditions under which $d(A, B) = d(B, A)$. See Exercise 8.)

Although the distance $d(A, B)$ is not a metric, it does satisfy the metric property (i), which is that $d(A, B) \geq 0$. Also, it satisfies part of property (ii), as we see next.

Theorem 6.12. If $A = B$, then $d(A, B) = 0$ and $d(B, A) = 0$.

Proof. This is obvious from the definition of $d(A, B)$ and $d(B, A)$.   □

Theorem 6.13. If $d(A, B) = 0$ then $A \subseteq B$.

Proof. If $d(A, B) = 0$, then $d(\mathbf{a}, B) = 0$ for each $\mathbf{a}$ in A. Consider an arbitrary such $\mathbf{a}$. By definition, if $d(\mathbf{a}, B) = 0$ then there is a sequence $(\mathbf{b}_n)_{n=1}^{\infty}$ in B such that $\lim_{n\to\infty} d(\mathbf{a}, \mathbf{b}_n) = 0$. Since B is compact, there is a subsequence converging to an element **b** of B. Evidently $d(\mathbf{a}, \mathbf{b}) = 0$, that is, $\mathbf{a} = \mathbf{b}$, so that **a** is in B. $\qquad\square$

Next, we prove that $d(A, B)$ satisfies property (iv) of metrics, which is the triangle inequality.

Theorem 6.14. For any A, B, and C in $\mathcal{K}$,

$$d(A, B) \leq d(A, C) + d(C, B)$$

Proof. Let A and B be fixed compact sets, and let **a** be arbitrary in A. Then since d satisfies the triangle inequality for single elements,

$$d(\mathbf{a}, B) \leq d(\mathbf{a}, \mathbf{b}) \leq d(\mathbf{a}, \mathbf{c}) + d(\mathbf{c}, \mathbf{b}) \text{ for all } \mathbf{b} \text{ in } B \text{ and all } \mathbf{c} \text{ in } C$$

By the definition of $d(\mathbf{c}, B)$ and $d(C, B)$, this means that

$$d(\mathbf{a}, B) \leq d(\mathbf{a}, \mathbf{c}) + d(\mathbf{c}, B) \leq d(\mathbf{a}, \mathbf{c}) + d(C, B) \text{ for all } \mathbf{c} \text{ in } C$$

Continuing in the same way, we find that

$$d(\mathbf{a}, B) \leq d(\mathbf{a}, C) + d(C, B) \leq d(A, C) + d(C, B)$$

Since **a** is arbitrary in A, we conclude that $d(A, B) \leq d(A, C) + d(C, B)$. $\qquad\square$

Because $d(A, B)$ lacks only the symmetry property, we can use it to create a metric on $\mathcal{K}$, named after the 20th century mathematician Felix Hausdorff.

Definition 6.15. Let A and B be members of $\mathcal{K}$. The **Hausdorff metric** on $\mathcal{K}$ is defined by the formula

$$D(A, B) = \text{the maximum value of } d(A, B) \text{ and } d(B, A)$$

To show that $D(A, B)$ yields a metric, we note the following:

i. $D(A, B) \geq 0$ for all A and B in K. This is true because D is derived from d, which is always nonnegative.

ii. $D(A, B) = 0$ if and only if $A = B$: If $d(A, B) = 0$, then $A \subseteq B$ by Theorem 6.13. Next, if $d(B, A) = 0$, then $B \subseteq A$ by Theorem 6.13 again. Thus if $D(A, B) = 0$, which means that $d(A, B) = 0 = d(B, A)$, then $A = B$. For the converse, observe that if $A = B$, then $D(A, B) = 0$, by Theorem 6.12.

iii. $D(A, B) = D(B, A)$. This is true by the definition of D.

iv. $D(A, C) \leq D(A, B) + D(B, C)$. We can see this by referring to the definition of D and Theorem 6.14. Indeed, notice that by definition, $D(A, B)$ = either $d(A, B)$ or $d(B, A)$, whichever is the larger. On the one hand, if $D(A, B) = d(A, B)$, then by Theorem 6.14,

$$D(A, B) = d(A, B) \leq d(A, C) + d(C, B) \leq D(A, C) + D(C, B)$$

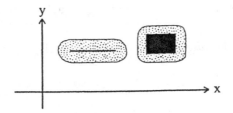

FIGURE 6.5
Corona of a line segment and of a rectangle

On the other hand, if $D(A, B) = d(B, A)$, then again by Theorem 6.14, along with property (iii) above,

$$D(A, B) = d(B, A) \leq d(B, C) + d(C, A) \leq D(B, C) + D(C, A)$$
$$= D(A, C) + D(C, B)$$

Thus (iv) is proved.

In summary, we have the following theorem.

Theorem 6.16. The Hausdorff distance D is a metric on $\mathcal{K}$.

For example, let A and B be as in Figure 6.4. Since $d(A, B) = 3$ and $d(B, A) = \sqrt{19 - 6\sqrt{2}} \approx 3.24$, it follows from the definition of the Hausdorff metric that $D(A, B) = \sqrt{19 - 6\sqrt{2}}$, which is the larger of the two distances $d(A, B)$ and $d(B, A)$.

Next, let A be a member of $\mathcal{K}$, and r a positive number. In the following discussion we will use the notation $A * r$ for the set defined by

$$A * r = \text{the collection of } \mathbf{v} \text{ in } \mathbb{R}^2 \text{ such that } ||\mathbf{v} - \mathbf{a}|| \leq r \text{ for some } \mathbf{a} \text{ in } A$$

The set $A * r$ is composed of all points whose distance from some member of A is no greater than r. If r is small, then you might think of $A * r$ as a kind of "corona" of A. Figure 6.5 shows two examples of A and $A * r$: a line segment with its corona, and a rectangle with its shaded corona.

Theorem 6.17. If B is compact, then so is $B * r$ for any $r > 0$. (In other words, if B is in $\mathcal{K}$, then $B * r$ is also in $\mathcal{K}$, for any $r > 0$.)

Proof. On the one hand, B is compact, so is bounded. If B lies inside a disk of radius M about the origin in $\mathbb{R}^2$, then $B * r$ automatically lies inside a like disk of radius $M + r$. Therefore $B * r$ is bounded. On the other hand, to show that $B * r$ is closed, suppose that $(\mathbf{v}_n)_{n=1}^{\infty}$ is a sequence in $B * r$ that converges to a point $\mathbf{v}$ in $\mathbb{R}^2$. Then by the definition of $B * r$ there is a sequence $(\mathbf{b}_n)_{n=1}^{\infty}$

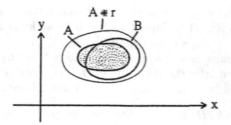

FIGURE 6.6
Relationship between A, B, and $A * r$

in B such that $\|\mathbf{v}_n - \mathbf{b}_n\| \leq r$. Since B is compact, there is a subsequence $(\mathbf{b}_{n_k})_{k=1}^{\infty}$ converging to an element $\mathbf{b}$ in B. For any $\epsilon > 0$, if k is large enough, then

$$\|\mathbf{v} - \mathbf{b}\| \leq \|\mathbf{v} - \mathbf{v}_{n_k}\| + \|\mathbf{v}_{n_k} - \mathbf{b}_{n_k}\| + \|\mathbf{b}_{n_k} - \mathbf{b}\| \leq \epsilon + r + \epsilon$$

As a result, $\|\mathbf{v} - \mathbf{b}\| \leq r$, so that $\mathbf{v}$ is in $B * r$, and thus $B * r$ is closed. Consequently $B * r$ is compact. $\qquad \square$

Next, we give a criterion in terms of the "corona" of a set that tells us when two sets A and B are close together in the Hausdorff metric. The criterion rests on the set inclusions $A \subseteq B * r$ and $B \subseteq A * r$ (see Figure 6.5).

Theorem 6.18. Let $r > 0$. Then $D(A, B) \leq r$ if and only if $A \subseteq B * r$ and $B \subseteq A * r$.

Proof. Assume that $D(A, B) \leq r$, and let $\mathbf{a}$ be in A. Since $d(A, B) \leq D(A, B) \leq r$, we know that $d(\mathbf{a}, B) \leq r$. Since B is compact, there is a point $\mathbf{b}_0$ in B such that $\|\mathbf{a} - \mathbf{b}_0\| \leq r$. Consequently $\mathbf{a}$ is in $B * r$. Therefore $A \subseteq B * r$. By the definition of the metric D, we have $D(A, B) = D(B, A)$, so we also have $B \subseteq A * r$.

Conversely, assume that $A \subseteq B * r$ and $B \subseteq A * r$. If $\mathbf{a}$ is any element of A, then automatically $\mathbf{a}$ is in $B * r$, so that $d(\mathbf{a}, B) \leq r$. Since $\mathbf{a}$ is arbitrary in A, $D(A, B) \leq r$. Similarly, $d(\mathbf{b}, A) \leq r$ for any $\mathbf{b}$ in B, so that $D(B, A) \leq r$. $\square$

Figure 6.6 shows the relationship between sets A, $A * r$, and B, in which $B \subseteq A * r$.

Finally, we prove that the set $\mathcal{K}$ of all compact subsets of $\mathbb{R}^2$ is a *complete* metric space under the Hausdorff metric D. The proof is technical, but gives us some insights into the distances between compact subsets of $\mathbb{R}^2$.

To begin with, let the pair $(\mathcal{K}, D)$ denote the set $\mathcal{K}$ of compact sets in $\mathbb{R}^2$, endowed with the Hausdorff metric D. Before we prove the completeness of the metric space $(\mathcal{K}, D)$, let us determine what it means for a sequence of compact sets to be a Cauchy sequence in $(\mathcal{K}, D)$. Suppose that $(A_n)_{n=1}^\infty$ is a sequence of compact sets in $\mathbb{R}^2$. Then from Theorem 6.18 we know that $D(A_n, A_m) < \epsilon$ provided that $A_n \subseteq A_m * \epsilon$, and similarly, $A_m \subseteq A_n * \epsilon$. In other words, the sets A_n and A_m must nearly coincide. It follows that the sequence is Cauchy provided that eventually the members of the sequence nearly coincide.

Now we are ready to prove that $\mathcal{K}$ is a complete metric space under D.

Theorem 6.19. $(\mathcal{K}, D)$ is a complete metric space.

Proof. Let $(A_n)_{n=1}^\infty$ be a sequence of compact sets that is Cauchy in the Hausdorff metric. Define A by

$A =$ the collection of **a** that have the following property: there is a sequence $(\mathbf{a}_n)_{n=1}^\infty$ in R^2 such that $\mathbf{a}_n$ is in A_n for all n and a subsequence of $(\mathbf{a}_n)_{n=1}^\infty$ converges to **a**.

To show that A is nonempty: Let $(\mathbf{a}_n)_{n=1}^\infty$ be a sequence such that $\mathbf{a}_n$ is in A_n for each n. Since $(A_n)_{n=1}^\infty$ is a Cauchy sequence, it follows that for any $r > 0$, there is an N such that $\mathbf{a}_n$ is in $A_N * r$ for all $n \geq N$. However, $A_N * r$ is compact by Theorem 6.17, so there is a subsequence of $(\mathbf{a}_n)_{n=1}^\infty$ converging to a point **a**. By the definition of A, **a** is in A, so A is not empty.

To show that A is compact: A is bounded since it lies inside $A_N * r$ for any positive r and some large enough N depending on r, and since A_N is compact. Next, A is closed because if a sequence $(\mathbf{v}_n)_{n=1}^\infty$ in A converges to some **v** in $\mathbb{R}^2$, then first one can find a sequence $(\mathbf{a}_{n_k})_{k=1}^\infty$ such that $\mathbf{a}_{n_k}$ lies in A_{n_k} and also $||\mathbf{v}_{n_k} - \mathbf{a}_{n_k}||$ approaches 0 as k increases. Since $(\mathbf{a}_{n_k})_{k=1}^\infty$ lies in a compact set, it has a convergent subsequence, which can be shown also to converge to **v**. Thus **v** is in A.

To show that $D(A_n, A)$ converges to 0 as n increases without bound: Let ϵ be an arbitrary positive number. It suffices to show that there is an N such that $A \subseteq A_n * \epsilon$ and $A_n \subseteq A * \epsilon$ for all $n \geq N$. The former inclusion is immediate because A is defined by means of sequences from A_n, and $\{A_n\}_{n=1}^\infty$ is a Cauchy sequence. To show that there is an N such that $A_n \subseteq A * \epsilon$ for $n \geq N$, recall that $(A_n)_{n=1}^\infty$ is a Cauchy sequence. As a result, we can find an N so large that $D(A_n, A_m) < \epsilon/3$ whenever $n \geq N$ and $m \geq N$. Now let **v** be in A_n for such an n. By the definition of $D(A_n, A_m)$, for each $m \geq N$ there is an $\mathbf{a}_m$ in A_m such that $||\mathbf{v} - \mathbf{a}_m|| < \epsilon/3$. Since the sequence $(\mathbf{a}_m)_{m=N+1}^\infty$ lies in the compact set $A_n + \epsilon/3$, there is a subsequence that converges to some **a** in $A_n * \epsilon/3$. Thus $||\mathbf{v} - \mathbf{a}|| < \epsilon$, so that **v** is in $A * \epsilon$. We conclude that $A_n \subseteq A * \epsilon$. $\qquad\square$

We have now completed the proof that the collection $\mathcal{K}$ of compact subsets in $\mathbb{R}^2$ with the metric D is a complete metric space. This is an important ingredient for the analysis of the fractal images we can obtain on the computer.

6.2.1 Section 6.2 Exercises

Exercise 1. Let B be in $\mathcal{K}$. Prove that $d(\mathbf{v}, B) = 0$ if and only if $\mathbf{v}$ is in B.

Exercise 2. Let A and B be in $\mathcal{K}$. Show that $d(A, B)$ exists.

Exercise 3. Show that the distance $D(A, B)$ between A and B in Figure 6.4 is $\sqrt{19 - 6\sqrt{2}}$.

Exercise 4. Let B be the circle with center $(3, 0)$ and radius 2. Find the distance from the point $(-2, 1)$ to B.

Exercise 5. Suppose that A and B are members of $\mathcal{K}$, and assume that $B \subseteq A$. Show that $d(\mathbf{v}, A) \le d(\mathbf{v}, B)$ for all $\mathbf{v}$ in $\mathbb{R}^2$.

Exercise 6. Prove that for any compact sets A and B in $\mathbb{R}^2$, the maximum of the numbers $d(\mathbf{v}, B)$, for all $\mathbf{v}$ in A, actually exists.

Exercise 7. Give an example to show that if A is not necessarily compact, then the minimum value of $\|\mathbf{v} - \mathbf{w}\|$ for distinct vectors $\mathbf{v}$ and $\mathbf{w}$ in A need not exist.

Exercise 8.

(a) Let A be the segment $[0, 1]$ on the x-axis, and let B be the line segment from $(0, 0)$ to $(1, 1)$ in the plane. Is $d(A, B) = d(B, A)$? Give reasons for your answer.

(b) Let A and B be congruent sets in $\mathbb{R}^2$ with the same orientation (i.e., not rotated!). Show that $d(A, B) = d(B, A)$.

Exercise 9. Constructs two compact subsets A and B (different from Exercise 8) of $\mathbb{R}^2$ such that $d(A, B) = d(B, A)$. Explain why this condition is not always required, even though it is part of the definition of a metric.

6.3 Contractions and Affine Functions

The notion of metric space, and in particular complete metric space, will be very important in the recipe for creating fractal images on the computer. An indispensable tool will be iterates of two special kinds of functions: contractions on metric spaces and affine functions on $\mathbb{R}^2$. This section is devoted to those two types of functions.

6.3.1   Contractions

As you might expect, a contraction shrinks the distance between points.

Definition 6.20. Let (S, d) be a metric space. A function $T : S \to S$ is a **contraction** if there is a number s such that $0 < s < 1$ and such that $d(T(x), T(y)) \le s\, d(x, y)$ for all x and y in S. In this case, s is called a **contraction constant**, or **contraction factor**, for T.

If $T : S \to S$ has the property that $d(T(x), T(y)) = sd(x, y)$ for all x and y in S and some $s > 0$, then T is a similarity, as defined in Section 5.1, and is a contraction if $0 < s < 1$. However, there are contractions that are not similarities. For example, if $f(x) = x^2$ for $0 \le x \le 1/3$, then

$$|f(x) - f(y)| = |x^2 - y^2| = |x + y||x - y| \le \frac{2}{3}|x - y|, \text{ for all } x, y \text{ in } [0, 1/3]$$

so that f is a contraction. However, since

$$\left| f(\tfrac{1}{3}) - f(0) \right| = \frac{1}{9} - 0 = \frac{1}{9} = \frac{1}{3}\left| \frac{1}{3} - 0 \right| \quad \text{and}$$

$$\left| f(\tfrac{1}{4}) - f(0) \right| = \frac{1}{16} - 0 = \frac{1}{16} = \frac{1}{4}\left| \frac{1}{4} - 0 \right|$$

it follows that f is not a similarity. In any case, we will be interested in the *existence* of a contraction constant for a given contraction.

Many of the real-valued functions we are familiar with are contractions. For example, if $f(x) = ax + b$ with $|a| < 1$, then f is a contraction, since

$$|f(x) - f(y)| = |(ax + b) - (ay + b)| = |a|\,|x - y|$$

Thus $|a|$ is a contraction constant for f. More generally, $f : R \to R$ is any differentiable real-valued function such that $|f'(x)| \le s < 1$ for all x in the domain of f, then f is a contraction. (See Exercise 2.)

It might appear that the sine function is a contraction because $|\sin x - \sin y| < |x - y|$ for all distinct real x and y. However, the sine function is *not* a contraction since there is no value of $s < 1$ such that $|\sin x - \sin y| \le s|x - y|$ for all x and y. (See Exercise 3.)

Next, consider the linear function L on $\mathbb{R}^2$, and assume that the eigenvalues λ and μ have the property that $|\mu| \le |\lambda| < 1$. Then although the iterates $L^{[n]}(\mathbf{v})$ converge to $\mathbf{0}$ for each vector $\mathbf{v}$ (thanks to Theorem 3.14), it is *not* true that L is automatically a contraction (see Exercise 4). However, in the event that the matrix $A_L = \begin{pmatrix} \lambda & 0 \\ 0 & \mu \end{pmatrix}$, then it is straightforward to show that L is a contraction. To that end, let $\mathbf{v} = \begin{pmatrix} x \\ y \end{pmatrix}$. Then because $|\mu| \le |\lambda| < 1$,

$$\|L(\mathbf{v})\| = \left\| \begin{pmatrix} \lambda & 0 \\ 0 & \mu \end{pmatrix} \begin{pmatrix} x \\ y \end{pmatrix} \right\| = \left\| \begin{pmatrix} \lambda x \\ \mu y \end{pmatrix} \right\| = \sqrt{(\lambda x)^2 + (\mu x)^2}$$

$$\le \sqrt{\lambda^2(x^2 + y^2)} = |\lambda|\sqrt{x^2 + y^2} = |\lambda|\|\mathbf{v}\|$$

Therefore L is a contraction with contraction constant λ.

A by-product of the definition of contraction is that a contraction on any metric space is a continuous function.

Theorem 6.21. Let T be a contraction on the metric space (S, d). Then T is a continuous function.

Proof. Let s be a contraction constant for T, and let x be an arbitrary element of S. Suppose that $\{x_n\}_{n=1}^{\infty}$ is a sequence in S that converges to x in S. By the definition of convergence, for any arbitrary $\epsilon > 0$ there is an N such that if $n > N$, then $d(x_n, x) < \epsilon$. Since s is a contraction constant, for such n we have

$$d(T(x_n), T(x)) \leq s\, d(x_n, x) < s\epsilon < \epsilon$$

Consequently T is continuous at x, so T is a continuous function. $\qquad\square$

We are now ready for the Contraction Mapping Theorem, which is very important not only in the study of fractals but also in other areas of mathematics such as differential equations. The theorem is valid for any complete metric space.

Theorem 6.22 (Contraction Mapping Theorem). Let (S, d) be a complete metric space, and let $T : S \to S$ be a contraction, with contraction constant s. Then T has exactly one fixed point p. Moreover, for *every* x in S, $\{T^{[n]}(x)\}_{n=1}^{\infty}$ converges to p as n increases without bound.

Proof. Let x be any element of S. We will show that $\{T^{[n]}(x)\}_{n=1}^{\infty}$ is a Cauchy sequence. To that end, let $m > n$. Using the triangle inequality several times with respect to the metric d, we find that

$$d(x, T^{[m-n]}(x))$$
$$\leq d(x, T(x)) + d(T(x), T^{[2]}(x)) + \cdots + d(T^{[m-n-1]}(x), T^{[m-n]}(x))$$
$$\leq d(x, T(x)) + s\, d(x, T(x)) + s^2\, d(x, T(x)) + \cdots + s^{m-n-1} d(x, T(x))$$
$$= d(x, T(x)) \sum_{k=0}^{m-n-1} s^k \leq d(x, T(x)) \sum_{k=0}^{\infty} s^k$$
$$= d(x, T(x)) \frac{1}{1-s}$$

Since $m > n$ by assumption, the preceding inequalities and the fact that T is a contraction imply that

$$d(T^{[n]}(x), T^{[m]}(x)) \leq s\, d(T^{[n-1]}(x), T^{[m-1]}(x)) \leq \cdots \leq s^n\, d(x, T^{[m-n]}(x))$$
$$\leq \frac{s^n}{1-s} d(x, T(x))$$

which converges to 0 as n increases without bound because $0 < s < 1$. Thus $\{T^{[n]}(x)\}_{n=1}^{\infty}$ is a Cauchy sequence. Since (S, d) is complete, it follows that the sequence converges to an element p of S.

Next, we will prove that p is a fixed point of T. To that end, we use the fact that $\{T^{[n]}(x)\}_{n=1}^{\infty}$ converges to p and T is continuous. Therefore $T^{n+1}(x) \to p$ and $T^{n+1}(x) = T(T^{[n]})(x) \to T(p)$ as n increases without bound. Consequently $T(p) = p$.

Finally, we will show that $\{T^{[n]}(z)\}_{n=1}^{\infty}$ converges to p for every z in S. Indeed, because $p = T^{[n]}(p)$, and by the second set of calculations in the proof,

$$d(p, T^{[n]}(z)) = d(T^{[n]}(p), T^{[n]}(z)) \leq s^n d(p, z) \to 0$$

as n increases without bound. That completes the proof of the theorem.   □

The Contraction Mapping Theorem is powerful, and from the perspective of geometry, the result is simply that the long-term iterates of a single contraction converge to a unique fixed point of the function. For example, consider the quadratic function Q_μ studied in Section 1.5, with $0 < \mu < 1$. Since $0 \leq Q_\mu(x) = \mu x(1 - x) < \mu x$ for all x in $(0, 1]$ and since $Q_\mu(0) = 0$, it follows that Q_μ is a contraction with contraction constant μ. We proved that all iterates of each x in the domain of Q_μ converge to the lone fixed point 0. That is consistent with the result of the Contraction Mapping Theorem.

Note: The Contraction Mapping Theorem is frequently referred to as the Banach Fixed Point Theorem. The first statement of the theorem was from Stefan Banach (1922). The history of the work done by Banach and his contemporaries is illustrated in *The Scottish Book* (originally published by Ulam and annotated by Mauldin), a fascinating account of the early stages of functional analysis.

6.3.2   Affine Functions

In Chapter 3 we discussed linear functions defined on $\mathbb{R}^2$. Although iterates of linear functions are quite tame, a close relative of linear functions, called affine functions, will play a major role in the creation of intricate fractals on the computer.

Definition 6.23. A function $T : \mathbb{R}^2 \to \mathbb{R}^2$ is **affine** if there are real constants a, b, c, d, e, and f such that

$$T \begin{pmatrix} x \\ y \end{pmatrix} = \begin{pmatrix} a & b \\ c & d \end{pmatrix} \begin{pmatrix} x \\ y \end{pmatrix} + \begin{pmatrix} e \\ f \end{pmatrix} \qquad \text{for all } x \text{ and } y \text{ in } R$$

In other words, an affine function T is a linear function that is followed by a translation (i.e., a shift).

Evidently the linear portion of the affine function is represented by $\begin{pmatrix} a & b \\ c & d \end{pmatrix}$, and the translation is represented by the vector $\begin{pmatrix} e \\ f \end{pmatrix}$. Thus to each affine function there is a unique associated linear function and a unique translation vector. The translation vector is the image of the origin: $T\begin{pmatrix} 0 \\ 0 \end{pmatrix} = \begin{pmatrix} e \\ f \end{pmatrix}$. In addition, images of line segments under an affine function are either line segments or points (in the degenerate case) (see Exercise 6). It follows that the image of a triangle under an affine function is either a triangle, a line segment, or a point.

Because an affine function is obtained from a linear function by a translation, it follows from Corollary 3.13 and the comment following it that every affine function is continuous. (See Exercise 7.) Moreover, an affine function is a contraction if and only if the associated linear function is a contraction.

Example 6.3.1. Let T be defined by $T\begin{pmatrix} x \\ y \end{pmatrix} = \begin{pmatrix} 2x + 3y - 1 \\ -y + 2 \end{pmatrix}$. Show that T is an affine function.

Solution. We find that

$$T\begin{pmatrix} x \\ y \end{pmatrix} = \begin{pmatrix} 2x + 3y - 1 \\ -y + 2 \end{pmatrix} = \begin{pmatrix} 2x + 3y \\ -y \end{pmatrix} + \begin{pmatrix} -1 \\ 2 \end{pmatrix}$$

$$= \begin{pmatrix} 2 & 3 \\ 0 & -1 \end{pmatrix} \begin{pmatrix} x \\ y \end{pmatrix} + \begin{pmatrix} -1 \\ 2 \end{pmatrix}$$

Thus T has the form of an affine function. $\square$

Example 6.3.2. Let $T : \mathbb{R}^2 \to \mathbb{R}^2$ be an affine function that satisfies the following 3 equations:

$$T\begin{pmatrix} 0 \\ 0 \end{pmatrix} = \begin{pmatrix} -1 \\ -1 \end{pmatrix}, \quad T\begin{pmatrix} 1 \\ 0 \end{pmatrix} = \begin{pmatrix} 0 \\ 0 \end{pmatrix}, \quad \text{and} \quad T\begin{pmatrix} 0 \\ 1 \end{pmatrix} = \begin{pmatrix} -1 \\ 2 \end{pmatrix}$$

Find a formula for $T\begin{pmatrix} x \\ y \end{pmatrix}$, for all x and y in R.

Solution. From the first equation in the hypotheses we find that

$$\begin{pmatrix} -1 \\ -1 \end{pmatrix} = T\begin{pmatrix} 0 \\ 0 \end{pmatrix} = \begin{pmatrix} a & b \\ c & d \end{pmatrix} \begin{pmatrix} 0 \\ 0 \end{pmatrix} + \begin{pmatrix} e \\ f \end{pmatrix} = \begin{pmatrix} e \\ f \end{pmatrix}$$

so that $e = -1 = f$. Next, we use the information in the second equation in the hypotheses, along with the fact that $e = -1 = f$. We conclude that

$$\begin{pmatrix} 0 \\ 0 \end{pmatrix} = T\begin{pmatrix} 1 \\ 0 \end{pmatrix} = \begin{pmatrix} a & b \\ c & d \end{pmatrix} \begin{pmatrix} 1 \\ 0 \end{pmatrix} + \begin{pmatrix} -1 \\ -1 \end{pmatrix} = \begin{pmatrix} a - 1 \\ c - 1 \end{pmatrix}$$

which means that $a = 1 = c$. Now we use the information in the third equation in the hypotheses, along with the information on a, c, e, and f that we have gained:

$$\begin{pmatrix} -1 \\ 2 \end{pmatrix} = T\begin{pmatrix} 0 \\ 1 \end{pmatrix} = \begin{pmatrix} 1 & b \\ 1 & d \end{pmatrix} \begin{pmatrix} 0 \\ 1 \end{pmatrix} + \begin{pmatrix} -1 \\ -1 \end{pmatrix} = \begin{pmatrix} b - 1 \\ d - 1 \end{pmatrix}$$

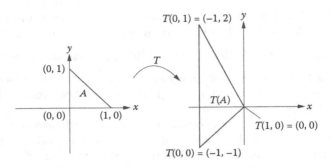

FIGURE 6.7
Applying an affine map to a triangle

The result is that $-1 = b - 1$ and $2 = d - 1$, so that $b = 0$ and $d = 3$. We conclude that

$$T\begin{pmatrix} x \\ y \end{pmatrix} = \begin{pmatrix} 1 & 0 \\ 1 & 3 \end{pmatrix} \begin{pmatrix} x \\ y \end{pmatrix} + \begin{pmatrix} -1 \\ -1 \end{pmatrix}$$

which has the form of an affine function.   □

Figure 6.7 indicates what happens to the triangle with vertices $(0, 0)$, $(0, 1)$, and $(1, 0)$ under the mapping T.

6.3.3   Isometries

A special kind of affine function on $\mathbb{R}^2$ is called an isometry.

Definition 6.24. An affine function $T : \mathbb{R}^2 \to \mathbb{R}^2$ is an **isometry** provided that

$$||T(\mathbf{v}) - T(\mathbf{w})|| = ||\mathbf{v} - \mathbf{w}|| \text{ for all } \mathbf{v} \text{ and } \mathbf{w} \text{ in } \mathbb{R}^2$$

By definition, an isometry preserves the distance between two vectors. This means, for instance, that the image of a triangle is congruent to the triangle, and similarly for any polygon. In other words, an isometry represents rigid motion in the plane. Moreover, an isometry is automatically one-to-one and is *never* a contraction.

Although an isometry need not have a fixed point, each isometry is closely related to an isometry with **0** as fixed point.

Theorem 6.25. Suppose that T is an isometry, and let T_0 be defined by $T_0(\mathbf{v}) = T(\mathbf{v}) - T(\mathbf{0})$, for all $\mathbf{v}$ in $\mathbb{R}^2$. Then T_0 is an isometry with $T_0(\mathbf{0}) = \mathbf{0}$.

Proof. On the one hand, $T_0(\mathbf{0}) = T(\mathbf{0}) - T(\mathbf{0}) = \mathbf{0}$. On the other hand,

$$||T_0(\mathbf{v}) - T_0(\mathbf{w})|| = ||[T(\mathbf{v}) - T(\mathbf{0})] - [T(\mathbf{w}) - T(\mathbf{0})]|| = ||T(\mathbf{v}) - T(\mathbf{w})||$$
$$= ||\mathbf{v} - \mathbf{w}||$$

since T is by hypothesis an isometry. Thus T_0 is also an isometry.   □

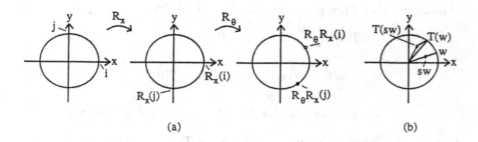

(a)   (b)

FIGURE 6.8
Reflection composed with rotation

Suppose that the isometry $T : \mathbb{R}^2 \to \mathbb{R}^2$ has fixed point $\mathbf{0}$. In this case, because $\|T(\mathbf{v}) - T(\mathbf{0})\| = \|\mathbf{v} - \mathbf{0}\|$, evidently $\|T(\mathbf{v})\| = \|\mathbf{v}\|$ for all $\mathbf{v}$ in $\mathbb{R}^2$. This means, for example, that the image of the circle C of radius 1 centered at the origin is a subset of C. Actually we can say more. Let $\mathbf{u}$ and $\mathbf{v}$ represent any two points on the circle C. Then $\|T(\mathbf{u}) - T(\mathbf{v})\| = \|\mathbf{u} - \mathbf{v}\|$. Thus the distance between $\mathbf{u}$ and $\mathbf{v}$ is the same as the distance between $T(\mathbf{u})$ and $T(\mathbf{v})$. It should be reasonable to conclude that there are two possibilities for T:

(a) T is a rotation in the counterclockwise direction by an angle θ around the origin, which rotation we will denote by R_θ. Thus $T(\mathbf{v}) = R_\theta(\mathbf{v})$.

(b) T is composed of the reflection across the x-axis, which we will denote by R_x, and a rotation by an angle θ. Thus $T(\mathbf{v}) = R_\theta(R_x(\mathbf{v}))$.

The three figures that comprise Figure 6.8(a) give an example describing the formula $T(\mathbf{v}) = R_\theta(R_x(\mathbf{v}))$, with a reflection through the x-axis followed by a counterclockwise rotation.

Now we are ready for the next theorem, which we state without proof.

Theorem 6.26. Let $T : \mathbb{R}^2 \to \mathbb{R}^2$ be an isometry such that $T(\mathbf{0}) = \mathbf{0}$. Then T is either a rotation about the origin or the reflection through the x-axis followed by a rotation about the origin.

Example 6.3.3. Let T be an isometry on $\mathbb{R}^2$, with $T(\mathbf{0}) = \mathbf{0}$. Show that T sends each line emanating from the origin onto a line emanating from the origin.

Solution. Suppose that $\mathbf{w}$ is a nonzero point in the plane, and let $s\mathbf{w}$ be on the line from $\mathbf{0}$ to $\mathbf{w}$, with $0 < s < 1$ (Figure 6.8(b)). The following formulas demonstrate that $T(s\mathbf{w})$ is on the line from $\mathbf{0}$ to $T(\mathbf{w})$:

$$||\mathbf{w}|| = ||T(\mathbf{w})|| = ||T(\mathbf{w}) - T(\mathbf{0})||$$

$$\leq ||T(\mathbf{w}) - T(s\mathbf{w})|| + ||T(s\mathbf{w}) - T(\mathbf{0})||$$

$$\leq ||\mathbf{w} - s\mathbf{w}|| + ||s\mathbf{w}|| = (1 - s)||\mathbf{w}|| + s||\mathbf{w}|| = ||\mathbf{w}||$$

As a result, $||T(\mathbf{w}) - T(s\mathbf{w})|| + ||T(s\mathbf{w}) - T(\mathbf{0})|| = ||\mathbf{w}|| = ||T(\mathbf{w})|| = ||T(\mathbf{w}) - T(\mathbf{0})||$. Thus the sum of the lengths of the vector from $T(\mathbf{0})$ to $T(s\mathbf{w})$ and the vector from $T(s\mathbf{w})$ to $T(\mathbf{w})$ equals the length of the vector from $T(\mathbf{0})$ to $T(\mathbf{w})$. Consequently $T(s\mathbf{w})$ lies on the line from $T(\mathbf{0})$, which is $\mathbf{0}$, to $T(\mathbf{w})$. We conclude that the image of the line through the origin and $\mathbf{w}$ is mapped by T onto the line through the origin and $T(\mathbf{w})$. $\square$

The following two results follow readily.

Theorem 6.27. Let $T : \mathbb{R}^2 \to \mathbb{R}^2$ be an isometry such that $T(\mathbf{0}) = \mathbf{0}$. Then T is a linear function.

Proof. The result follows directly from Theorem 6.26 and the fact that rotations and reflections are linear functions. $\square$

Corollary 6.28. Let $T : \mathbb{R}^2 \to \mathbb{R}^2$ be an isometry. Then T is an affine function.

Proof. This is left as Exercise 15. $\square$

6.3.4 Section 6.3 Exercises

Exercise 1. Let $f(x) = x^2$ for $0 \leq x \leq 1/3$. Show that for any number a with $0 < a < 2/3$, there is x in $[0, 1/3)$ such that $|f(x) - f(1/3)| > a|x - 1/3|$.

Exercise 2. Let $f : R \to R$ be a function, let $0 < s < 1$, and assume that $|f'(x)| \leq s < 1$ for all x in the domain of f. Show that f is a contraction.

Exercise 3. Show that the sine function (with domain R) is *not* a contraction.

Exercise 4. Let the linear function L have associated matrix $A_L = \begin{pmatrix} 1/2 & 1 \\ 0 & 1/2 \end{pmatrix}$. Show that L is *not* a contraction, although the eigenvalues λ and μ satisfy $0 < \lambda, \mu < 1$.

Exercise 5. Let the linear function L have associated matrix $A_L = \begin{pmatrix} 1/2 & b \\ b & 1/2 \end{pmatrix}$, with $0 < b < 1/2$. Show that L is a contraction.

Exercise 6. Prove that if T is an affine function on $\mathbb{R}^2$, then the image of a line segment is either a line segment or a point.

Exercise 7. Show that if T is an affine function on $\mathbb{R}^2$, then T is a continuous function.

Exercise 8. Let S be the unit square with vertices $(0, 0)$, $(1, 0)$, $(1, 1)$, and $(0, 1)$. Find the image of S under the affine function given by

(a) $T\begin{pmatrix} x \\ y \end{pmatrix} = \begin{pmatrix} 1/2 & 0 \\ 0 & 1/2 \end{pmatrix}\begin{pmatrix} x \\ y \end{pmatrix} + \begin{pmatrix} 1/4 \\ 1/2 \end{pmatrix}$

(b) $T\begin{pmatrix} x \\ y \end{pmatrix} = \begin{pmatrix} 0 & 1 \\ -1 & 0 \end{pmatrix}\begin{pmatrix} x \\ y \end{pmatrix} + \begin{pmatrix} 2 \\ 3 \end{pmatrix}$

Exercise 9. Find a formula for the affine function T that maps the unit circle $x^2 + y^2 = 1$ onto the ellipse $4x^2 + 9y^2 = 1$.

Exercise 10. In (a) – (c), find a formula for the affine function T defined by the given equations.

(a) $T\begin{pmatrix} 0 \\ 0 \end{pmatrix} = \begin{pmatrix} 2 \\ 0 \end{pmatrix}$, $T\begin{pmatrix} 1 \\ 0 \end{pmatrix} = \begin{pmatrix} 2 \\ -1 \end{pmatrix}$, $T\begin{pmatrix} 0 \\ 1 \end{pmatrix} = \begin{pmatrix} 1 \\ 0 \end{pmatrix}$

(b) $T\begin{pmatrix} 0 \\ 0 \end{pmatrix} = \begin{pmatrix} -1 \\ -2 \end{pmatrix}$, $T\begin{pmatrix} 1 \\ 0 \end{pmatrix} = \begin{pmatrix} 0 \\ -2 \end{pmatrix}$, $T\begin{pmatrix} 0 \\ 1 \end{pmatrix} = \begin{pmatrix} -1 \\ -3/2 \end{pmatrix}$

(c) $T\begin{pmatrix} 1 \\ 0 \end{pmatrix} = \begin{pmatrix} 1/2 \\ 1 \end{pmatrix}$, $T\begin{pmatrix} 0 \\ 1 \end{pmatrix} = \begin{pmatrix} 1 \\ 3/2 \end{pmatrix}$, $T\begin{pmatrix} 1 \\ 1 \end{pmatrix} = \begin{pmatrix} 3/2 \\ 3/2 \end{pmatrix}$

Exercise 11. Let e and f be fixed real numbers, and let $T\begin{pmatrix} x \\ y \end{pmatrix} = \begin{pmatrix} 1/2 & 0 \\ 0 & 1/2 \end{pmatrix}\begin{pmatrix} x \\ y \end{pmatrix} + \begin{pmatrix} e \\ f \end{pmatrix}$. Find the fixed point of T.

Exercise 12. Show that the reflection R_x through the x-axis is a **linear isometry**, that is, an isometry that is a linear function.

Exercise 13. Show that the counterclockwise rotation R_θ through an angle of θ radians about the origin is a linear isometry.

Exercise 14. Show that the reflection R_y through the y-axis is the composite of the reflection R_x and an appropriate rotation.

Exercise 15. Prove Corollary 6.28.

6.4   Iterated Function Systems

In Section 6.3 we proved the Contraction Mapping Theorem (Theorem 6.22). For that theorem, we assumed that (S, d) was a complete metric space, and $T : S \to S$ a contraction. Then the iterates of any element z in S converge to a unique fixed point p of T. As a consequence, the resulting long-term iterates are not very exciting geometrically, since they lie within ever smaller neighborhoods of p.

The iterates can approach a very different kind of set if we combine contractions in a suitable way.

Definition 6.29. Let $T_1, T_2, \ldots, T_n$ be a finite set of contractions on R^2. Define the function $F : \mathbb{R}^2 \to \mathbb{R}^2$ by

$$F(A) = T_1(A) \cup T_2(A) \cup \cdots \cup T_n(A), \text{ for all compact sets } A \text{ in } R^2$$

Then F is called the **union** of the functions $T_1, T_2, \ldots, T_n$. When we want to emphasize the finite collection of functions $T_1, T_2, \ldots, T_n$ that describes F, we call F, or the collection $T_1, T_2, \ldots, T_n$, an **iterated function system**. The expression "iterated function system" is usually abbreviated **IFS**.

Sometimes the union F is called a **Hutchinson map**, after the American mathematician John Hutchinson, who investigated such maps in the early 1980s. (See Hutchinson, 1981.) We should also recall that $\mathcal{K}$ is the set of all compact subsets of R^2. With this terminology, the union T is defined on the elements of $\mathcal{K}$.

Because the functions that comprise an IFS are contractions by definition, we will prove in Theorem 6.30 that the IFS is a contraction in the complete metric space $(\mathcal{K}, D)$ of compact sets on $\mathbb{R}^2$ with the Hausdorff metric. Before we can prove that result, we have three lemmas.

Lemma 6.1. Let F be the union of the contractions $T_1, T_2, \ldots, T_n$, and let A be a compact set in R^2. Then $F(A)$ is also compact in R^2.

Proof. Notice first that for each $k = 1, 2, \ldots, n$, T_k is continuous by Theorem 6.21. This means by Theorem 6.8(b) that $T_k(A)$ is compact. Since

$$F(A) = T_1(A) \cup T_2(A) \cup \cdots \cup T_n(A)$$

and since the finite union of compact sets is compact by Theorem 6.8(a), $F(A)$ is automatically compact. $\square$

In preparation for Lemma 6.2, recall that the Hausdorff distance between compact subsets A and B of $\mathbb{R}^2$ is denoted by $D(A, B)$, and the distance *from* A *to* B is $d(A, B)$.

Lemma 6.2. For any compact sets A, B, C, and E in R^2, we have

$$D(A \cup B, C \cup E) \leq \text{maximum of } D(A, C) \text{ and } D(B, E)$$

Proof. Assume without loss of generality that $D(A \cup B, C \cup E) = d(A \cup B, C \cup E)$. By the comment following the definition of the distance d, there is a vector $\mathbf{v}$ in $A \cup B$ such that

$$d(A \cup B, C \cup E) = d(\mathbf{v}, C \cup E)$$

Now if $\mathbf{v}$ is in A, then $d(\mathbf{v}, C \cup E) \leq d(\mathbf{v}, C) \leq d(A, C)$, whereas if $\mathbf{v}$ is in B, then $d(\mathbf{v}, C \cup E) \leq d(\mathbf{v}, E) \leq d(B, E)$. Either way, we deduce that

$$d(A \cup B, C \cup E) \leq \text{maximum of } d(A, C) \text{ and } d(B, E)$$
$$\leq \text{maximum of } D(A, C) \text{ and } D(B, E)$$

We conclude that

$$D(A \cup B, C \cup E) \leq \text{maximum of } D(A, C) \text{ and } D(B, E)$$

This finishes the proof of Lemma 6.2. □

For our final preparation for Theorem 6.30, Lemma 6.3 shows that if T is a contraction on R^2, then with respect to distances between compact sets in R^2, T satisfies a contraction type of inequality with respect to the Hausdorff metric on $\mathcal{K}$.

Lemma 6.3. Let $T : R^2 \to R^2$ be a contraction with contraction constant $s < 1$. For any compact sets A and B in R^2, we have $D(T(A), T(B)) \leq s\, D(A, B)$.

Proof. Assume without loss of generality that $D(T(A), T(B)) = d(T(A), T(B))$. Let $\mathbf{v}$ be an arbitrary vector in A, and let $\mathbf{w}$ be a vector in B such that $d(\mathbf{v}, \mathbf{w}) = d(\mathbf{v}, B)$. Since T is a contraction with contraction constant s, it follows from the definition of $\mathbf{w}$ and Definition 6.9 that

$$d(T(\mathbf{v}), T(B)) \leq d(T(\mathbf{v}), T(\mathbf{w})) \leq s\, d(\mathbf{v}, \mathbf{w}) = s\, d(\mathbf{v}, B) \leq s\, d(A, B)$$
$$\leq s\, D(A, B)$$

Since this set of inequalities is valid for the arbitrary vector $\mathbf{v}$ of A, we conclude that

$$D(T(A), T(B)) = d(T(A), T(B)) \leq s\, D(A, B)$$

This completes the proof. □

Now we are ready to show that because the constituent functions of an IFS are contractions, then the IFS is a contraction also.

Theorem 6.30. Let F be the union of the contractions $T_1, T_2, \ldots, T_n$, with contraction constants $s_1, s_2, \ldots, s_n$, respectively. Let $s = $ the maximum of $s_1, s_2, \ldots, s_n$. Then F is a contraction with contraction constant no greater than s.

Proof. By Lemma 6.1, if A is any compact subset of R^2, then $F(A)$ is also a compact subset of R^2. Next, let $n = 2$, and assume that A and B are compact subsets of R^2. Then by Lemma 6.2 (with $T_1(A), T_2(A), T_1(B), T_2(B)$ replacing A, B, C, E, respectively), and then by Lemma 6.3,

$$D(F(A), F(B)) = D(T_1(A) \cup T_2(A), \ T_1(B) \cup T_2(B))$$

$$\leq [\text{maximum of } D(T_1(A), T_1(B)) \text{ and } D(T_2(A), T_2(B))]$$

$$\leq sD(A, B)$$

This proves the theorem for $n = 2$. To complete the proof, one can let $1 < k < n$, and use the induction hypothesis that the union of $T_1, T_2, \ldots, T_k$ is a contraction with appropriate similarity constant. The argument above for $n = 2$ then proves that the IFS composed of $T_1, T_2, \ldots, T_{k+1}$ is also a contraction. Consequently, the Law of Induction yields the conclusion of the theorem. $\square$

The Contraction Mapping Theorem, which applies to any contraction defined on any complete metric space, tells us that the iterates of any point in the domain of a contraction converge to the unique fixed point of the contraction. The same result is valid for iterated function systems (which we just proved are contractions) with respect to the complete metric space $\mathcal{K}$ of all compact subsets of R^2, endowed with the Hausdorff metric. This yields the main theorem that we have been preparing for in Chapter 6.

Theorem 6.31. Let F be the union of $T_1, T_2, \ldots, T_n$, each of which is a contraction on $\mathbb{R}^2$. Then there is a unique compact subset A_F in R^2 such that for *any* element B of $\mathcal{K}$, the sequence of iterates $\{F^{[n]}(B)\}_{n=1}^\infty$ of B converge in the Hausdorff metric to A_F.

Proof. First, we know that F is a contraction on $\mathcal{K}$, by Theorem 6.30. Second, $(\mathcal{K}, D)$ is a complete metric space, by Theorem 6.19. Third, the Contraction Mapping Theorem (Theorem 6.22) implies that there is a unique attracting fixed point for F. More precisely, there is a compact set A_F such that the sequence $\{F^{[n]}(B)\}_{n=1}^\infty$ converges to A_F in the Hausdorff metric for each B in $\mathcal{K}$. Thus the proof is complete. $\square$

This Main Theorem appeared as Theorem 7.1 in Barnsley's book (1988). The set A_F is so important that it is given a name.

Definition 6.32. The compact set A_F is the **attractor** for the IFS F.

The attractor A_F is unique (as the Contraction Mapping Theorem asserts), compact, and frequently a very interesting planar set that can be profoundly robust.

An observation that will help to identify A_F for various IFS F is the following: Given an IFS F, if we can find a compact set in R^2 such that

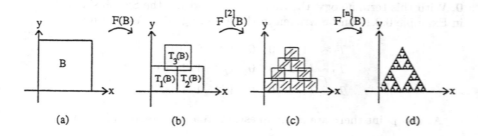

FIGURE 6.9
Attractor for an IFS

$F(B) = B$, then $B = A_F$. The reason is that A_F is a fixed point of F, and A_F is unique since the iterates of each compact set in $\mathbb{R}^2$ converge to A_F by the Main Theorem.

Example 6.4.1. Let F be the union of $T_1, T_2,$ and T_3, where $T_1(\mathbf{v}) = \frac{1}{2}\mathbf{v}$, $T_2(\mathbf{v}) = \frac{1}{2}\mathbf{v} + \begin{pmatrix} 1/2 \\ 0 \end{pmatrix}$, and $T_3(\mathbf{v}) = \frac{1}{2}\mathbf{v} + \begin{pmatrix} 1/4 \\ \sqrt{3}/4 \end{pmatrix}$, for all $\mathbf{v}$ in $\mathbb{R}^2$. Find the attractor A_F.

Solution. The three functions $T_1, T_2,$ and T_3 are evidently contractions, so F is also a contraction by Theorem 6.30. To get an idea of what A_F might look like, let us consider the initial set B to be the unit square in Figure 6.9(a). Then $F(B)$ consists of the three half-sized squares in Figure 6.9(b), and $F^{[2]}(B)$ consists of the nine quarter-sized squares in Figure 6.9(c). This leads us to conjecture that A_F should be the Sierpiński gasket, which emerges after a large number of iterates (Figure 6.9(d)). The conjecture is valid! □

Because the attractor in Example 6.3.1 is a Sierpiński gasket, we will refer to it as the **Sierpiński gasket attractor**.

As you can see from Figure 6.9, the attractor of an IFS can be highly nontrivial. Even when the IFS F is composed of affine functions, as is the case in Example 6.3.1, there is an endless variety of shapes for possible attractors.

The IFS's that are most easily created by computers and calculators are those corresponding to affine functions. Let the affine function T be defined by

$$T\begin{pmatrix} x \\ y \end{pmatrix} = \begin{pmatrix} a & b \\ c & d \end{pmatrix}\begin{pmatrix} x \\ y \end{pmatrix} + \begin{pmatrix} e \\ f \end{pmatrix}$$

For simplicity we will associate T with a 6-letter "code":

$$T \sim (a \quad b \quad c \quad d \quad e \quad f)$$

The first four letters $a, b, c,$ and d identify the linear portion of the affine function, and the last two letters e and f identify the shift of the fixed point

0. With this terminology, the functions comprising the Sierpiński gasket IFS in Example 6.3.1 can be written in the following condensed form:

$$T_1 \sim (1/2 \quad 0 \quad 0 \quad 1/2 \quad 0 \quad 0)$$
$$T_2 \sim (1/2 \quad 0 \quad 0 \quad 1/2 \quad 1/2 \quad 0)$$
$$T_3 \sim (1/2 \quad 0 \quad 0 \quad 1/2 \quad 1/4 \quad \sqrt{3}/4)$$

At this point there are many questions that are reasonable to ask:

1. If we are given an IFS, how can we predict the attractor for it?

2. If we are given a shape in the plane, under what conditions can we obtain it as the attractor of an IFS?

3. How many iterates of a given compact set for an IFS does it take before we can be sure that successive iterates would appear identical, or virtually identical, on the computer screen?

4. What is the efficiency of the process, that is, how many computer calculations are needed in order to obtain a shape that looks essentially like the attractor of a given IFS?

To answer Question 1, we observe that there are computer programs that produce iterates of an IFS F. Since the attractor is independent of the initial compact set B whose iterates for F are evaluated, we can consider B to be, for example, a unit square, or circle, or triangle, or even a single point in R^2, and determine what happens for the first dozen or so iterates. This is what we did in Example 6.3.1 in letting B be a unit square. You might consider the initial set in Example 6.3.1 to be the circle centered at the origin with radius $1/2$ (Exercise 2), and see what develops in taking higher iterates.

We can give a reasonable response to Question 2 if the given shape is self-similar. The reason is that in this case we can determine how to obtain each miniature from the shape. For example, in Figure 6.10, consider the entire shape, which we will call A.

Notice that A is composed of three miniatures: bottom right, top left, and top right. If we consider A as having exterior vertices located at $(1, 0)$, $(0, 1)$, and $(1, 1)$, then we can deduce that the first miniature (bottom right) is obtained from A by first shrinking A by half (so it has exterior vertices $(1/2, 0)$, $(0, 1/2)$, and $(1/2, 1/2)$), and then shifting the resulting figure to the right by $1/2$ unit. Likewise, the second miniature (top left) is obtained from A by first shrinking A as before, and then shifting the resulting figure upward by $1/2$ unit. Finally, the third miniature (top right) can be obtained from A by first shrinking A to one-fourth its size in the same manner, and then shifting the resulting figure to the right by $3/4$ unit and upward by $3/4$ unit. From these observations we conclude that A is the union of $T_1(A)$, $T_2(A)$, and $T_3(A)$,

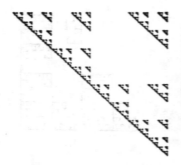

FIGURE 6.10
Attractor of an IFS

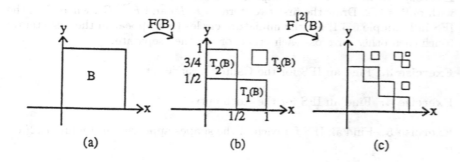

FIGURE 6.11
Demonstrating that the union of sets leads to Figure 6.10

where T_1, T_2, and T_3 are given by the following codes:

$$T_1 \sim (1/2 \quad 0 \quad 0 \quad 1/2 \quad 1/2 \quad 0)$$
$$T_2 \sim (1/2 \quad 0 \quad 0 \quad 1/2 \quad 0 \quad 1/2)$$
$$T_3 \sim (1/4 \quad 0 \quad 0 \quad 1/4 \quad 3/4 \quad 3/4)$$

A glance at Figure 6.11 will substantiate our claim that the union of T_1, T_2 and T_3 produces the shape A in Figure 6.10.

We will address questions 3 and 4 in the next, final section of Chapter 6.

6.4.1   Section 6.4 Exercises

Exercise 1. Suppose that the IFS F is the "union" of just one function T. Show that A_F consists of a single point in the plane. (*Hint:* Is the Contraction Mapping Theorem relevant?)

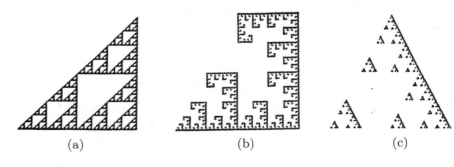

(a)   (b)   (c)

FIGURE 6.12
Illustrations for Exercise 5 of Section 6.4

Exercise 2. Let the initial compact set B be the circle centered at the origin with radius $1/2$. Draw the first two iterates $F(B)$ and $F^{[2]}(B)$, where F is the IFS in Example 6.3.1. Do the miniature circles that appear in the first iterate touch each other, or cross each other, or are they separate?

Exercise 3. Find an IFS for the Cantor ternary set.

Exercise 4. Find an IFS for the Koch curve.

Exercise 5. Find an IFS for each of the shapes appearing in Figure 6.12(a)–(c).

Exercise 6. Let F be the union of T_1, T_2, and T_3, where
$$T_1 \sim (1/2 \quad 0 \quad 0 \quad 1/2 \quad 0 \quad 0)$$

$$T_2 \sim (1/2 \quad 0 \quad 0 \quad 1/2 \quad 0 \quad 1/2)$$

$$T_3 \sim (1/2 \quad 0 \quad 0 \quad 1/2 \quad 1/2 \quad 0)$$
Find A_F. (*Hint:* Start with the unit square with vertices $(0, 0)$, $(1, 0)$, $(0, 1)$, and $(1, 1)$.)
Let F be the union of T_1, T_2, and T_3, where

$$T_1 \sim (1/2 \quad 0 \quad 0 \quad 1/2 \quad 0 \quad 0)$$

$$T_2 \sim (1/2 \quad 0 \quad 0 \quad 1/2 \quad 1/2 \quad 0)$$

$$T_3 \sim (1/2 \quad 0 \quad 0 \quad 1/2 \quad 1/2 \quad 1/2)$$

$$T_4 \sim (.45 \quad 0 \quad 0 \quad .45 \quad 0 \quad .5)$$
Find A_F. (*Hint:* Start with the unit square with vertices $(0, 0)$, $(1, 0)$, $(0, 1)$, and $(1, 1)$.)

6.5   Algorithms for Drawing Fractals

Among the questions concerning the attractors of IFS that we asked in Section 6.4 were two that we still need to address:

3. How many *iterates* of a given compact set for an IFS F does it take before we can be sure that successive iterates would appear identical, or virtually identical, on the computer screen?

4. What is the efficiency of the process, that is, how many *computer calculations* are needed in order to obtain a shape that looks essentially like the attractor of a given IFS F?

In this section we will address these two questions in relation to two algorithms that are utilized in producing attractors on the computer: the complete iteration algorithm and the random iteration algorithm.

6.5.1   The Complete Iteration Algorithm

For a given IFS F, the **complete iteration algorithm**, abbreviated **CIA**, involves the selection of an initial compact set B in R^2, and then the calculation of the iterates $F(B)$, $F^{[2]}(B)$, ... By the Main Theorem (Theorem 6.31) we know that the iterates of B approach the fixed point A_F of F in the Hausdorff metric.

Turning to Question 3, we let F be an IFS with contraction constant s. If B is any compact subset of R^2, then from the proof of Theorem 6.30 and the fact that $F(A_F) = A_F$ we deduce that

$$D(F(B), A_F) = D(F(B), F(A_F)) \leq s\,D(B, A_F)$$

By induction we obtain the formula

$$D(F^{[n]}(B), A_F) \leq sD(F^{[n-1]}(B), A_F) \leq \cdots \leq s^n\,D(B, A_F) \qquad (6.1)$$

for every positive integer n. In particular, the distance between A_F and the iterates of B shrinks by at least a factor of s with each successive iterate of B for F.

Let us assume that if images on the computer screen are closer together than 0.02 inch (in the Hausdorff metric), then they are virtually indistinguishable to the eye. It then follows from (6.1) that the nth iterate of B is virtually indistinguishable to the eye if

$$D(F^{[n]}(B), A_F) \leq s^n\,D(B, A_F) < \frac{1}{50} = 0.02$$

In order to see how (6.1) and the calculation above apply to the computer creation of a given attractor, consider the iterated function system G that produces the Sierpiński gasket attractor A_G. In this case, $s = 1/2$. By (6.1),

$$D(G^{[n]}(B), A_G) \leq \frac{1}{2^n} D(B, A_G) \tag{6.2}$$

How many iterates would it take in order to have successive iterates of B be indistinguishable from the real attractor, that is, less than 0.02 inch apart? By (6.2) this means that n need only satisfy the inequalities

$$D(G^{[n]}(B), A_G) \leq \frac{1}{2^n} D(B, A_G) < \frac{1}{50} = 0.02 \tag{6.3}$$

In the event that the Hausdorff distance between B and A_G is no greater than 20 inches, that is, $D(B, A_G) < 20$ (inches), then (6.3) tells us that we can guarantee that A_G and the nth iterate of B are within 0.02 inch provided that

$$\frac{1}{2^n}(20) < \frac{1}{50} = 0.02 \text{ (inches)}$$

which occurs if $n \geq 10$. Requiring both B and A_G to be on the computer screen simultaneously and less than 20 inches apart in the Hausdorff metric is generally not difficult to achieve.

In Figure 6.13 you can see why iterates of the unit circle converge rapidly to the Sierpiński gasket.

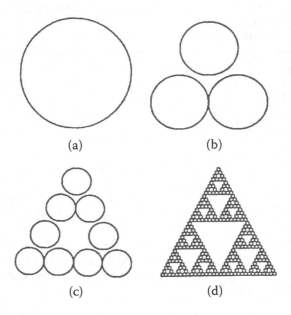

(a)   (b)

(c)   (d)

FIGURE 6.13
A union of circles converging to a Sierpiński gasket.

How many iterates are required in order for the iterates of a compact set to be guaranteed to appear just like a given attractor depends on how precise you wish the accuracy to be, how many pixels per square inch there are on the computer, and where the initial compact set B is placed. At any rate, we have now given a reasonable answer to Question 3 with respect to the CIA.

To answer Question 4 for the CIA, we start with a given compact set B and a given IFS F, and then we must determine how many calculations it takes to produce an iterate of B that is essentially indistinguishable from A_F. If there are k contractions comprising F, then the first iterate of B contains k copies of B, and requires k computations for each point in B. The second iterate contains k^2 copies of B. Thus to obtain the first two iterates of B, in theory it requires $k + k^2$ computations for each point in B. Of course, the larger the number of points in B is, the larger is the number of computations needed in deriving $F^{[n]}(B)$. The simplest case occurs when B consists of a single point. In that case, for the CIA the number N of computations performed in obtaining the nth iterate of B is expected to be

$$N = k + k^2 + \cdots k^n = \frac{k^{n+1} - 1}{k - 1} - 1 = \frac{k^{n+1} - k}{k - 1} \qquad (6.4)$$

unless some iterates are duplicates.

For example, if B consists of a single point, and the number k of functions is 3 (as with the Sierpiński gasket), and if the number n of iterates of F is 10 (the number it might take to obtain an approximation to A_F if $D(B, A_G)$ is less than 20 inches), then by (6.4) we expect that

$$N = \frac{3^{11} - 3}{2} = 88{,}572$$

This means that under the hypothesis that B is a single point and is less than 20 inches (in the Hausdorff metric) from the Sierpiński gasket on the computer screen, it would take no more than approximately 89,000 calculations to produce an iterate of B that appears virtually identical to the actual Sierpiński gasket. This is true if B is but a single point. If B were a square, or a circle, or another figure with many points, then it would take vastly more calculations.

To see how N can vary if F has more constituent contractions and/or the contraction constants of the constituent contractions might be larger than $1/2$, consider the fern in Figure 6.14, made famous by Michael Barnsley in the mid 1980s. Denote the IFS by F. There are 4 constituent contractions, and the largest contraction constant of those contractions is approximately 0.85. Thus we let $s = 0.85$. If once again we assume that B is a single point and that the Hausdorff distance between B and A_F is less than 20 inches, then by (6.1),

$$D(F^{[n]}(B), A_F) \le (0.85)^n D(B, A_F) < (0.85)^n (20)$$

We can guarantee that $D(F^{[n]}(B), A_F) < 0.02$ provided that $(0.85)^n(20) < 0.02$, which occurs if $n \ge 43$. How many calculations would

FIGURE 6.14
Barnsley's fern

it take to compose the 43th iterate of a *single* point set B for F with respect
to the CIA? Letting $k = 4$ in (6.4), we find that

$$N = \frac{4^{44} - 4}{3} \approx 1.032 \times 10^{26}$$

calculations. This number is astronomical! Indeed, if the computer is capable
of performing one million calculations per second, how long would it take for
the computer to make 10^{26} calculations? Fortunately there is another, more
economical algorithm for producing a rendition of the fern on the computer.
We turn to it next.

6.5.2   The Random Iteration Algorithm

The **random iteration algorithm**, abbreviated **RIA**, was first described
by Michael Barnsley. It utilizes the ability of the computer to produce essentially "random" numbers between 0 and 1. For the IFS F with n constituent
contractions $T_1, T_2, \ldots, T_n$, the algorithm proceeds as follows:

STEP 1.   Choose an arbitrary initial point $\mathbf{v}$ in R^2.

STEP 2.   Generate a random number r in $(0, 1)$. Then rn is a "random" number in the interval $(0, n)$.

STEP 3.   If $k - 1 < rn \leq k$, then plot the point $T_k(\mathbf{v})$.

STEP 4.   Let $T_k(\mathbf{v})$ be designated the new $\mathbf{v}$.

STEP 5.   With the new point $\mathbf{v}$, repeat Steps 2–4, and continue repeating
the process as often as needed in order to generate a reasonable
representation of the attractor A_F.

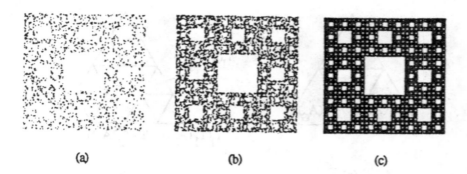

(a)　　　　　　　　　　(b)　　　　　　　　　　(c)

FIGURE 6.15
The RIA for the Sierpiński carpet

As one executes the algorithm, points in (or essentially in) the attractor begin appearing. Generally it takes many thousands of iterations to obtain a reasonable rendition of the attractor. For example, in Figure 6.15(a)–(c), we display the results for the Sierpiński carpet after 5,000, 20,000, and 50,000 iterates, respectively.

With the RIA, a basic question arises: Why should the points that are plotted with the help of the RIA eventually (essentially) fill out the attractor, or even reside (essentially) in the attractor?

To respond to these questions, let F be an IFS. On the one hand, the Main Theorem (Theorem 6.31) tells us that if $\mathbf{v}$ is any point in the plane, then its iterates for F converge in the Hausdorff metric to A_F, so that if n is large enough, $F^{[n]}(\mathbf{v})$ is essentially A_F. On the other hand, the nth point that is plotted by means of the RIA is *not* the nth iterate of F, but rather one point in the nth iterate of F. Nevertheless these two statements together imply that if n is very large, then the nth point plotted by means of the RIA is either in A_F or would appear to be in A_F.

In order to show that the attractor of an IFS F is (essentially) filled out if we apply the RIA enough times, we will first identify an "address" for each point in A_F. To begin this process, recall that if A_F is the attractor of F, then $F(A_F) = A_F$. Next, assume that F is the union of the contractions $T_1, T_2, \ldots, T_n$. Since A_F is the fixed point of F, we know that

$$A_F = F(A_F) = T_1(A_F) \cup T_2(A_F) \cup \cdots \cup T_n(A_F)$$

so that A_F is the union of n miniature images of itself. Likewise,

$$T_1(A_F) = T_1(T_1(A_F)) \cup T_1(T_2(A_F)) \cup \cdots \cup T_1(T_n(A_F))$$

so that again, $T_1(A_F)$ is the union of n miniature images of itself. The same holds for $T_2(A_F), T_3(A_F), \ldots, T_n(A_F)$.

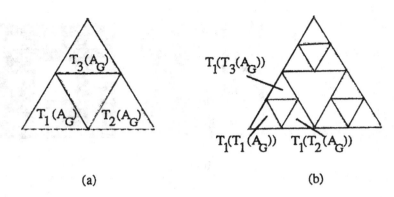

(a)   (b)

FIGURE 6.16
Sierpiński subgaskets

For an example we turn again to the Sierpiński gasket IFS G and its attractor A_G, with constituent maps T_1, T_2, and T_3. To assist in our identifications, we split the Sierpiński gasket into three subregions, and label them as in Figure 6.16(a):

1. $T_1(A_G)$ is the bottom left subregion of A_G.

2. $T_2(A_G)$ is the bottom right subregion of A_G.

3. $T_3(A_G)$ is the upper subregion of A_G.

This identifies the three first-stage images when G is applied to A_G. We will call these images the first-stage subgaskets. In Figure 6.16(b) we identify three of the nine second-stage subgaskets when G is applied to $G(A_G)$, that is, when $G^{[2]}$ is applied to A_G. The process can be continued to obtain the 3rd stage subgaskets, etc.

Now we are ready to assign a numerical address to each point in the Sierpiński gasket. To each point in $T_i(A_G)$ we assign i to be the first digit in the address (see Figure 6.17(a)). Next, to each point in $T_i(T_j(A_G))$ we assign the first two digits in the address to be ij, as in Figure 6.17(b). In general, we assign $i_1 i_2 \cdots i_k$ to represent the first k digits in the address of $T_{i_1}(T_{i_2} \cdots (T_{i_k}(A_G)) \cdots))$. Because we write the composition of functions from right to left, the left-to-right order in the address is backwards from the order in which the composition is executed.

The way you can tell the first few digits in the address of a point $\mathbf{v}$ in A_G is to ascertain which first-stage subgasket $\mathbf{v}$ lies in, then the pair of numbers representing the secondary subgasket where $\mathbf{v}$ lies, and so on. For any $\mathbf{v}$ in the Sierpiński gasket, the infinite sequence of 1's, 2's, and 3's produced by continuing the procedure indefinitely is the **address** of $\mathbf{v}$. In Figure 6.17(c) we identify a point $\mathbf{v}$ in A_G whose address has the four initial numbers 2122.

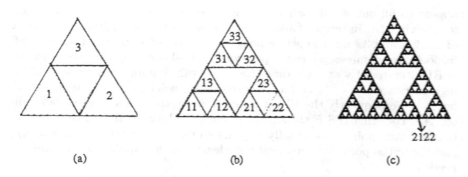

(a)   (b)   (c)

FIGURE 6.17
Addresses for the Sierpiński gasket

As described above, each point in A_G has an address. Nevertheless, it may not yet be clear that we can identify the location of any point $\mathbf{v}$ of A_G precisely by its address. However, notice that because $\mathbf{v}$ lies in one of the first-stage subgaskets of A_G (Figure 6.17(a)), each of which has side length $1/2$, the first digit in the address of $\mathbf{v}$ describes the location of $\mathbf{v}$ to within a distance of $1/2$. Similarly, the first two digits describe the location of $\mathbf{v}$ to within a distance of $1/4$, since the second-stage subgaskets have side length $1/4$ (Figure 6.17(b)). In general, the first n digits of the address describe the location of $\mathbf{v}$ to within a distance $1/2^n$. This implies that the address of $\mathbf{v}$ (which is an infinite sequence of 1's, 2's, and 3's) identifies precisely the location of $\mathbf{v}$.

You might have observed that certain points in A_G have two addresses, much as the ternary expansions are not unique for numbers in the Cantor ternary set. (See Exercise 2.) What is important is that each point in A_G *has* an address that uniquely defines its location in the Sierpiński attactor.

The RIA iterates of an arbitrary initial point $\mathbf{v}$ will appear to fill out A_G only if they (eventually) visit each subgasket of A_G that is large enough to detect. What is the probability of that happening? Let us assume, as we did earlier, that the naked eye can only easily discern differences in images on the computer screen if they are more than 0.02 inch apart. Since $\mathbf{v}$ can be *any* point in the plane, and since its iterates approach A_G by the Main Theorem, we may need to begin by taking enough preliminary iterates, say m iterates, so that $G^{[m]}(\mathbf{v})$ is within 0.02 of A_G. If, for example, $\mathbf{v}$ were 20 inches away from A_G, then it would take 10 iterates to guarantee an nth iterate within 0.02 of A_G. In this way the particular selection of an initial point $\mathbf{v}$ is generally nearly irrelevant to the number of iterates needed in order to render a reasonable image of A_G by the RIA. For simplicity let us assume henceforth that $\mathbf{v}$ itself is within 0.02 of A_G.

If A_G is 16 inches on an outer side, then because the contraction constant for G is $1/2$, then Equation (6.2) assures us that iterates of $\mathbf{v}$ via RIA will appear to fill out A_G if each 10th stage subgasket is visited by at least one iterate via RIA. In terms of addresses, this means that each 10-digit sequence of 1's, 2's, and 3's must appear among the iterates of **v**. However, by the definition of randomness, we can expect with virtual certainty that the sequence of RIA iterates of **v** at some time or another will contain *every* prescribed 10-digit sequence of 1's, 2's, and 3's. Consequently such iterates will appear to fill out A_G. Consequently the RIA iterates of any vector **v** for G essentially fills out A_G. For other IFS F the answer is similar: The collection of RIA iterates of any initial point **v** eventually appears to fill out the entire attractor A_F, and the initial point **v** is practically irrelevant to the number of RIA iterates needed.

Next, we turn to Question 4 at the outset of the section, which concerns the efficiency of the RIA for an IFS F. Again we will assume that the initial point **v** lies (essentially) in the attractor A_F. The question we will address is a little more precise than Question 4, since it refers to the specific IFS G. The question is: How many iterates would guarantee that with a probability of 0.99 each 10th stage subregion of A_G is visited? We have chosen the 10th stage subregion because $D(G^{[10]}(B), A_G) \leq 0.02$ if $D(B, A_G) \leq 20$ (inches). The estimates below are rough, and markedly exceed the minimum such number of iterates that would be needed.

As before, let us concentrate on the Sierpiński gasket, with outer side length of 16 inches. For other attractors the analysis is similar. Evidently there are 3^{10} 10-tuples of 1's, 2's, and 3's, so the probability of *not* obtaining a given 10-tuple during the first 10 iterates is no more than $1 - 1/3^{10}$. Similarly, the probability of *not* obtaining a given 10-tuple during the first 20 iterates is no more than $(1 - 1/3^{10})^2$, which is the probability that the 10-tuple does not appear during either the first 10 iterates or during the next 10 iterates. (The probability turns out to be much less than that. Why? See Exercise 3.) Continuing along this line of reasoning, we find that for each positive integer n, the probability of *not* obtaining a given 10-tuple in the first $10n$ iterates is no greater than $(1 - 1/3^{10})^n$. Thus we will be assured with at least a probability of 0.99 that a given 10th stage subgasket is visited during the first $10n$ iterates if n satisfies

$$(1 - \frac{1}{3^{10}})^n < \frac{1}{100}, \text{ or equivalently, } n > \frac{\ln 0.01}{\ln(1 - 1/3^{10})}$$

which is approximately 271,929. This is an upper bound for n. However, what we have shown is that with a little patience (as the computer calculates!) we can virtually guarantee a very accurate image for the attractor A_G if we take at least $10n = (271,929)(10) = 2,719,290$ RIA iterates. Figure 6.17(c) was created by taking 50,000 RIA iterates. If you obtain the image after, say, 500,000 RIA iterates, you can determine whether there is any demonstrable difference in the image.

How does the RIA succeed for Barnsley's fern attractor, A_F? Before we make our calculations, let us recall that in the CIA we determined an upper bound of 10^{26} calculations to render a good approximation to A_F, based on $n = 4$ and the largest contraction constant $s = 0.85$. With the RIA we first perform a few preliminary iterates, if necessary, to make sure that a certain RIA iterate is essentially in the attractor. The fern IFS F has 4 contractions, and we found that we could guarantee the nth iterate to be within 0.02 inch of A_F if $n = 43$. This means that n must be large enough that

$$(1 - \frac{1}{4^{43}})^n < \frac{1}{100}, \text{ or equivalently, } n > \frac{\ln 0.01}{\ln(1 - 1/4^{43})}$$

which most hand-held calculators cannot evaluate. We know that this is a very rough, upper bound for the number of calculations that would be necessary. Though this number is very large, it is nothing like the corresponding number for the CIA!

In addition to the randomness of choices of iterates, one can assign probabilities to the choices. More particularly, instead of having the RIA choose $T_1, T_2, \ldots, T_n$ randomly with equal probability, we can choose them randomly so that on the average, T_1 is chosen the fraction p_1 of the time, T_2 is chosen the fraction p_2 of the time, etc., where $p_1 + p_2 + \cdots p_n = 1$. Interestingly enough, it is by choosing probabilities wisely that Barnsley was able to produce a reasonable image of his fern attractor by using approximately 50,000 RIA iterations.

6.5.3   The Chaos Game Revisited

Recall from Section 5.2 that the chaos game, introduced by Michael Barnsley, consists of the following setup:

Let B, C, and D be three non-collinear points arbitrarily placed in the plane, and let $\mathbf{v}_0$ be any point whatsoever in the plane, as in Figure 6.18(a). Next, assign each side of a fair die one of the three letters B, C, or D, so

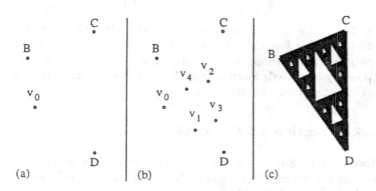

FIGURE 6.18
Results of the chaos game

that each letter appears twice on the die. Then roll the die. Let the face that shows up be denoted by A_1 (so that A_1 is B, C, or D, depending on the roll). Define $\mathbf{v}_1$ to be the midpoint of the line segment joining $\mathbf{v}_0$ and the point corresponding to A_1. Proceed inductively: For $n > 1$, if the nth roll of the die produces the face denoted A_n, then $\mathbf{v}_n$ is the midpoint of the line segment joining $\mathbf{v}_{n-1}$ and the point corresponding to A_n. Figure 6.18(b) shows the first four steps for a particular selection of $\mathbf{v}_0, B, C$, and D. Repeating the procedure a vast number of times yields a figure that has the features of a Sierpiński gasket (Figure 6.18(c)).

Armed with IFS's, we are ready to show how this "game" is related to the attractor of a suitable IFS. To that end, let

$$\mathbf{v}_0 = \begin{pmatrix} v_1 \\ v_2 \end{pmatrix}, B = \begin{pmatrix} b_1 \\ b_2 \end{pmatrix}, C = \begin{pmatrix} c_1 \\ c_2 \end{pmatrix}, \text{ and } D = \begin{pmatrix} d_1 \\ d_2 \end{pmatrix}.$$

The midpoint of the line segment joining $\mathbf{v}_0$ and the point corresponding to B is

$$\frac{1}{2} \left(\begin{pmatrix} v_1 \\ v_2 \end{pmatrix} + \begin{pmatrix} b_1 \\ b_2 \end{pmatrix} \right) = \frac{1}{2} \begin{pmatrix} v_1 \\ v_2 \end{pmatrix} + \begin{pmatrix} b_1/2 \\ b_2/2 \end{pmatrix}$$

This reminds us of the result of an affine function, with shift $\begin{pmatrix} b_1/2 \\ b_2/2 \end{pmatrix}$. Similar observations apply to C and D. Thus we define the three affine functions T_B, T_C, and T_D by the equations

$$T_B \begin{pmatrix} x \\ y \end{pmatrix} = \begin{pmatrix} 1/2 & 0 \\ 0 & 1/2 \end{pmatrix} \begin{pmatrix} x \\ y \end{pmatrix} + \begin{pmatrix} b_1/2 \\ b_2/2 \end{pmatrix}$$

$$T_C \begin{pmatrix} x \\ y \end{pmatrix} = \begin{pmatrix} 1/2 & 0 \\ 0 & 1/2 \end{pmatrix} \begin{pmatrix} x \\ y \end{pmatrix} + \begin{pmatrix} c_1/2 \\ c_2/2 \end{pmatrix}$$

$$T_D \begin{pmatrix} x \\ y \end{pmatrix} = \begin{pmatrix} 1/2 & 0 \\ 0 & 1/2 \end{pmatrix} \begin{pmatrix} x \\ y \end{pmatrix} + \begin{pmatrix} d_1/2 \\ d_2/2 \end{pmatrix}$$

The chaos game is equivalent to letting the IFS F be the IFS with constituent affine functions T_B, T_C, and T_D, and employing the RIA beginning with the initial compact set consisting of the single point $\mathbf{v}_0$. The result is a (possibly distorted) Sierpiński gasket, whatever noncollinear points B, C, D are selected, and whatever $\mathbf{v}_0$ is chosen.

6.5.4   Section 6.5 Exercises

Exercise 1. Suppose a computer is capable of performing one million calculations per second. In that case, how long would it take for the computer to make 10^{25} calculations?

Exercise 2.

(a) Show that the point whose address is $1\overline{2}$ (that is, $1222\cdots$) in the Sierpiński gasket has a second address.

(b) Determine the points in the Sierpiński gasket that have a second address. Can any points in the Sierpiński gasket have three distinct addresses? Explain why or why not.

Exercise 3. In the discussion of the efficiency of the RIA for the Sierpiński gasket, why is the probability significantly less than $(1 - 1/3^{10})^2$ that a given 10-tuple does not appear during the first 20 iterates?

Exercise 4. Assess the difference in the resulting image when the Sierpiński gasket is approximated by the RIA with initial point **0** by taking 50,000 iterates, vs. 500,000 iterates.

For the remaining exercises, you may wish to use the ITERATED FUNCTION SYSTEM code provided in the Appendix. The Sierpinski gasket and Barnsley fern are provided there, but we encourage you to extend this to the other examples in the section.

(a) Consider the Sierpiński gasket once again, but with the points representing the 3 exterior vertices shifted around in the plane (so we have a skewed Sierpiński gasket!) but still with maximum side length 16 inches. Under these conditions, could the answer to Questions 3 and 4 have been significantly altered from what we gave in the text? Explain your answer.

(b) For the Barnsley fern, the IFS is

$$
\begin{array}{llllll}
T_1 \sim & (0 & 0 & 0 & .16 & 0 & 0) \\
T_2 \sim & (.85 & .04 & -.04 & .85 & 0 & 1.6) \\
T_3 \sim & (.2 & -.26 & .23 & .22 & 0 & 1.6) \\
T_4 \sim & (-.15 & .28 & .26 & .24 & 0 & .44)
\end{array}
$$

Assess the difference in the resulting image when the Barnsley fern is approximated by the RIA with initial point **0** by taking 50,000 iterates, vs. 500,000 iterates.

(c) Determine the difference in the appearance of the Barnsley Fern attractor when 50,000 RIA iterates are taken with equal probability, and with the special probabilities: $p(T_1) \sim .01$, $p(T_2) \sim .85$, $p(T_3) \sim .07$, $p(T_4) \sim .07$.

(d) Suppose that the IFS F is comprised of four constituent contractions, and the contraction factor is s, with $0 < s < 1$. Using the analysis in this section, determine how many iterates would guarantee that with a probability of 0.99 that each 10th level subattractor is visited?

7

Complex Fractals: Julia Sets and the Mandelbrot Set

Many of the beautiful, exotic sets that appear on calendars and covers of magazines are sets that can be described in terms of iterates of complex functions. In Section 7.1 we give a brief introduction to complex numbers, the complex plane, and complex functions. We devote Section 7.2 to the fascinating collection of sets called Julia sets in the complex plane, and devote the final section to the famous, beautiful Mandelbrot set, which is also in the complex plane.

In what has become a theme of this text, you will find terms below that were previously defined. We extend our earlier definitions to include this extended number set, and encourage you to return to early chapters to convince yourself of the consistency between definitions.

7.1 Complex Numbers and Functions

We begin with the definition of a complex number.

Definition 7.1. A **complex number** has the form $x + yi$, where x and y are real numbers and $i = \sqrt{-1}$. We frequently interchange the i and y, and write $x + iy$ for $x + yi$. Also, customarily we will denote complex numbers by z or w. Thus $z = x + yi$.

It follows from the definition of complex number that $2 - \sqrt{-3}$ is a complex number that can be written as $2 - i\sqrt{3}$ or as $2 - \sqrt{3}\,i$. Next, the sum and product of two complex numbers $z = x + yi$ and $w = u + vi$ are given, respectively, by

$$z + w = (x + u) + (y + v)i \qquad \text{and} \qquad zw = (xu - yv) + (xv + yu)i$$

In particular,

$$z^2 = (x + yi)(x + yi) = x^2 + 2xyi + y^2 i^2 = (x^2 - y^2) + 2xyi$$

A complex number $x + yi$ can be considered either as an ordered pair (x, y) of real numbers or as a point (x, y) in the plane. The total collection of complex numbers is referred to as the **complex plane**. A real number x

DOI: 10.1201/9781032678757-7

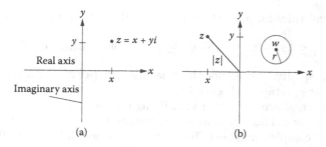

FIGURE 7.1
The complex plane

is identified with the complex number $x + 0i$, and the collection of all real numbers is the **real axis**. Similarly, the collection of all numbers of the form $0 + yi$, or equivalently, yi, comprises the **imaginary axis**, and consists of **pure imaginary numbers**. (See Figure 7.1(a).)

Next, let $z = x + iy$. Because of the identification of complex numbers with points in the plane, we denote the **absolute value** or **modulus** of z by $|z|$, where

$$|z| = \sqrt{x^2 + y^2}$$

Thus $|z|$ is the distance between the complex number z and the origin. Similarly, $|z - w|$ is the distance between the complex numbers z and w. The open disk about w with radius r consists of all z such that $|z - w| < r$ (Figure 7.1(b)). A set S in the complex plane is **bounded** if $S \subseteq U$ for some open disk U about 0 with finite radius r. A bounded set of special interest is the unit circle centered at the origin. It consists of all z such that $|z| = 1$ (Figure 7.2(a)).

A sequence $\{z_n\}_{n=1}^{\infty}$ **converges** to a complex number w if and only if $|z_n - w| \to 0$ as n increases without bound. If S contains the limit point of each convergent sequence in S, then S is **closed**.

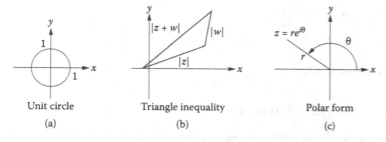

FIGURE 7.2
Geometry in the complex plane

The usual rules for absolute values are valid for complex numbers:

$$|z + w| \leq |z| + |w|, \quad |z - w| \geq ||z| - |w||, \quad \text{and} \quad |zw| = |z||w|$$

The first inequality represents the **triangle property**, with the associated geometric interpretation (Figure 7.2(b)).

There is an alternate way of identifying complex numbers that uses polar coordinate values. This alternate identification will assist us in finding powers and roots of complex numbers. To start, let θ be any angle, with $0 \leq \theta < 2\pi$. The complex number $\cos \theta + i \sin \theta$ has modulus 1, since $\cos^2 \theta + \sin^2 \theta = 1$. If $r \geq 0$, then $r(\cos \theta + i \sin \theta)$ represents the point in the plane r units from the origin and at an angle θ (in the counterclockwise direction) from the real axis (Figure 7.2(c)).

If $z = r(\cos \theta + i \sin \theta)$, then we can write z in an alternate **polar form**: $z = re^{i\theta}$. Thus

$$z = r(\cos \theta + i \sin \theta) = re^{i\theta}$$

By definition, r is the **modulus** of $re^{i\theta}$, and θ is called the (principal) **argument** of z.

For a simple example, consider $z = 3i$, which lies in the upper part of the imaginary axis. Since $r = 3$ and the argument is $\pi/2$, the polar form of $3i$ is

$$3e^{i\pi/2}, \quad \text{or equivalently,} \quad 3\left(\cos \frac{\pi}{2} + i \sin \frac{\pi}{2}\right)$$

Also, we mention that the usual properties of the exponential function hold: If z and w are complex numbers, and n is a positive integer, then $e^{z+w} = e^z e^w$ and $e^{nz} = (e^z)^n$.

For later use we remark that the unit circle $r = 1$ centered at the origin consists of all complex numbers with polar form $e^{i\theta}$, where θ is any real number. Indeed,

$$|e^{i\theta}| = |\cos \theta + i \sin \theta| = \sqrt{\cos^2 \theta + \sin^2 \theta} = \sqrt{1} = 1$$

Example 7.1.1. Show that $e^{2n\pi i} = 1$ for all integers n.

Solution. Let $n = 1$. Then

$$e^{2\pi i} = 1(\cos 2\pi + i \sin 2\pi) = 1(1 + i \cdot 0) = 1$$

More generally,

$$e^{2n\pi i} = (e^{2\pi i})^n = 1^n = 1, \forall n$$

This completes the solution. $\square$

Products of complex numbers have a particularly simple form in polar notation. More specifically, let

$$z_1 = r_1(\cos \theta_1 + i \sin \theta_1) \quad \text{and} \quad z_2 = r_2(\cos \theta_2 + i \sin \theta_2)$$

By the trigonometric sum formulas for $\sin(\theta_1 + \theta_2)$ and $\cos(\theta_1 + \theta_2)$, we find that

$$z_1 z_2 = [r_1(\cos\theta_1 + i\sin\theta_1)]\,[r_2(\cos\theta_2 + i\sin\theta_2)]$$

$$= r_1 r_2[(\cos\theta_1\cos\theta_2 - \sin\theta_1\sin\theta_2) + i(\cos\theta_1\sin\theta_2 + \sin\theta_1\cos\theta_2)]$$

$$= r_1 r_2[\cos(\theta_1 + \theta_2) + i\sin(\theta_1 + \theta_2)]$$

Therefore

$$z_1 z_2 = r_1 r_2[\cos(\theta_1 + \theta_2) + i\sin(\theta_1 + \theta_2)] \tag{7.1}$$

Alternatively,

$$z_1 z_2 = r_1 r_2\, e^{i(\theta_1 + \theta_2)}$$

Note: Formula (7.1) says that the modulus of a product is the product of the moduli, and the argument of the product is the sum of the arguments. An application of (7.1) and the Law of Induction yield De Moivre's Theorem, named after one of the early 17th-century founders of probability theory, Abraham De Moivre.

De Moivre was a relatively early contributor to the field of complex analysis. While Cardano first formalized the idea of a complex number in 1545, De Moivre's formula as we know it was described in 1707. De Moivre received credit for this in the case that n in a positive integer, but, as is often the case, Euler extended this formula for any real n in 1749.

Theorem 7.2 (De Moivre's Theorem). Let $z = r(\cos\theta + i\sin\theta) = re^{i\theta}$. Then

$$z^n = r^n(\cos n\theta + i\sin n\theta) = r^n e^{in\theta}, \quad \text{for any integer } n \geq 2 \tag{7.2}$$

Proof. We proceed by induction. For $n = 2$ the formula is valid by (7.1) if we let $z_1 = z_2 = z$. Now assume that the formula in (7.2) is true for a given $n \geq 2$. Then by the induction hypothesis and by another application of (7.1) we obtain

$$z^{n+1} = z^n z = [r^n(\cos n\theta + i\sin n\theta)]\,[r(\cos\theta + i\sin\theta)]$$
$$= r^{n+1}[\cos(n+1)\theta + i\sin(n+1)\theta]$$

Thus also $z^{n+1} = r^{n+1}e^{i(n+1)\theta}$. By the Law of Induction, (7.2) is valid for all $n \geq 2$. $\qquad\square$

De Moivre's Theorem helps us to determine powers and roots of complex numbers.

Example 7.1.2. Let $z = 1 + i$. Find the 2nd, 3rd, 4th, and 6th powers of z.

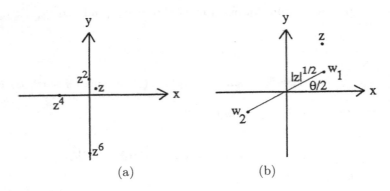

(a)   (b)

FIGURE 7.3
Powers and roots of z

Solution. Notice that $|z| = \sqrt{1^2 + 1^2} = \sqrt{2}$, so that the modulus of z is $\sqrt{2}$. Also, the argument is $\pi/4$ because z lies on the line $x = y$. Thus in polar form, $1 + i = \sqrt{2}\left(\cos \pi/4 + i \sin \pi/4\right)$. Therefore by De Moivre's Theorem,

$$z^2 = 2\left(\cos \pi/2 + i \sin \pi/2\right) = 2i$$
$$z^3 = 2^{3/2}\left(\cos 3\pi/4 + i \sin 3\pi/4\right) = 2^{3/2}(-\sqrt{2}/2 + i\sqrt{2}/2) = -2 + 2i$$
$$z^4 = 4\left(\cos \pi + i \sin \pi\right) = -4, \text{ which also equals } (z^2)^2$$
$$z^6 = 8(\cos 3\pi/2 + i \sin 3\pi/2) = -8i, \text{ which also equals } (z^2)^3$$

Figure 7.3(a) displays z, z^2, z^4 and z^6.   □

Next, let $z = re^{i\theta}$. A **square root** of z is any (complex) number w such that $w^2 = z$. If $z \neq 0$, then De Moivre's Theorem yields the distinct two square roots of z:

$$w_1 = \sqrt{r}\, e^{i\theta/2} \quad \text{and} \quad w_2 = \sqrt{r}\, e^{i(\theta/2+\pi)}$$

Indeed,

$$w_1^2 = re^{i\theta} = z \quad \text{and} \quad w_2^2 = re^{i(\theta+2\pi)} = re^{i\theta}e^{2\pi i} = z$$

We conclude that the two square roots of z lie directly opposite one another across the origin (Figure 7.3(b)), and the argument of one of the square roots is half the argument of z. In fact, this happens for the two square roots of any complex number $z \neq 0$: the square roots are directly opposite one another across the origin.

Example 7.1.3. Let $z = 3i$. Find the two square roots of z.

Solution. First, we write $3i$ in polar form:

$$3i = 3\left(\cos \frac{\pi}{2} + i \sin \frac{\pi}{2}\right)$$

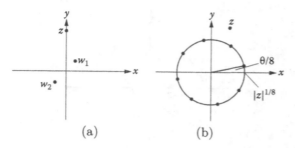

(a)   (b)

FIGURE 7.4
Square roots and 8th roots of z

Then we take the square root of 3 and half the argument $\pi/2$, which yields

$$w_1 = \sqrt{3}\left(\cos\frac{\pi}{4} + i\sin\frac{\pi}{4}\right) = \sqrt{3}\left(\frac{1}{2}\sqrt{2} + \frac{1}{2}\sqrt{2}\,i\right) = \frac{\sqrt{6}}{2}(1+i)$$

The other square root, symmetric with respect to the origin, is $w_2 = -\sqrt{6}\,(1+i)/2$ (Figure 7.4(a)). $\square$

In general, the n nth roots of a nonzero complex number z lie equi-spaced on the circle of radius $|z|^{1/n}$ about the origin. In particular, if $z = |z|e^{i\theta}$, then the nth roots of z are equally spaced around the circle, with one located at an angle θ/n with respect to the positive real axis (Figure 7.4(b)).

Example 7.1.4. Find the three 3rd roots of $z = -8$.

Solution. In polar form we have

$$z = -8 = 8\,(\cos\pi + i\sin\pi)$$

By De Moivre's Theorem, one 3rd root w_1 of z is given by

$$w_1 = 2\,(\cos\pi/3 + i\sin\pi/3) = 2\left(\frac{1}{2} + i\frac{\sqrt{3}}{2}\right) = 1 + \sqrt{3}\,i$$

Note that w_1 lies on the circle of radius 2 centered at the origin. The other two 3rd roots of z are thus placed on the same circle so that the three 3rd roots are equi-spaced. Since the argument of w_1 is $\pi/3$, this means that the argument of w_2 is $\pi/3 + 2\pi/3 = \pi$, and the argument of w_3 is $\pi/3 + 4\pi/3 = 5\pi/3$. Consequently

$$w_2 = 2\,(\cos\pi + i\sin\pi) = 2\,(-1+0) = -2$$

$$w_3 = 2\,[\cos(5\pi/3) + i\sin(5\pi/3)] = 2\left(\frac{1}{2} - i\frac{\sqrt{3}}{2}\right) = 1 - \sqrt{3}\,i$$

Figure 7.5 exhibits z as well as its three 3rd roots. $\square$

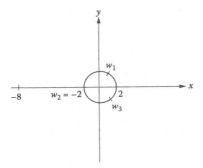

FIGURE 7.5
Three 3rd roots of -8

7.1.1   Complex Functions

A **complex function** is a function whose domain and range consist of complex numbers. (The range may consist only of real numbers, however.) Complex polynomial functions are written in the usual way, but with z's instead of x's:

$$f(z) = a_n z^n + a_{n-1} z^{n-1} + \cdots + a_2 z^2 + a_1 z + a_0 \qquad (7.3)$$

where $a_n, a_{n-1}, ..., a_1, a_0$ are complex constants, and $a_n \neq 0$.

Limits and derivatives of complex functions are defined as they are for real-valued functions of a real variable. In particular,

$$f'(z_0) = \lim_{z \to z_0} \frac{f(z) - f(z_0)}{z - z_0}$$

provided that the limit exists. Thus if f is the polynomial in (7.3), then

$$f'(z) = na_n z^{n-1} + (n-1)a_{n-1} z^{n-2} + \cdots + 2a_2 z + a_1$$

For example, if $f(z) = z^3 + \pi z^2 - 3z - 1$, then $f'(z) = 3z^2 + 2\pi z - 3$.

7.1.2   Zeros and Fixed Points of Complex Functions

Next, we turn to the relationship between zeros of a complex function and fixed points for the function. We begin with the Fundamental Theorem of Algebra, whose first complete proof was given in 1797 by the great German mathematician Carl Friedrich Gauss (1777–1855) when he was 20 years old! The proof of the Fundamental Theorem of Algebra appears in complex variables textbooks.

Theorem 7.3 (Fundamental Theorem of Algebra). Let f be a non-constant (complex) polynomial function. Then f has a zero, that is, there is a complex number z_0 such that $f(z_0) = 0$.

We observe that z_0 is a zero of a polynomial function f if and only if $z - z_0$ is a factor of $f(z)$. Thus if $f(z) = a_n z^n + a_{n-1} z^{n-1} + \cdots + a_1 z + a_0$, with $a_n, \ldots, a_0$ complex numbers, and if z_0 is a zero of f, then

$$f(z) = (z - z_0)g(z)$$

where $g(z)$ is a polynomial function. It then follows from the Fundamental Theorem of Algebra that if f is a nonconstant polynomial function of degree n (as in (7.3)), then f has precisely n zeros – though they need not be distinct. However, the Fundamental Theorem gives no indication as to the values of the zeros; the theorem only guarantees their existence.

In the event that f is a second-degree polynomial, one can use the quadratic formula to determine the zeros. For example, let $f(z) = az^2 + bz + c$ with $a \neq 0$. Then the zeros are given by

$$z = -\frac{b}{2a} \pm \frac{1}{2a}\sqrt{b^2 - 4ac} \tag{7.4}$$

Let a, b, and c be real numbers. On the one hand, if $b^2 - 4ac \geq 0$, then the zeros are real. On the other hand, if $b^2 - 4ac < 0$, then the zeros are complex. For an example, consider $f(z) = z^2 - z + 1$. It follows from (7.4) with $a = 1, b = -1, c = 1$ that the roots are

$$z_0 = \frac{1}{2} \pm \frac{1}{2}\sqrt{-3} = \frac{1}{2} \pm \frac{1}{2}\sqrt{3}\,i$$

As with real functions, a complex number p is a **fixed point** of f if $f(p) = p$. A fixed point of f corresponds to a zero of the function g, where $g(z) = f(z) - z$. That is,

$$f(p) = p \quad \text{if and only if} \quad g(p) = f(p) - p = 0$$

Using this information, along with the Fundamental Theorem of Algebra, we deduce that there are n (not necessarily distinct) fixed points for any nonconstant complex polynomials of degree n. Finding a fixed point of f is easy if f is the function defined by $f(z) = z^2 + c$ with c a complex constant, because z is a fixed point of f provided that $z = f(z) = z^2 + c$, or equivalently, z is a solution of the equation $z^2 - z + c = 0$. By the quadratic formula, the fixed points of f are given by

$$z = \frac{1}{2} \pm \frac{1}{2}\sqrt{1 - 4c} \tag{7.5}$$

Consequently if $f(z) = z^2 + c$, then f has two fixed points unless $c = 1/4$, in which case f has the unique fixed point $1/2$.

7.1.3 Periodic Points

We will be analyzing iterates of quadratic functions $f(z) = z^2 + c$, where c is a complex (or real) constant. We will be interested in not only the fixed points of such functions, but also the periodic points.

Definition 7.4. A complex number z_0 has **period n** for a (complex) function f if the iterates $z_0, f(z_0), f^{[2]}(z_0), \ldots, f^{[n-1]}(z_0)$ are distinct, and if $f^{[n]}(z_0) = z_0$. In that case, the collection

$$\{z_0, f(z_0), f^{[2]}(z_0), \ldots, f^{[n-1]}(z_0)\}$$

is called an **n-cycle** for f.

If z_0 has period n, then the iterates of z_0 "periodically" return to z_0; in fact, they return every nth iterate. Notice that a fixed point is a periodic point with period 1. Normally when we speak of a periodic point, however, we will mean that the period is greater than 1.

It is easy to see that the pair $\{0, 1\}$ of real numbers is a 2-cycle for the function $f(z) = 1 - z^2$, because $f(0) = 1$ and $f(1) = 0$. The next example may not be quite as easy to resolve by eye.

Example 7.1.5. Let $g(z) = z^2 + i$. Show that $\{-1 + i, -i\}$ is a 2-cycle.

Solution. We notice that

$$g(-1+i) = (-1+i)^2 + i = (1-2i-1) + i = -i \quad \text{and} \quad g(-i) = (-i)^2 + i = -1+i$$

Thus $\{-1 + i, -i\}$ is a 2-cycle for g.   □

Next, we will find a period-3 point of $f(z) = z^2$.

Example 7.1.6. Let $f(z) = z^2$. Show that $e^{2\pi i/7}$ has period 3 for f.

Solution. First, we observe that

$$f(e^{i\theta}) = (e^{i\theta})^2 = e^{2i\theta}$$

Using this fact, we calculate that

$$f(e^{2\pi i/7}) = e^{4\pi i/7}, \quad f^{[2]}(e^{2\pi i/7}) = e^{8\pi i/7}, \quad \text{and} \quad f^{[3]}(e^{2\pi i/7}) = e^{16\pi i/7}$$

However,

$$e^{16\pi i/7} = e^{2\pi i/7 + 2\pi i} = e^{2\pi i/7} e^{2\pi i} = e^{2\pi i/7}(1) = e^{2\pi i/7}$$

Thus $f^{[3]}(e^{2\pi i/7}) = e^{2\pi i/7}$, so that $e^{2\pi i/7}$ has period 3, and its 3-cycle is $\{e^{2\pi i/7}, e^{4\pi i/7}, e^{8\pi i/7}\}$.   □

We comment that if p is a positive integer and $p > 1$, then the following n points are distinct, placed equidistantly on the unit circle:

$$e^{2\pi i(0/p)}, e^{2\pi i(1/p)}, e^{2\pi i(2/p)}, \ldots, e^{2\pi i((p-1)/p)}$$

Since $e^{2\pi i} = 1$ and hence $e^{2\pi i m} = (e^{2\pi i})^m = 1$, we find that for any angle θ and any integer m,

$$e^{2\pi i \theta} = e^{2\pi i \theta} e^{2\pi i m} = e^{2\pi i(\theta+m)}$$

Thus if θ and ϕ differ by an integer, so that $\phi = \theta + m$ for some integer m, then $e^{2\pi i\theta} = e^{2\pi i\phi}$. It follows that there are precisely p distinct points on the unit circle of the form $e^{2\pi i(k/p)}$, where p is a positive integer and k is any nonnegative integer.

Now we are ready to exhibit lots of periodic points on the unit circle for the function $f(z) = z^2$. We will use the fact that if $e^{2\pi i\theta}$ is on the unit circle, then $f(e^{2\pi i\theta}) = e^{2\pi i(2\theta)}$.

Theorem 7.5. Let $f(z) = z^2$, so that $f(e^{2\pi i\theta}) = e^{2\pi i(2\theta)}$, for all real θ. Let p and k be positive numbers that are relatively prime to each other, with p odd and $k < p$. Then $e^{2\pi ik/p}$ is a periodic point for f.

Proof. First, notice that $f^{[n]}(e^{2\pi ik/p}) = e^{2\pi i(2^n)k/p}$, for each positive integer n. From the observation just prior to the theorem, there are at most p different points in the complex plane represented in the form $e^{2\pi i(2^n)k/p}$, for nonnegative integers n. This means that there are positive integers m and n that are relatively prime with $n > m$, such that

$$e^{2\pi i(2^n)k/p} = f^{[n]}(e^{2\pi ik/p}) = f^{[m]}(e^{2\pi ik/p}) = e^{2\pi i(2^m)k/p}$$

Therefore there is a positive integer s such that $(2^n k)/p = [(2^m k)/p] + s$, which means that $[(2^n - 2^m)k]/p = s$. Since s is an integer, it follows that p divides $k(2^n - 2^m)$, or equivalently, p divides $k2^m(2^{n-m} - 1)$. Since k and p are relatively prime and p is odd, we deduce that p divides $2^{n-m} - 1$. Thus there is a positive integer r such that $pr = 2^{n-m} - 1$, or equivalently, $2^{n-m} = pr + 1$. Consequently

$$f^{[n-m]}(e^{2\pi ik/p}) = e^{2\pi i(2^{n-m})k/p} = e^{2\pi i(pr+1)k/p} = e^{2\pi ikr}e^{2\pi ik/p} = e^{2\pi ik/p}$$

We conclude that $e^{2\pi ik/p}$ is periodic, with period that divides $n - m$. $\square$

Theorem 7.5 implies that numbers like $e^{2\pi i(2/5)} = e^{4\pi i/5}$ and $e^{2\pi i(4/9)} = e^{8\pi i/9}$ are periodic, though the theorem does not indicate the period. In fact, Theorem 7.5 implies that there are infinitely many periodic points for $f(z) = z^2$, and one can show that for each positive integer n, there are periodic points of period n (see Exercise 8).

7.1.4 Attracting and Repelling Periodic Points

The definitions of attracting and repelling fixed points for complex functions are virtually identical to their counterparts for real-valued functions.

Definition 7.6. Assume that p is a fixed point of a complex function f. Then p is **attracting** provided that there is a disk U in the plane centered at p such that if z is in the domain of f and in U, then $|f^{[n]}(z) - p| \to 0$ as n increases without bound. By contrast, a fixed point p is **repelling** if there is a disk U centered at p such that if z is in the domain of f and in U and $z \neq p$, then $|f(z) - p| > |z - p|$.

For examples of an attracting fixed point and a repelling fixed point, let us again consider $f(z) = z^2$.

Example 7.1.7. Let $f(z) = z^2$. Show that 0 is an attracting fixed point and 1 is a repelling fixed point.

Solution. First, we will show that 0 is an attracting fixed point. To that end, notice that

$$f(z) = z^2, \quad f^{[2]}(z) = z^4, \quad \text{and in general,} \quad f^{[n]}(z) = z^{(2^n)}$$

so that

$$\text{if } |z| < 1, \text{ then } |f(z)| < |z| < 1 \tag{7.6}$$

Letting U be the open disk of all z with $|z| < 1$, we note that the powers of z approach 0, so that 0 is an attracting fixed point. Next, we will show that 1 is a repelling fixed point. To begin with, we notice that if $z \neq 1$ but $|z - 1| < 1$, then $|z + 1| = |z - (-1)| > 1$. Consequently for such z,

$$|f(z) - 1| = |z^2 - 1| = |z - 1||z + 1| > |z - 1|$$

which means that 1 is a repelling fixed point. $\qquad \square$

The following theorem can expedite the proof that a fixed point of f is attracting or is repelling. The proof is almost identical for complex functions and for real functions, so we omit the proof.

Theorem 7.7. Let f be a differentiable complex function with fixed point p.

    a. If $|f'(p)| < 1$, then p is attracting.

    b. If $|f'(p)| > 1$, then p is repelling.

    c. If $|f'(p)| = 1$, then p may be attracting, repelling, or neither.

In the case of the function $f(z) = z^2$, we showed in Example 7.1.7 that 0 is an attracting fixed point and 1 is a repelling fixed point. Since $f'(z) = 2z$, Theorem 7.7 implies directly that 0 is attracting, whereas 1 is repelling.

Next, a point z_0 is an **attracting period-n point** (or a **repelling period-n point**) of the complex function f if z_0 is an attracting (or repelling) fixed point of $f^{[n]}$. In determining whether a periodic point is attracting or repelling, the Chain Rule plays a crucial role. Indeed, let $\{w, z\}$ be a 2-cycle for f. The fact that $f(w) = z$ and $f(z) = w$, along with the Chain Rule, together yield

$$(f^{[2]})'(w) = (f \circ f)'(w) = [f'(f(w))]\,[f'(w)] = f'(z)\,f'(w) \tag{7.7}$$

By symmetry, $(f^{[2]})'(z) = f'(w)\,f'(z)$. It follows from Theorem 7.7 applied to period-2 points that

$$\text{if } |f'(w)|\,|f'(z)| < 1, \quad \text{then the 2-cycle } \{w, z\} \text{ is attracting}$$

and

$$\text{if } |f'(w)|\,|f'(z)| > 1, \quad \text{then the 2-cycle } \{w, z\} \text{ is repelling}$$

From the symmetry we see that if $\{w, z\}$ is a 2-cycle, then w is attracting (repelling) if and only if z is attracting (repelling).

Example 7.1.8. Let $f(z) = z^2 + i$. Show that $\{-1 + i, -i\}$ is a repelling cycle.

Solution. In Example 7.1.5 we showed that the pair $\{-1 + i, -i\}$ is a 2-cycle. Next, we use the Chain Rule to find that

$$(f^{[2]})'(-1+i) = f'(-1+i)\,f'(-i) = [2(-1+i)]\,[2(-i)] = 4(1+i) = (f^{[2]})'(-i)$$

Since $|4(1+i)| = 4\sqrt{2} > 1$, we know from the comments preceding the example that the cycle is repelling. $\qquad\square$

More generally, let g be differentiable, with n-cycle $\{z_0, z_1, \ldots, z_{n-1}\}$. One member of the n-cycle is attracting (repelling) if and only if all the remaining members of the n-cycle are attracting (repelling). Moreover, by an extension of the Chain Rule,

$$|(g^{[n]})'(z_0)| = |g'(z_0)|\,|g'(z_1)| \cdots |g'(z_{n-1})|$$

Let us see how this applies to periodic points for the function $f(z) = z^2$, discussed in Theorem 7.5.

Example 7.1.9. Let $f(z) = z^2$, and let $z_0 = e^{2\pi i k/p}$, with k and p relatively prime positive integers and p odd. Then z_0 is a repelling periodic point.

Solution. By Theorem 7.5, z_0 is a periodic point. Suppose that z_0 has period n, with corresponding n-cycle $\{z_0, z_1, \ldots, z_{n-1}\}$. Noting that $f'(z) = 2z$ for all z, and using the extended Chain Rule above, we find that

$$|(f^{[n]})'(z_0)| = |f'(z_0)|\,|f'(z_1)| \cdots |f'(z_{n-1})| = 2^n > 1$$

Thus z_0 is a repelling period-n point. $\qquad\square$

The upshot of Example 7.1.9 is that the set of *repelling* periodic points on the unit circle for the function $f(z) = z^2$ is dense in the unit circle! In the next two sections repelling periodic points will play a key role in the discussion of Julia sets and the Mandelbrot set.

7.1.5   Section 7.1 Exercises

Exercise 1. Find the four 4th roots of -1.

Exercise 2. Find the three 3rd roots of $8i$.

Exercise 3. Let $f(z) = z^2$. Find the period-4 points on the unit circle.

Exercise 4. Let $f(z) = z^2 + i/4$. Find all fixed points of f.

Exercise 5. Show that the repelling periodic points of $f(z) = z^2$ form a dense set of the unit circle.

Exercise 6. Let $g_1(z) = z^2 + 1$. Determine the period-2 points for g_1.

Exercise 7. Let $g_{1/4}(z) = z^2 + 1/4$. Find the fixed point of $g_{1/4}$, and determine whether it is attracting, repelling, or neither.

Exercise 8. Let $f(z) = z^2$. Show that for each positive integer n, there are periodic points of period n.

Exercise 9. Let $f(z) = z^2$. Show that if $z = e^{2\pi i \theta}$, where in reduced form θ is not of the form k/p with p odd, then z cannot be a periodic point for f.

Exercise 10. Describe the regions in the complex plane given by each of the equations below:

- $|z| = 1$
- $|z| \geq 1$
- $|z - 1| = 2$
- $|z - i| = 2$
- $|z| = |z - 1|$

7.2   Julia Sets

In Section 7.1 we introduced the complex plane and complex functions, and discussed attracting and repelling periodic points. In the present section and the next section we will focus on famous sets that arise from attracting and repelling periodic points for members of the family $\{g_c\}$ of complex functions defined by

$$g_c(z) = z^2 + c, \quad \text{for all complex numbers } z$$

where c is an arbitrary fixed complex parameter for each function g_c. Sometimes we refer to the family as the g_c-**family**. Since this family consists of quadratic functions, the quadratic formula yields fixed points. More precisely,

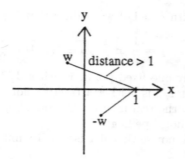

FIGURE 7.6
The distance between 1 and w or $-w$ is greater than 1

p is a fixed point of g_c provided that $p = g_c(p) = p^2 + c$, or equivalently, $p^2 - p + c = 0$. By the quadratic formula, this means that

$$p = \frac{1}{2} \pm \frac{1}{2} \sqrt{1 - 4c} \qquad (7.8)$$

Notice that g_c has two fixed points precisely when $1 - 4c \neq 0$, that is, when $c \neq 1/4$. In Theorem 7.8 will use (7.8) and Theorem 7.7 to determine the nature of the fixed points of g_c when $c \neq 1/4$.

Theorem 7.8. Let $c \neq 1/4$. Then at least one fixed point of g_c is repelling.

Proof. By (7.8), any fixed point p of g_c has the form $p = \frac{1}{2} \pm \frac{1}{2} \sqrt{1 - 4c}$. Since $g_c'(z) = 2z$, it follows that

$$g_c'(p) = g_c'(\frac{1}{2} \pm \frac{1}{2} \sqrt{1 - 4c}) = 1 \pm \sqrt{1 - 4c}$$

If $w = \sqrt{1 - 4c}$, then $w \neq 0$ since $c \neq 1/4$ by hypothesis. Moreover,

$$g_c'(p) = 1 \pm w$$

Since $w \neq 0$, either w or $-w$ lies on or to the left of the imaginary axis and is not the origin (Figure 7.6), so that the distance between 1 and either w and $-w$ (or both) is greater than 1. Therefore

$$\text{either} \quad |g_c'(p)| = |1 - w| > 1 \quad \text{or} \quad |g_c'(p)| = |1 + w| > 1$$

The fixed point corresponding to that w is necessarily repelling by Theorem 7.7. We conclude that g_c has a repelling fixed point. $\qquad \square$

Next, we study the case in which $c = 1/4$.

Lemma 7.1. Show that $g_{1/4}(z) = z^2 + \frac{1}{4}$ has exactly one fixed point, which is neither attracting nor repelling.

Proof. From (7.8) with $c = 1/4$ we deduce that if p is a fixed point of $g_{1/4}$, then

$$p = \frac{1}{2} \pm \frac{1}{2}\sqrt{1 - 4(1/4)} = \frac{1}{2}$$

Thus $g_{1/4}$ has exactly one fixed point, which is $1/2$. To complete the proof, we need to show that $1/2$ is neither attracting nor repelling. We will do that by restricting $g_{1/4}$ to the real numbers R, and showing that even on R, $g_{1/4}$ is neither attracting nor repelling.

To begin with, we notice that if x is a real number and $x \neq 1/2$, then

$$g_{1/4}(x) = x^2 + \frac{1}{4} > x \quad \text{if and only if} \quad (x - \frac{1}{2})^2 = x^2 - x + \frac{1}{4} > 0$$

which is valid whenever $x \neq 1/2$. Thus g_c is an increasing function. Next, assume that $0 < x < 1/2$. Then $g_{1/4}(x) = x^2 + 1/4 < (1/2)^2 + 1/4 = 1/2$. Consequently $\{g_{1/4}^{[n]}(x)\}_{n=1}^{\infty}$ is not only an increasing positive sequence, but is bounded above by $1/2$. Therefore the sequence converges, so by Theorem 1.3 it must converge to a fixed point, namely $1/2$. That means that the fixed point $1/2$ cannot be repelling. By contrast, let $x > 1/2$. Then $g_{1/4}(x) = x^2 + 1/4 > 1/4 + 1/4 = 1/2$. This means that the sequence $\{g_{1/4}^{[n]}(x)\}_{n=1}^{\infty}$ is increasing, with all members greater than $1/2$, and thus is receding from $1/2$. Therefore $1/2$ cannot be attracting. The conclusion is that $1/2$ is neither attracting nor repelling. $\square$

Lemma 7.2. Show that $g_{1/4}(z) = z^2 + \dfrac{1}{4}$ has a repelling 2-cycle.

Proof. To find period-2 points for $g_{1/4}$ we need to solve the equation $z = g_{1/4}^{[2]}(z)$, or equivalently,

$$z = g_{1/4}^{[2]}(z) = g_{1/4}(z^2 + \frac{1}{4}) = (z^2 + \frac{1}{4})^2 + \frac{1}{4} = z^4 + \frac{1}{2}z^2 + \frac{5}{16}$$

Thus $z^4 + z^2/2 - z + 5/16 = 0$.

Since $1/2$ is a fixed point of $g_{1/4}$, we know that $1/2$ is a fixed point of $g_{1/4}^{[2]}$, so that $z - 1/2$ divides $g_{1/4}^{[2]}(z) - z = z^4 + z^2/2 - z + 5/16$. It turns out that $1/2$ is a double root of $z^4 + z^2/2 - z + 5/16$, which leads to the following factorization:

$$z^4 + \frac{1}{2}z^2 - z + \frac{5}{16} = (z - \frac{1}{2})^2 (z^2 + z + \frac{5}{4})$$

By the quadratic formula,

$$z^2 + z + \frac{5}{4} = 0 \quad \text{if and only if} \quad z = -\frac{1}{2} \pm i$$

Therefore both $-1/2 + i$ and $-1/2 - i$ are fixed points of $g_{1/4}^{[2]}$. Since neither $-1/2 + i$ nor $-1/2 - i$ is a fixed point of $g_{1/4}$ but each is a fixed point of $g_{1/4}^{[2]}$, we conclude that $g_{1/4}$ has the 2-cycle $\{-\frac{1}{2} + i, -\frac{1}{2} - i\}$.

To show that the 2-cycle is repelling, we need only use (7.8) in Section 7.1:

$$|(g_{1/4}^{[2]})'(-\frac{1}{2} + i)| = |g_{1/4}'(-\frac{1}{2} + i)\, g_{1/4}'(-\frac{1}{2} - i)| = [2(-\frac{1}{2} + i)]\,[2(-\frac{1}{2} - i)]| = 5$$

Therefore the 2-cycle $\{-\frac{1}{2} + i, -\frac{1}{2} - i\}$ is repelling. □

Theorem 7.8 and Lemmas 7.1 and 7.2 together yield the following theorem.

Theorem 7.9. If $c \neq 1/4$, then g_c has a repelling fixed point. If $c = 1/4$, then g_c has a repelling 2-cycle.

Let us recall the special function $g_0(z) = z^2$, whose fixed points and periodic points we studied in Section 7.1. By Theorem 7.5, g_0 has an infinite number of periodic points on the unit circle $r = 1$: every point of the form $e^{2\pi i k/p}$, where k and p are relatively prime positive integers with p odd and $k < p$. Since $g_0'(z) = 2z$, so that $|g_0'(e^{2\pi i k/p})| = 2|e^{2\pi i k/p}| = 2$, Theorem 7.7 implies that all of the periodic points on the unit circle are repelling. Consequently the collection of repelling periodic points of g_0 forms a dense subset of the unit circle in the complex plane. This feature leads us to the following definition that applies to all functions in the g_c-family.

Definition 7.10. Let c be any complex number. The smallest closed set in the complex plane that contains all repelling fixed points and all repelling periodic points of g_c is called the **Julia set** of g and is denoted by J_c.

Note: Julia sets are named for the French mathematician Gaston Julia (1893–1978), who made profound discoveries about iterates of complex rational functions in the early 20th century. His study of iterated polynomial functions, along with similar work by Pierre Fatou, laid the foundation for Mandelbrot's work in the 1950's. Without this work, the notion of a fractal would not have been formalized 80 years later. This is an important example of mathematics that was done for its own sake, but which leads to vital technological advances in the next century. We will investigate the relationship between a Juila set and the "Fatou dust" in a later exercise.

By the definition of Julia sets and by our discussion of the repelling points of g_0, the Julia set J_0 of g_0 is the unit circle. For the remaining functions g_c the Julia set is much more complicated. Figure 7.7 exhibits four Julia sets, with accompanying values of c.

The next set of propositions demonstrates several salient properties of Julia sets.

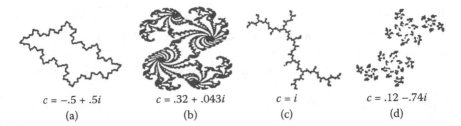

$c = -.5 + .5i$ $c = .32 + .043i$ $c = i$ $c = .12 - .74i$

(a)   (b)   (c)   (d)

FIGURE 7.7
Four Julia sets

Theorem 7.11. For every complex number c, the Julia set J_c is nonempty.

Proof. By Theorems 7.8 and 7.9, g_c has a repelling fixed point or a repelling period-2 cycle, so J_c is nonempty for each complex number c.   □

You might observe that we have merely shown that there *exists* an element in J_c. The question of the size of J_c is more difficult to answer. Note that J_0 = the unit circle in the complex plane, so it has an infinite number of points. It turns out that J_c always has an infinite number of points, though we will not prove that here.

Presently we continue with properties related to J_c.

Theorem 7.12. If $|z| > |c| + 1$, then the orbit of z for g_c is unbounded.

Proof. If $c = 0$, then $g_c(z) = z^2$, and the inequality $|z| > |c| + 1$ becomes $|z| > 1$. It follows that the iterates of z are just powers of z, the collection of which is an unbounded set. So henceforth we will assume that $c \neq 0$, and let $|z| > |c| + 1$. Then

$$|g_c(z)| = |z^2 + c| = |z| \left| z + \frac{c}{z} \right| \geq |z| \left| |z| - \frac{|c|}{|z|} \right| > |z| \left| |c| + 1 - \frac{|c|}{|c| + 1} \right|$$

Let

$$r = |c| + 1 - \frac{|c|}{|c| + 1} = 1 + (|c| - \frac{|c|}{|c| + 1}), \quad \text{so that } r > 1$$

Then by our calculations, $|g_c(z)| > r|z|$. Since $r > 1$, it follows that $|g_c^{[n]}(z)| > r^n|z| \to \infty$ as n increases without bound. Therefore if $|z| > |c| + 1$, then the iterates of z form an increasing, unbounded sequence, so that the orbit of z is unbounded.   □

Theorem 7.13. If z is in J_c, then $|z| \leq |c| + 1$, so that J_c is a bounded subset of the complex plane.

Proof. If z is a periodic point for g_c, then necessarily the orbit z is bounded, so by Theorem 7.12, $|z| \leq |c| + 1$. Since J_c is the smallest closed set containing all repelling periodic points, any z in J_c also has the property that $|z| \leq |c| + 1$. Consequently J_c is bounded.   □

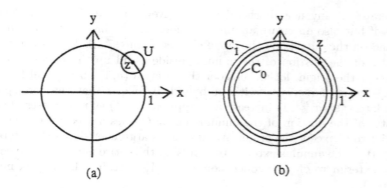

FIGURE 7.8
Iterates of points in U

Recall that a set S is **invariant** under a function f if $f(S) \subseteq S$. In other words, S is invariant if its image under f is contained in S.

Theorem 7.14. J_c is invariant under g_c. In fact, $J_c = g_c(J_c)$.

Proof. If S is the set of all the repelling periodic points for g_c, then since iterates of repelling periodic points are themselves repelling periodic points, it follows that $S = g_c(S)$. The closures of S and $g_c(S)$ are, respectively, J_c and $g_c(J_c)$, which are thus equal. □

Theorem 7.15. If z is in J_c, then the orbit of z is bounded.

Proof. Let z be in J_c. Then by Theorem 7.13, $|z| \leq |c| + 1$. However, the iterates of J_c for g_c are precisely J_c by Theorem 7.14. Therefore, the iterates of z also are bounded by $|c| + 1$. □

Next, we turn to the iterates of points in a neighborhood of an element of J_c. To that end, let z be in J_c, and let U be any disk centered at z. We will denote by V_c the set defined by

$$V_c = \bigcup \{g_c^{[n]}(U) : n = 1, 2, \ldots\} = \text{all iterates of all elements of } U$$

For example, let $c = 0$, and let z be in J_0, so that z lies on the unit circle. In addition, let U be a disk centered at z. Then part of U lies inside the unit circle and part lies outside the circle (Figure 7.8(a)). This leads us to Theorem 7.16, which describes V_0.

Theorem 7.16. Let $c = 0$, and let U be any neighborhood of any point z on the unit circle C. Then V_0 contains all complex numbers except 0.

Proof. Since the angle subtended by any nontrivial arc of C is doubled by g_0, it is apparent that the union of an appropriate finite number of iterates of any arc of C contains all of C. In this same vein, an appropriate finite number of iterates of any neighborhood U of a given point in C contains not only C itself but also an entire annulus bounded on the inside of C by the circle C_0 and on the outside by the circle C_1 (see Figure 7.8(b)). To show that the iterates of the portion of the annulus inside C fill up the region interior to C except the origin, let us consider the interval $(a, 1)$ on the real line. Since $a^{(2^n)} \to 0$ as n increases without bound, it follows that the corresponding open intervals $(a^{(2^n)}, 1)$ increase and approach $(0, 1)$ as n increases. Thus the iterates of the portion of the annulus inside C does indeed fill up the region interior to C except the origin. An analogous argument shows that iterates of the part of the annulus exterior to C fills up the entire portion of the complex plane exterior to C. The conclusion is that V_0 contains all complex numbers except 0. $\qquad\square$

If $c \neq 0$, then a corresponding result holds for V_c, but the proof is much more technical, in part because J_c is not so tractable. The result is essentially a corollary of a famous, 90-year-old theorem by the French mathematician Paul Montel. Its proof can be found in books on complex analysis.

Theorem 7.17 (Montel's Theorem). Let c be any complex number, and suppose that z is an arbitrary element of J_c. Then V_c contains all complex numbers except at most one complex number.

Theorem 7.16, Montel's Theorem, and a little additional argument yields the following slight extension of Theorem 7.17.

Theorem 7.18. Let z be any element of J_c, and let U be any disk centered at z. Let V_c be associated with z and U, as defined above.

    a. If $c = 0$, then V_c consists of all complex numbers except 0.

    b. If $c \neq 0$, then V_c consists of all complex numbers.

Proof. Theorem 7.16 yields (a). For (b), let $c \neq 0$, and assume that V_c does not contain all the complex numbers. In that case, Montel's Theorem assures us that V_c contains all but a single point s. We will obtain a contradiction by proving that $c = 0$.

First, we use the fact that any complex number has a square root. So let $b = \sqrt{s - c}$, that is, assume that b is one square root of $s - c$. Then $b^2 = s - c$, so that $g_c(b) = b^2 + c = s$. If $b \neq s$ then since V_c contains all complex numbers except s, it follows that b is in V_c. However, that means that $s = g_c(b)$ is also in V_c because $g_c(V_c) \subseteq V_c$ by the definition of V_c. This contradicts the assumption that s is *not* in V_c. Thus $b = s$. Therefore b is uniquely defined as s, so that $s - c$ has only one square root, namely $b = s$. The only complex number with a single square root is 0. Consequently $s - c = 0$, which means that $s = c$. As a result, $b = s = c$. Thus $g_c(b) = b^2 + c = s$ becomes $g_c(b) = c^2 + c = c$, which means that $c = 0$. This contradicts the assumption in (b) that $c \neq 0$. We conclude that V_c contains all complex numbers. Part (b) is thus proved. $\square$

Recall that the Julia set J_0 for $g_0(z) = z^2$ is the unit circle. The unit circle has no interior, that is, it contains no open disks about any of its points. This property is characteristic of all Julia sets, as we now prove.

Theorem 7.19. For each complex number c, J_c has empty interior.

Proof. Let z be in J_c, and suppose that a nonempty open disk U centered at z also is in J_c. By Theorem 7.18, the corresponding set V_c with respect to z and U contains all complex numbers except possibly 0. Thus there is a w in U such that some iterate $|g_c^{[n]}(w)| > |c| + 1$. However, Theorem 7.15 implies that such a w cannot be in J_c. We conclude that there is no disk about z contained in J_c. In other words, J_c has empty interior. □

A second characterization of the Julia set J_c is this:

J_c is the boundary B_c of the set of all z such that the orbit of z for g_c is unbounded.

The proof is nontrivial, and we won't give it here. However, you might be able to imagine that $J_c \subseteq B_c$. Indeed, let z be in J_c. Then by Theorem 7.18, in any given open disk about z there are points w whose orbits are unbounded. Since the iterates of z are bounded by Theorem 7.15, we conclude that z is in B_c.

We will use the new characterization of J_c to help determine properties of symmetry for J_c.

Theorem 7.20. J_c is symmetric with respect to the origin, that is, z is in J_c if and only if $-z$ is in J_c.

Proof. Let z be in J_c. On the one hand, the orbit of z is bounded by Theorem 7.15. Since

$$g_c(-z) = (-z)^2 + c = z^2 + c = g_c(z)$$

it follows that $-z$ also has a bounded orbit. On the other hand, by the alternate characterization of J_c, in any small neighborhood U of z there is a w such that the orbit of w is unbounded. Because $g_c(-w) = g_c(w)$, it follows that $-w$, which lies in the small neighborhood $-U$ about $-z$, also has an unbounded orbit. Consequently $-z$ is in J_c. We conclude that J_c is symmetric with respect to the origin. □

Before we discuss symmetry with respect to the axes, we define the **conjugate** $\overline{z}$ of z:

$$\text{if} \quad z = x + yi, \quad \text{then} \quad \overline{z} = x - yi$$

Thus $\overline{z}$ is the point symmetric to z with respect to the x-axis. You can show that $\overline{z + w} = \overline{z} + \overline{w}$ and that $\overline{zw} = \overline{z}\,\overline{w}$ for all complex numbers z and w. Moreover, it is apparent that if z is real, then $\overline{z} = z$.

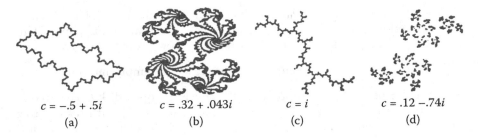

$$c = -.5 + .5i \qquad c = .32 + .043i \qquad c = i \qquad c = .12 - .74i$$

(a)   (b)   (c)   (d)

FIGURE 7.9
Varied Julia sets from a simple closed curve to a totally disconnected set

Theorem 7.21. If c is real, then J_c is symmetric with respect to both the x and the y axes.

Proof. Since c is real by hypothesis, it follows that for any complex number z, we have $\overline{z^2} + c = \overline{z^2 + c}$. Therefore

$$g_c(\overline{z}) = (\overline{z})^2 + c = \overline{z^2} + c = \overline{z^2 + c} = \overline{g_c(z)}$$

By induction, $g_c^{[k]}(\overline{z}) = \overline{g_c^{[k]}(z)}$ for all positive integers k. In particular, if z has period n for g_c, then

$$\overline{z} = \overline{g_c^{[n]}(z)} = g_c^{[n]}(\overline{z})$$

Thus $\overline{z}$ also has period n. Because of the reflection properties of conjugates, z is a *repelling* periodic point if and only if $\overline{z}$ is a *repelling* periodic point. Therefore the set of repelling periodic points is symmetric with respect to the x-axis. The closure of that set, which is J_c, inherits the symmetry property as well. Finally, symmetry with respect to the y-axis follows from symmetry with respect to the origin and with respect to the x-axis (see Exercise 7). $\square$

Julia sets can have widely varied shapes, as Figure 7.7, reproduced here as Figure 7.9, suggests. During the past several decades these sets have been studied very carefully, and many of their properties are known. Here we list five major classes of Julia sets.

1. J_0 is the unit circle.

2. J_c is a simple closed curve that is nowhere differentiable, provided that g_c has an attracting fixed point and $c \neq 0$ (which occurs if, for example, $c < 1/4$). (See Figure 7.9(a).)

3. J_c is connected and encloses a figure with interior, but is a more complicated shape, provided that g_c has an attracting cycle that is not a fixed point. (See Figure 7.9(b).)

4. J_c is connected but does not enclose a figure with interior, provided that some iterate of 0 is a fixed point or is periodic. Such a Julia set is called a **dendrite**. (See Figure 7.9(c).)

5. J_c is a totally disconnected, two-dimensional Cantor-like set, provided that the orbit of 0 is unbounded. Such a Julia set is called a **Fatou dust**, after the French mathematician Pierre Fatou

---

Note: Fatou dust is illustrated in Figure 7.9(d), though computer renditions make the figure look like lots of little petals, rather than a totally disconnected point set. A fascinating, if not slightly cheeky, paper illustrating the limits of computer graphics is *Drawing the Pseudo-arc* by Lewis and Minc (2010).

How can one derive an approximation of a given Julia set J_c on the computer screen? More particularly, if we select any z_0 in the complex plane, what process could lead to an approximation of J_c? We will describe, without proof, a relatively short process that does the job.

First, we notice that if $z^2 + c = z_0$, then $z_0 - c = z^2$, so that $\sqrt{z_0 - c}$ is either z or $-z$. In other words, the pre-image $\sqrt{z_0 - c}$ of z_0 is either z or $-z$.

Next, we use the fact that the repelling periodic points of g_c are dense in J_c, along with the result of Theorem 7.18 that iterates of neighborhoods of points in J_c fill up the entire plane (except possibly 0). Together these observations direct us to the following **Julia set algorithm** for approximating J_c on the computer:

1. Choose any complex number z_0 that is *not* a periodic point of g_c.

2. Choose randomly one of the two pre-images of z_0 of g_c: $\sqrt{z_0 - c}$ or $-\sqrt{z_0 - c}$. Call the chosen pre-image z_1, and plot z_1.

3. Continue the selection process, plotting $z_1, z_2, z_3, \ldots$.

The Julia set algorithm process might remind you of the Random Iteration Algorithm in Section 6.5. In the present case, the random selection derives from two functions!

We conclude with a description of one Julia set other than J_0 that we can draw and analyze.

Example 7.2.1. Show that J_{-2} is the interval $[-2, 2]$ on the real line.

Partial Solution. The function $g_{-2}(z) = z^2 - 2$, pictured in Figure 7.10(a), has the following notable properties:

1. $g_{-2}(-2) = (-2)^2 - 2 = 2 = g_{-2}(2)$

2. $g_{-2}(0) = -2$

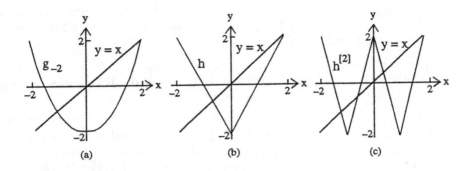

FIGURE 7.10

g_2, a related function h, and its iterates

3. g_{-2} is decreasing on $[-2, 0]$ and increasing on $[0, 2]$.

4. $g_{-2}^{[n]}$, which is a polynomial of degree 2^n, has *precisely* 2^n zeros by the Fundamental Theorem of Algebra, and hence g_{-2} has precisely 2^n periodic points with period dividing n.

It is possible to show that g_{-2} is related to the function h whose graph is a V and appears in Figure 7.10(b). In particular, one can show that g_{-2} has the same number of period-n points as does h, and moreover those period-n points are spread out fairly evenly over the interval $[-2, 2]$. Calculating the period-n points of h is straightforward. Indeed, from Figure 7.10(b) we see that h has two fixed points, and from Figure 7.10(c) that $h^{[2]}$ has 2 additional fixed points, i.e., h has a period-2 cycle. More generally, the graph of the nth iterate of h has n figure V's, each of which crosses the line $y = x$ twice. Therefore h has 2^n periodic points with period dividing n. Our comment just above implies that g_{-2} also has 2^n such periodic points, which are more-or-less evenly spread out over the interval $[-2, 2]$. Since g_{-2} has precisely 2^n such periodic points, we deduce that all the periodic points of g_{-2} lie on the interval $[-2, 2]$, and are spread out more-or-less evenly. The conclusion is that the periodic points of g_{-2} form a dense subset of the interval $[-2, 2]$, so that $J_{-2} = [-2, 2]$.

There are many other features of Julia sets, such as "quasi"-self-similarity for certain such sets, and the actual behavior of iterates of 0 in relation to J_c. However, these are beyond the scope of this text.

7.2.1   Section 7.2 Exercises

Exercise 1. Show that g_{-1} has no attracting fixed points.

Exercise 2. Show that 0 is an attracting period-2 point for g_{-1}.

Exercise 3. Determine whether $g_{1/4}$ has a 3-cycle, giving reasons.

Exercise 4. Show that if $|c| < 1/4$, then g_c has an attracting fixed point.

Exercise 5. Let c be a real number. Give conditions under which g_c has two repelling fixed points.

Exercise 6. Let $z = x + iy$ and $w = u + iv$. Show that $\overline{zw} = \overline{z}\,\overline{w}$.

Exercise 7. Show that if a set S in the complex plane is symmetric with respect to the origin and with respect to the x-axis, then it is also symmetric with respect to the y-axis.

Exercise 8. Suppose that $z_0 = 10$ and that we use the Julia set algorithm to approximate the Julia set J_0, which is the unit circle. What is the minimum integer n such that z_n is within $1/10$ of the Julia set?

7.3 The Mandelbrot Set

We turn to one of the most celebrated sets in all of mathematics: the Mandelbrot set. Popular on dorm room walls for decades, we'll see how this set has important mathematical consequences, beyond its attractiveness.

Definition 7.22. The collection of complex numbers c for which the orbit of 0 for g_c is bounded is called the **Mandelbrot set** and is denoted by M.

The Mandelbrot set is named for the mathematician Benoit Mandelbrot (1924–2010), who in the late 1970s brought the set wide acclaim with the help of images on a "high-speed" computer.

> **Note:** Mandelbrot's first ideas in this field came from studying "noise" in signal processing. He was working as a researcher in 1961, helping IBM identify how to correct the errors that occurred in transmitting computer data over phone lines. While it took until nearly 1980 for computing power to allow him to iterate this function effectively, it was his insight that some behavior was self-similar that would eventually lead to coining the term "fractal," and the mathematics underlying to chapters 5–7.

How are Julia sets and the Mandelbrot set related? On the one hand, for each complex number c there is a Julia set: the smallest closed set containing the repelling periodic points for g_c. On the other hand, the Mandelbrot set is defined by the behavior of 0 for the various functions in the g_c-family. In other

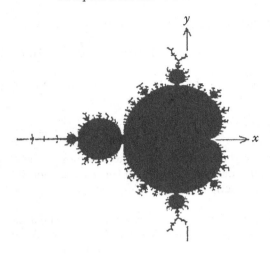

FIGURE 7.11
The Mandelbrot set

words, there are infinitely many Julia sets (one for each c), whereas there is only one Mandelbrot set. Figure 7.11 gives a rendition of the Mandelbrot set, including its ultra-complicated boundary.

Our first goal is to learn about what is surely *in* M, and what is surely *not* in M.

Theorem 7.23. The Mandelbrot set has the following major features:

    a. M contains all complex numbers c such that $|c| \leq 1/4$.

    b. If c is in M, then $|c| \leq 2$. Equivalently, if $|c| > 2$, then c is not in M.

Proof. To prove (a), let $|c| \leq 1/4$, and recall that $g_c(z) = z^2 + c$, for all z. Then $|g_c(0)| = |c| \leq 1/4$. In addition,

$$|g_c^{[2]}(0)| = |g_c(c)| = |c^2 + c| \leq |c|^2 + |c| \leq (\frac{1}{4})^2 + \frac{1}{4} < \frac{1}{2}$$

To apply induction, we assume that $|g_c^{[n]}(0)| \leq 1/2$. Then

$$|g_c^{[n+1]}(0)| = |g_c(g_c^{[n]}(0))| = |[g_c^{[n]}(0)]^2 + c| \leq |[g_c^{[n]}(0)]^2| + |c| \leq \frac{1}{4} + \frac{1}{4} = \frac{1}{2}$$

Therefore the Law of Induction implies that $|g_c^{[n]}(0)| \leq 1/2$ for all positive n, so that c is in M. Thus part (a) is proved.

To prove (b), suppose that $|c| > 2$, and let $r = |c| - 1$, so that $r > 1$. If $|z| \geq |c|$, then

$$|g_c(z)| = |z^2 + c| = |z|\,|z + \frac{c}{z}| \geq |z|\,(|z| - \frac{|c|}{|z|}) \geq |z|\,(|c| - 1) = |z|r \quad (7.9)$$

Since $|g_c(0)| = |c|$, it follows that $|g_c^{[2]}(0)| = |g_c(c)| \geq |c|r$. In order to apply induction, assume that $|g_c^{[n]}(0)| \geq |c|r^{n-1}$. Then by (7.9) and the induction hypothesis,

$$|g_c^{[n+1]}(0)| = |g_c(g_c^{[n]}(0))| \geq |g_c^{[n]}(0)|r \geq (|c|r^{n-1})r = |c|r^n$$

Therefore by the Law of Induction, $|g_c^{[n]}(0)| \geq |c|r^{n-1}$ for all positive integers n. Since $r > 1$, the iterates of 0 are unbounded, so that c is not in M. This completes the proof of (b). $\square$

Next, we will use a complex version of Singer's Theorem (Theorem 1.23); the complex version was proved by Julia nearly a century ago. Its proof, which is beyond the scope of this book, appears in Devaney (1989).

Theorem 7.24. Assume that p is an attracting periodic point of a complex polynomial function f. Then there is a critical point that lies in the basin of attraction of p.

Corollary 7.25. If g_c has an attracting periodic point, then c is in the Mandelbrot set M.

Proof. Since $g_c'(z) = 2z$, 0 is the lone critical point of g_c. If g_c has an attracting periodic point p, then by Theorem 7.24 the iterates of the critical point 0 for g_c converge to p and its iterates. Hence the iterates of 0 are bounded, so that c is in M. $\square$

It is time to describe those complex numbers c that comprise the large heart-like, or cardioid-shaped region that is located on the right side of the Mandelbrot set. That region is called the **body** of the Mandelbrot set and is designated B. To set the stage, let us recall from (7.9) in Section 7.2 that the fixed points of g_c are $(1 \pm \sqrt{1-4c})/2$. Since

$$g_c'(\tfrac{1}{2}(1 \pm \sqrt{1-4c})) = 1 \pm \sqrt{1-4c}$$

it follows that g_c has an attracting fixed point only if

$$\text{either} \quad |1 - \sqrt{1-4c}| \leq 1 \quad \text{or} \quad |1 + \sqrt{1-4c}| \leq 1$$

Theorem 7.26. The set of all complex numbers c such that g_c has an attracting fixed point occupies the region bounded by the cardioid that represents the body B of the Mandelbrot set.

Proof. First, we notice that if g_c has an attracting fixed point, then c is in M, by Corollary 7.25. Next, let $w = \sqrt{1-4c}$. We will determine all complex numbers w such that $|1-w| \leq 1$ or $|1+w| \leq 1$. We will show that the complex number c is an element of the cardioid if and only if either $|1 - w| = 1$ or $|1 + w| = 1$. Then we will analyze those w, next analyze the associated w^2, and after that the associated $(1 - w^2)/4$, finally relating the end result with the values of c.

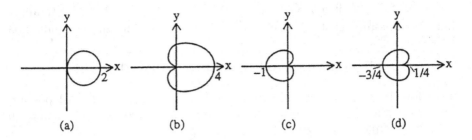

(a)   (b)   (c)   (d)

FIGURE 7.12
Analysis of the body of the Mandelbrot set

The collection of all w such that $|1 - w| = 1$ is the circle of radius 1 centered at the point $(1, 0)$ (Figure 7.12(a)), which in polar coordinates is given by $r = 2\cos\theta$. In other words, w can be written $w = (r, \theta) = (2\cos\theta, \theta)$. Then $w^2 = (4\cos^2\theta, 2\theta)$. By the trigonometric half-angle formula,

$$2\cos^2\theta = 1 + \cos 2\theta, \quad \text{so that} \quad 4\cos^2\theta = 2 + 2\cos 2\theta$$

Thus in polar coordinates, $w^2 = (2 + 2\cos 2\theta, 2\theta)$, or equivalently (by substituting θ for 2θ), $w^2 = (2 + 2\cos\theta, \theta)$. It follows that w^2 has polar coordinates (r, θ) satisfying $r = 2 + 2\cos\theta$. However, this is an equation of the cardioid represented in (Figure 7.12(b)). A similar conclusion is drawn if we consider w such that $|1 + w| = 1$ (instead of $|1 - w| = 1$).

We have shown thus far that the collection of all w^2 such that $|1 - w| = 1$ or $|1 + w| = 1$ forms the cardioid $r = 2 + 2\cos\theta$. All that remains is to identify the boundary of M with an appropriately repositioned cardioid. To that end, we note that if $w = \sqrt{1 - 4c}$, then $c = (1 - w^2)/4$. Now if the collection of w^2 as discussed in the paragraph above represents the boundary of the cardioid in Figure 7.12(b), then the points $-w^2/4$ form the cardioid in Figure 7.12(c) obtained by reflecting across the origin and contracting by a factor of $1/4$. Finally, $(1 - w^2)/4$ is obtained by adding $1/4$ (Figure 7.12(d)). Since $w = \sqrt{1 - 4c}$ yields $c = (1 - w^2)/4$, we conclude that c is on the cardioid of Figure 7.12(d). This means that the complex numbers c such that $|1 - \sqrt{1 - 4c}| \le 1$ or $|1 + \sqrt{1 - 4c}| \le 1$ form the cardioid in Figure 7.12(d) and its interior. We conclude that the complex numbers c for which g_c has an attracting fixed point comprises the region bounded by the cardioid represented by the body of M. $\square$

As we prove next, the region just to the left of the body in the Mandelbrot set is a bona-fide disk, called the **head** of the Mandelbrot set.

Theorem 7.27. The set of complex numbers c such that g_c has an attracting 2-cycle is the region bounded by the disk H centered at the point $(-1, 0)$ and with radius $1/4$.

Proof. Again from (7.8) in Section 7.2, we know that the fixed points p of g_c are given by $p = (1 \pm \sqrt{1 - 4c})/2$. Let us denote them by p_1 and p_2. (If $c = 1/4$, then $p = p_1 = p_2$, and p serves as a double fixed point.) Next, the fixed points of $g_c^{[2]}$ include the fixed points of g_c and the period-2 points of g_c. Now $z = g_c^{[2]}(z)$ if and only if

$$z = g_c(g_c(z)) = [g_c(z)]^2 + c = (z^2 + c)^2 + c = z^4 + 2cz^2 + c^2 + c$$

or equivalently,

$$z^4 + 2cz^2 - z + c^2 + c = 0$$

Since p_1 and p_2 are fixed points of g_c and hence of $g_c^{[2]}$, it follows that $z - p_1$ and $z - p_2$ are divisors of the polynomial $z^4 + 2cz^2 - z + c^2 + c$, yielding:

$$z^4 + 2cz^2 - z + c^2 + c = (z - p_1)(z - p_2)(z^2 + z + c + 1)$$

You can check that the solutions of $z^2 + z + c + 1 = 0$ are given by

$$u = -\frac{1}{2} + \frac{1}{2}\sqrt{-3 - 4c} \quad \text{and} \quad v = -\frac{1}{2} - \frac{1}{2}\sqrt{-3 - 4c}$$

They are also fixed points of $g_c^{[2]}$, and hence they constitute a period-2 cycle. (Why?) In order to determine conditions under which the cycle $\{u, v\}$ is attracting, let us first calculate and reduce the product uv:

$$uv = \frac{1}{4} - \frac{1}{4}(-3 - 4c) = 1 + c$$

Using this formula, the fact that $\{u, v\}$ is a 2-cycle, and the Chain Rule, we find that

$$|(g_c^{[2]})'(u)| = |g_c'(g_c(u))| \, |g_c'(u)| = |g_c'(v)| \, |g_c'(u)|$$
$$= |2v| \, |2u| = 4|1 + c| = |(g_c^{[2]})'(v)|$$

It follows that $\{u, v\}$ is attracting provided that $|4(1 + c)| < 1$, or equivalently, $|c - (-1)| < 1/4$ (or potentially it can be attracting if equality holds). This inequality describes the (open) disk centered at -1 with radius $1/4$. We conclude that the open disk just described, with its boundary, is the head of the Mandelbrot set. $\square$

There is no way to sketch precisely the Mandelbrot set, or even describe it decently, without a computer. Nevertheless, a simple program produces a facsimile of the set on the computer screen. The principle behind the program is that if $|c| > 2$, then by Theorem 7.23 we know that c is not in the Mandelbrot set.

To obtain a reasonable image of the Mandelbrot set M, we select a relatively dense collection S of c's inside the disk of radius 2 centered at the origin in the complex plane. Then we look at the iterates of 0 for g_c for the values of c in S. If, say, the first 50 iterates of 0 for g_c are no more than 2 in absolute value, then we accept that c is (very likely to be) in M. Such a procedure gives a relatively accurate representation of M. More accuracy in the representation can be achieved by analyzing 100, or 200, or more iterates of 0 for the values of c in S.

In Theorems 7.26 and 7.27 we determined the body and the head of the Mandelbrot set. Now we will present results that provide insight into the various bulbs that are situated everywhere around the boundary of the Mandelbrot set. We omit the proofs because most of them are technically challenging.

Let

$$K = \text{all } c \text{ such that } g_c \text{ has an attracting } n\text{-cycle for some } n > 0$$

Notice that $K \subseteq M$, by Corollary 7.25. It can be shown that K is composed of an infinite number of (open) connected components. The interior W_B of the body B of M is one such component, consisting of complex numbers c such that g_c has an attracting fixed point. Similarly, the interior W_H of the head H of M is another component, consisting of complex numbers c such that g_c has an attracting 2-cycle. More generally, if W is any open connected component of K, then there is a unique positive integer n such that g_c has an attracting n-cycle for each c in W. We will refer to such a W as an n-**cycle component**.

Consider a fixed n-cycle component W of K. Let c be an arbitrary point in W, and designate by $\{z_{c1}, z_{c2}, \ldots, z_{cn}\}$ the (unique) attracting n-cycle of g_c. (There is a theorem like Corollary 1.24 that says that any g_c can have at most *one* attracting cycle.) By the Chain Rule, for each $k = 1, 2, \ldots, n$, we have

$$(g_c^{[n]})'(z_{ck}) = [g_c'(z_{c1})][g_c'(z_{c2})] \cdots [g_c'(z_{cn})] = (2z_{c1})(2z_{c2}) \cdots (2z_{cn})$$
$$= 2^n z_{c1} z_{c2} \cdots z_{cn}$$

This leads us to define the function ρ_W by the formula

$$\rho_W(c) = (g_c^{[n]})'(z_{ck}) = 2^n z_{c1} z_{c2} \cdots z_{cn} \quad \text{for all } c \text{ in } W$$

Since the cycle is attracting, we deduce that $|\rho_W(c)| = |(g_c^{[n]})'(z_{ck})| \leq 1$. If D denotes the open unit disk centered at the origin, then it follows that ρ_W is a function from W to D. The function ρ_W is called the **multiplier** of W.

Mathematicians Adrien Douady and John Hubbard together proved that ρ_W is a one-to-one correspondence between W and D. (See Douady and Hubbard, 1982.) For example, if W is the interior of the body of M, then ρ_W yields a correspondence between the body and the unit disk D. Generally, for each W there is a unique complex number c_W in W such that $\rho_W(c_W) = 0$. Since

$$0 = \rho_W(c_W) = 2^n z_{cw1} \cdots z_{cwn}$$

we conclude that $z_{cwk} = 0$ for some k. Therefore 0 is in the attracting cycle $\{z_{cw1}, z_{cw2}, \ldots, z_{cwn}\}$, as is c_W, because $g_{c_W}(0) = c_W$. It follows that

$g_{cw}^{[n]}(0) = 0$, and thus $g_{cw}^{[n-1]}(c_W) = 0$. Because c_W corresponds to 0 via the map ρ_W, c_W is called the **center** of W. Each component W has its own center that uniquely defines W.

Notice that $g_{cw}^{[n-1]}$ is a polynomial of degree 2^{n-1}, so by the Fundamental Theorem of Algebra the polynomial has 2^{n-1} solutions, which can be shown to be distinct. Each solution serves as the center of a connected component W associated with k-cycles, where k divides n. This gives us a method of identifying the centers of various connected components:

1. $n = 1$: $g_{cw}^{[n-1]}(c_W) = g_{cw}^{[n]}(0) = 0$. Since $n = 1$, the equations translate to $c_W = g_{cw}(0) = 0$. Thus 0 is the center of the body, W_B.

2. $n = 2$: $g_{cw}^{[n-1]}(c_W) = 0$, which means that $0 = g_{cw}(c_W) = c_W^2 + c_W = c_W(c_W + 1)$, so that $c_W = 0$ or $c_W = 1$. Since 0 is the center of the body, we conclude that -1 is the center of the head, W_H.

3. $n = 3$: $g_{cw}^{[n-1]}(c_W) = 0$, which means that $0 = g_{cw}^{[2]}(c_W) = (c_W^2 + c_W)^2 + c_W = c_W(c_W^3 + 2c_W^2 + c_W + 1)$. Since 0 is the center of the body, we conclude that there are three 3-cycles, whose centers are the roots of $c_W^3 + 2c_W^2 + c_W + 1$. You can calculate that the roots are approximately

$$-1.75488, \quad -0.122561 + 0.744862i, \quad \text{and} \quad -0.122561 - 0.744862i$$

The two reasonably large 3-cycle components above and below the body have chemically oriented names (thanks to Mandelbrot): **north radical** and **south radical**. The other 3-cycle component located to the left of the head is called the **little midget**.

4. $n = 4$: The same technique yields a polynomial of degree 8, two solutions of which are 0 and -1. Thus there are 6 centers of 4-cycle components located approximately at

$$-1.9408, \quad -1.3107, \quad -0.15652 \pm 1.03225i, \quad \text{and} \quad 0.282271 \pm 0.530061i$$

The largest of the 4-cycle components, which is just to the left of the head, is called the **secondary head**.

There is much, much more to the story of the Mandelbrot set. The fact that the Mandelbrot set is connected was proved by Douady and Hubbard, and the mathematician Mitsuhiro Shishikura proved that the dimension of the boundary of M is actually 2. The proofs of these two results are deep and very technical.

We continue our study, addressing bulbs attached to the body of M. Notice in Figure 7.13 that between the 2-cycle and 3-cycle components lies a 5-cycle component. Analogously, between the 7-cycle and 5-cycle components there is a 12-cycle. More generally, the largest bulb between an m-cycle component and

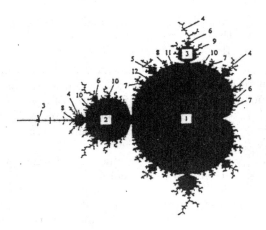

FIGURE 7.13
Identification of components of M: (1) body; (2) head; and (3) north radical

an n-cycle component has period $m+n$. This is the so-called **Farey sequence** pattern. The same pattern occurs for bulbs attached to the head of M, the components having cycles of order $2, 4, 6, \ldots$, arranged in the Farey sequence. Analogously, the same arrangement occurs for the miniature bulbs attached to any given bulb. Bulbs and cardioid-like mini-Mandelbrot sets adorn the boundary of M. Even when one zooms deeper and deeper into the boundary of M, one finds mini-Mandelbrot-like sets appearing forever!

You might wonder why the bulbs in the Mandelbrot set appear to be cardioids and disks. This is neither a coincidence nor an artifact of computer drawings. It can be shown that these bulbs are either disks or cardioids. Moreover, Douady and Hubbard showed that for any component W, ρ_W can be extended to be a one-to-one correspondence between the closure $\overline{W}$ of W and the closed unit disk $\overline{D}$ centered at the origin. Of course this means that there is a one-to-one correspondence between the boundary of W and the unit circle C centered at the origin. The element r_W of $\overline{W}$ corresponding to the point $(1, 0)$ on the unit circle is called the **root** of W. For cardioid-like components the root is the cusp. Thus the root of the body of M is the point $(1/4, 0)$ on the horizontal axis. By contrast, the root of a disk is the point of tangency with the larger component. For the head, the root is thus the point $(-3/4, 0)$ on the horizontal axis.

Until now we have focused our attention on bulbs of M, that is, components with interiors. However, tendrils and branches appear virtually everywhere on the boundary of M. In our final discussion of the Mandelbrot set, we attend to these tendrils. We say that a point is **eventually fixed** if it is *not* fixed but if some iterate *is* fixed. Analogously, a point is **eventually periodic** if it is *not* periodic but some iterate *is* periodic. Two simple examples are afforded by functions in the g_c-family:

$c = -2$: Iterates of 0 are -2, 2, 2, 2, ... Thus 0 is eventually a fixed point for g_{-2}.

$c = i$: Iterates of 0 are i, $-1 + i$, $-i$, $-1 + i$, $-i$, ... Thus 0 is eventually periodic for g_i.

The complex number c is a **Misiurewicz point** if 0 is eventually fixed or eventually periodic for g_c. These points are named after the modern Polish mathematician Michal Misiurewicz, who has studied the dynamics of one-dimensional maps. Such points have many interesting properties:

1. Each Misiurewicz point is in the Mandelbrot set, and the corresponding cycle is repelling.

2. Each Misiurewicz point lies on the boundary of the Mandelbrot set.

3. Misiurewicz points are dense in the boundary of the Mandelbrot set.

4. If c is a Misiurewicz point, then the portion of M in small neighborhoods of c is virtually self-similar and is nearly identical to the corresponding Julia set J_c. (This result is only recently proved, by the Chinese mathematician Tan Lei.)

Figure 7.14 illustrates Property 4 listed above, for the Misiurewicz point that is approximately $-0.1011 + 0.9563i$. The top portion shows the Julia set along with a zoom, and the bottom portion shows the location of c in the Mandelbrot set, along with a comparable zoom. Notice the great similarity of the two zooms.

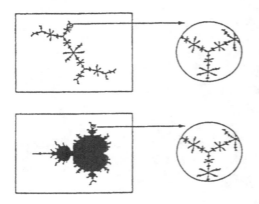

FIGURE 7.14
Misiurewicz point in M

Self-similarity is not restricted to purely dendritic regions of M, but can be seen in other parts of the boundary when magnified sufficiently. In this way one can identify many points in the boundary of M with corresponding Julia sets. The book by Peitgen, Jürgens and Saupe (1992) has a significant discussion concerning the Mandelbrot and its connections with Julia sets.

7.3.1   Section 7.3 Exercises

Exercise 1. Show that M is a closed set in the complex plane.

Exercise 2. Show that M is symmetric with respect to the x-axis.

Exercise 3. Show that i is in M.

Exercise 4. Find a complex number c such that c is in M but $-c$ is not in M.

Exercise 5. The iterates of 0 for g_i are $i, -1+i, -i$, etc. Does that mean that $-1+i$ is automatically in M? Explain why or why not.

Exercise 6. Why is each Misiurewicz point in M?

A

Appendix: Computer Programs

The following computer programs are written in Python. The # indicates a comment in Python. These should be copied along with the code when you run these programs. Keep in mind that the indentations must be copied exactly as they are for the code to run. Python requires white space indentation.

The programs may be downloaded at www.ibldynamics.com/ECAF

Program 1: ITERATE

```
# ITERATE computes and prints the n initial iterates of some function Q.
# We need to select an initial condition x, and number n of iterates to be
evaluated.
# Use the numpy library to change the given function. A few examples are
provided, including an example using a parameter m.
# When retyping these, remove the # from whichever function you want to use

import numpy as np

#If you choose to switch to another sample function, or to create your own,
#a parameter m is not required.
def Q(x, m=1):
    return m*x*(1-x)
    # return np.cos(x)
    # return 3*x - 3*x**2

def Iterate(x, n, Q, m=1):
    pts = [x]
    for i in range(0, n):
        pts.append(Q(pts[i]),m)
    print(pts)

#Change x to the initial condition, n to the number of iterates,
# and m to the parameter for Q if you used one.
#If your function doesn't have a parameter, stop after you enter Q.
Iterate(x,n,Q, m)
```

DOI: 10.1201/9781032678757-A

Program 2: NUMBER OF ITERATES

\# NUMITER computes the iterates of a function Q until an iterate is within d
\# of p. We need to select the initial point x, fixed point or
\# periodic point p, the maximum allowed distance d between the desired
\# iterate and p, and maximum number n of iterates to be evaluated.

```
import numpy as np

# As in the previous program, define your function.
# You may ignore the parameter if it is not used
def Q(x, m=1):
    return np.cos(x)
    # return m*x*(1-x)
    # return 3*x - 3*x**2

# If your function didn't have a parameter, you can skip a value for m
def Numiter(x, p, d, n, Q, m=1):
    iter = 0
    while np.abs(x-p) > d and iter <= n:
        x = Q(x, m)
        iter += 1
    if iter >= n:
        print("After" + str(n) + "iterations the distance is still"
            + str(np.abs(x-p)))
    else:
        print("After" + str(iter) + "iterations the distance is less than" +
str(d))

Numiter(0.5, 0.666, 0.001, 100000, Q, 4)
```

Program 3: PLOT

```
# PLOT plots the indicated function and the line y = x

import matplotlib.pyplot as plt
import numpy as np

plt.style.use('bmh')
# Change the values below the adjust the axes, depending on your function.
xmin = -0.1
xmax = 1.1
ymin = -0.1
ymax = 1.1

# Change this to indicate how often you want the ticks to show up on each
axis
ticks = 0.1

# As in the previous program, define your function. You may ignore the pa-
rameter if it is not used
def Q(x, m=0):
    return m*x*(1-x)
    # return np.cos(x)
    # return 3*x - 3*x**2
    # return (3 + np.sqrt(9 + 12*x))/(-6)

# defines 60 evenly spaced values along the x-axis
x = np.linspace(xmin, xmax, 60)
# Computes the corresponding y-values
y = Q(x, 3)
# Defines the area for the plot
fig, ax = plt.subplots()

ax.plot(x, y, linewidth=2.0, color='black') # plots the function
ax.plot(x, x, linewidth=2.0, color='red') # plots the diagonal

# sets the limits for the axes and tickmarks
ax.set(xlim=(xmin, xmax), xticks=np.arange(xmin + ticks, xmax, ticks),
ylim=(ymin, ymax), yticks=np.arange(ymin + ticks, ymax, ticks))

# Shows the graph in a new window
plt.show()
```

Program 4: GRAPHICAL ANALYSIS

```
# Plots the indicated function and the line y = x
# For a given initial condition, shows the graphical analysis for the orbit of
that point.

import matplotlib.pyplot as plt
import numpy as np

plt.style.use('bmh') #Feel free to look up other available styles
# Change the values below the adjust the axes, depending on your function.
xmin = -0.1
xmax = 1.1
ymin = -0.1
ymax = 1.1


# Change this to indicate how often you want the ticks to show up on each
axis
ticks = 0.1
# Choose the initial condition for your input
ic = 0.1
# Choose how many steps you want in your orbit
orbit = 20
# As in the previous program, define your function.
# You may ignore the parameter if it is not used. Some examples are included.
def Q(x, m=1):
    return m*x*(1-x)
    # return np.cos(x)
    # return 3*x - 3*x**2
    # return (3 + np.sqrt(9 + 12*x))/(-6)


# Generates the points to plot the graphical analysis for a given function and
initial condition.
def pts(Q, ic, orbit, m=1):
    xpts = [ic]
    ypts = [0]
    for i in range(0, orbit*2, 2):
        xpts.append(xpts[i])
        ypts.append(Q(xpts[i], m))
        xpts.append(ypts[i+1])
        ypts.append(ypts[i+1])
    return [xpts, ypts]


# defines 60 evenly spaced values along the x-axis
x = np.linspace(xmin, xmax, 60)
```

# Computes the correponding y-values. Change the parameter for *m* here if you wish
y = Q(x, 3.5)
# Defines the area for the plot
fig, ax = plt.subplots()

ax.plot(x, y, linewidth=2.0, color='black') # plots the function
ax.plot(x, x, linewidth=2.0, color='red') # plots the diagonal

xpts = pts(Q, ic, orbit, m)[0]
ypts = pts(Q, ic, orbit, m)[1]
ax.plot(xpts, ypts, linewidth=2.0) # plots orbit

Sets the limits for the axes and tickmarks
ax.set(xlim=(xmin, xmax), xticks=np.arange(xmin + ticks, xmax, ticks),
ylim=(ymin, ymax), yticks=np.arange(ymin + ticks, ymax, ticks))

# Shows the graph in a new window
plt.show()

Program 5: BIFURCATION

```
# BIFURCATION plots the bifurcation diagram of the quadratic family Qm
# for m in [p,q] by increments of 0.001. Select the min value and max
# value of the window, and the value for the parameter m

import numpy as np
import matplotlib.pyplot as plt
fig, ax = plt.subplots(1, 1)

# The bifurcation here is for a specific function, so leave this alone
def Q(x, m):
    return m*x*(1-x)

def Bifurcation(p, q): # For the interval [p,q] with function Qm.
    ax.set_xlim(p, q)
    ax.set_title("Bifurcation diagram")
    n = int((q-p)/0.001) # Number of points for the horizontal axis
    r = np.linspace(p, q, n)
    iterations = 1000
    last = 100 # Takes 1000 iterations, and only shows the last 100
    x = 0.01*np.ones(n) # defines a lit of initial conditions
    for i in range(iterations):
        x = Q(x, r)
        if i >= (iterations - last):
        ax.plot(r, x, ',k', alpha=0.25)
    plt.show()

Bifurcation(3, 4)
```

Program 6: SCHWARZ

```
# This computes the Schwarzian derivative for f(x) = ax^n e^{-bx} # and plots
its graph
import matplotlib.pyplot as plt
import numpy as np
import sympy as sym #This lets us use symbolic variables for derivatives.
from sympy.abc import a, b, n, x # defines variables
from sympy import diff, exp # exponential function

f = a*x**n*exp(-b*x) #Defines the intial function. The lines below take
derivatives.
f1 = diff(f, x)
f2 = diff(f1, x)
f3 = diff(f2, x)
S = f3/f1 - 1.5*(f2/f1)**2
S_eval = sym.lambdify(args=[x, a, b, n], expr=S) # Turns the variable S
into a function we can evaluate

# Change these values to alter the max and min for the x-axis. Exclude zero.
xmin = 0.01
xmax = 10
x = np.linspace(xmin, xmax, 60)

# Change the parameters a,b, and n below.
# These are renamed to separate them from the symbolic variables
sa = 1
sb = 1
sn = 2
y = S_eval(x, sa, sb, sn)

fig, ax = plt.subplots()
ax.plot(x, y)
plt.show()
```

Program 7: HENON

\# HENON draws the Henon attractor of $\{H_{a.3}\}$. It plots from the 51st iterate
\# of the initial point (x, y) to the 5000th iterate. We need to select the
\# parameter a and the initial point (x, y).
\# The parameter 0.3 is indicated in the comments and could be modified if
you wish.

```
import numpy as np
import matplotlib.pyplot as plt

fig, ax = plt.subplots(1, 1)

def plotHenon(a, x, y): # Parameter a and initial condition (x, y)
    ax.set_xlim(-2, 2)
    ax.set_ylim(-1.5, 1.5)
    ax.set_title("Henon Attractor")
    ax.plot(x, y, 'ro') # puts a red circle at the initial condition
    for i in range(1000):
        xt = 1 - a*x**2 + y
        yt = 0.3*x #Modify the parameter if you wish
        x = xt
        y = yt
        if i >= 51:
            ax.plot(x, y, '.k', alpha=0.25)
    plt.show()

plotHenon(1.4, 0.1, 0.1)
```

Program 8: PENDULUM

#PENDULUM takes in a second order differential equation of the form
$#mL^2\frac{d^2\theta}{dt^2} + cL\frac{d\theta}{dt} + mgL\sin(\theta) = 0.$
#The output is a graph of the x and y values as functions of time
using the substitution provided earlier in the text.

You'll notice a new package here, scipy, which is needed for
integration import numpy as np
from scipy.integrate import odeint
import matplotlib.pyplot as plt

# Defining the system of differential equations
def system(xy, t, c, g, L, m):
    x = xy[0]
    y = xy[1]
    dx_dt = y
    dy_dt = -(c/(m*L))*y-g/L*np.sin(x)
    dtheta_dt = [dx_dt, dy_dt]

    return dtheta_dt

Inputs, the notation is the same as in Chapter 4
Refer to the relevant exercises to see which term should be 0 in an un-
damped system.
c = 0.05
g = 9.81
L = 1
m = 1

# initial condition
xy_0 = [0, 3]

# Provides evenly spaced time intervals to compute our solutions
t = np.linspace(0, 20, 240)

# Solves the system of ODEs numerically
theta = odeint(system, xy_0, t, args=(c, g, L, m))

plt.figure()
# Changes to x (θ) are in blue, changes to y are in red.
plt.plot(t, theta[:, 0], 'b-')
plt.plot(t, theta[:, 1], 'r--')
plt.show()

Program 9: LORENZ

```
# Plots the Lorenz attractor.

import numpy as np
from scipy.integrate import solve_ivp
import matplotlib.pyplot as plt
from mpl_toolkits.mplot3d import Axes3D


# Since we're making a 3d image, it's important to set up
# the screen correctly. This sets the width and height of
# the window. The DPI determines the resolution.
# You can make it higher, but it might slow things down
WIDTH, HEIGHT, DPI = 1000, 750, 100


# Lorenz parameters and initial conditions. This follows the notation from
# equation (1) in section 4.4
sigma, b, r = 10, 2.667, 166
x0, y0, z0 = 0, 1, 1.05


# Assigns t as a set of n evenly spaced points from 0 to tmax
tmax, n = 100, 10000
t = np.linspace(0, tmax, n)


# t is the grid on which we plot this, V is the vector of the 3
variables # and the remaining terms are the constants for the system.
def lorenz(t, V, sigma, b, r):
    x, y, z = V
    xp = sigma*y - sigma*x
    yp = r*x - y - x*z
    zp = x*y-b*z
    return xp, yp, zp


# Integrate the Lorenz equations. This is the 3d version of the numerical
# solver used in PENDULUM.
soln = solve_ivp(lorenz, (0, tmax), (x0, y0, z0), args=(sigma, b, r),
    dense_output=True)


# Uses interpolation to smooth out the curve between the
# number of points we chose earlier
x, y, z = soln.sol(t)


# Plot the Lorenz attractor using a Matplotlib 3D projection.
fig = plt.figure(facecolor='k', figsize=(WIDTH/DPI, HEIGHT/DPI))
ax = fig.gca(projection='3d')
```

```
ax.set_facecolor('k')
fig.subplots_adjust(left=0, right=1, bottom=0, top=1)

# This is easier to see if we use multiple colors. We plot this in segments and
# change the colors on each segment for clarity.
s = 10
cmap = plt.cm.winter
for i in range(0, n-s, s):
ax.plot(x[i:i+s+1], y[i:i+s+1], z[i:i+s+1], color=cmap(i/n), alpha=0.4)
# Remove all the axis clutter, leaving just the curve.
ax.set_axis_off()

plt.show()
```

Program 10: CHAOS GAME

CHAOS GAME selects the midpoint of the line between a given point (x, y) and one
of the three given points B, C, D.
Which of A, B, or C is chosen is random. The process continues, and
eventually yields a (possibly deformed) Sierpinski gasket. The first 50
iterates in the process are ignored. We need to select the three points
A, B, and C, and the initial point (x, y), and the total number n of iterates
to be evaluated.

import matplotlib.pyplot as plt import random
# We need an auxilliary function to check if the points are colinear
# B = (b1,b2), C = (c1,c2), D = (d1,d2)
# This computes double the area of a triangle, given these 3 points
# If returns true if the area is zero, meaning the points are colinear.

def colinearCheck(b1, b2, c1, c2, d1, d2):
    a = b1*(c2-d2) + c1*(d2-b2) + d1*(b2-c2)
    return a == 0

```
# Checks that the point are not colinear, and iterates the function
# The first 50 are ignored, and this returns a list of points to plot

def chaosGame(x, y, b1, b2, c1, c2, d1, d2, n):
    xpts, ypts = [ ],[ ]
    if colinearCheck(b1, b2, c1, c2, d1, d2):
        print("Points are colinear, change one of them.")
        return
    else:
        for i in range(n):
            r = random.randint(1, 3)
            if r == 1:
                x = (x+b1)/2
                y = (y+b2)/2
            elif r == 2:
                x = (x+c1)/2
                y = (y+c2)/2
            else:
                x = (x+d1)/2
                y = (y+d2)/2
            if i>50:
                xpts.append(x)
                ypts.append(y)
    return(xpts, ypts)


# Initial conditions for all points and the number of iterations
x0 = 1
y0 = 1
b1 = 0
b2 = 0
c1 = 100
c2 = 0
d1 = 50
d2 = 100
iter = 10000

x, y = chaosGame(x0, y0, b1, b2, c1, c2, d1, d2, iter)
plt.figure(figsize=(5, 5))
plt.scatter(x, y, alpha=0.25, color='black', s=10)
plt.axis('equal')
plt.axis('off')
plt.show()
```

Program 11: ITERATED FUNCTION SYSTEM

\# ITERFUNCSYS plots the iterates in the Iterated Function System given by three
\# functions that are defined in the program.
\# For each k, one of the three given functions is chosen randomly and the kth iterate
\# of the initial point (x, y) is plotted. The resulting figure is the Sierpinski gasket.
\# We need to select the initial point (x_0, y_0), as well as the total number of n iterates
\# to be evaluated.

```
import matplotlib.pyplot as plt
import numpy as np
import random
```

\# We can define a function T for our iterated system. The code below differentiates each T
\# Input the values for the point you wish to transform
\# and the 6 parameters from section 6.4 of the text.

```
def T(x, y, a, b, c, d, e, f):
    return [a*x + b*y + e, c*x + d*y + f]

def IFS(x, y, n):
    xpts, ypts = [], []
    for i in range(n):
        r = random.randint(1, 3)
        if r == 1:
            x = T(x, y, 0.5, 0, 0, 0.5, 0, 0)[0]
            y = T(x, y, 0.5, 0, 0, 0.5, 0, 0)[1]
        elif r == 2:
            x = T(x, y, 0.5, 0, 0, 0.5, 0.5, 0)[0]
            y = T(x, y, 0.5, 0, 0, 0.5, 0.5, 0)[1]
        else:
            x = T(x, y, 0.5, 0, 0, 0.5, 0.25, np.sqrt(3)/4)[0]
            y = T(x, y, 0.5, 0, 0, 0.5, 0.25, np.sqrt(3)/4)[1]
        if i > 100:
            xpts.append(x)
            ypts.append(y)
    return(xpts, ypts)
```

\# Initial conditions and number of loops to run
x0 = 0

```
y0 = 0
iter = 1000

x, y = IFS(x0, y0, iter)
plt.figure(figsize=(5, 5))
plt.scatter(x, y, color='black', s=10)
plt.axis('equal')
plt.axis('off')
plt.show()
```

Program 12: FERN LEAF

```
# FERNLEAF plots the iterates given by the four functions given in
# the program. The probability of
# selecting each of the four functions is give. Notice that
# the percentages are not equal like they were in the last program. #Other
than the additional choice, the code is identical. #We need to select the initial
```

(x_0, y_0) and the total number of

```
# iterates iter to be evaluated.


import matplotlib.pyplot as plt
import numpy as np
import random
```

```
# We can define a function T for our iterated system. The code below differ-
```
entiates each T

```
# Input the values for the point you wish to transform
# and the 6 parameters from section 6.4 of the text.

def T(x, y, a, b, c, d, e, f):
    return [a*x + b*y + e, c*x + d*y + f]

def fern(x, y, n):
    xpts, ypts = [], []
    for i in range(n):
        r = random.randint(0, 101)
        if r < 85:
            x = T(x, y, 0.85, 0.04, -0.04, 0.85, 0, 3)[0]
            y = T(x, y, 0.85, 0.04, -0.04, 0.85, 0, 3)[1]
        elif r < 92:
            x = T(x, y, 0.2, -0.26, 0.23, 0.22, 0, 1.4)[0]
            y = T(x, y, 0.2, -0.26, 0.23, 0.22, 0, 1.4)[1]
        elif r < 99:
            x = T(x, y, 0.2, -0.26, 0.23, 0.22, 0, 1.4)[0]
            y = T(x, y, 0.2, -0.26, 0.23, 0.22, 0, 1.4)[1]
        else:
            x = T(x, y, 0.5, 0, 0, 0.5, 0.25, np.sqrt(3)/4)[0]
            y = T(x, y, 0.5, 0, 0, 0.5, 0.25, np.sqrt(3)/4)[1]
        if i > 100:
            xpts.append(x)
            ypts.append(y)
    return(xpts, ypts)


# Initial conditions and number of loops to run
x0 = 0
```

```
y0 = 0
iter = 10000

x, y = fern(x0, y0, iter)
plt.figure(figsize=(5, 5))
plt.scatter(x, y, color='black', s=5, alpha=0.5)
plt.axis('equal')
plt.axis('off')
plt.show()
```

Program 13: JULIA

JULIA plots the Julia set for $g_c(z) = z^2 + c$

```
import numpy as np
import matplotlib.pyplot as plt

#Change the values for c here. You may also adjust the number iter of iter-
ates
#The number of evenly spaced points on the complex axes N
#for the bounds on the complex plane X0.
def julia_set(c=-0.835 - 0.2321 * 1j,iter=50,
        N=1000, X0=np.array([-2, 2, -2, 2])):


    x0 = X0[0]
    x1 = X0[1]
    y0 = X0[2]
    y1 = X0[3]


    # This makes a grid array for the x and y values, and combines then into
a grid z of complex numbers.
    x, y = np.meshgrid(np.linspace(x0, x1, N), np.linspace(y0, y1, N) * 1j)
    z = x + y


    # F keeps track of which grid points are bounded
    # even after many iterations of z = z * *2 + c.
    # This is done by making an N × N array of zeros
    # and iterating our function. For each entry in the array z, if the absolute
    # value of z is finite, the corresponding entry in F is increased by 1.
    # These values are used to produce the colors in the eventual image.
    F = np.zeros([N, N])
    # Iterate through the operation z = z^2 + c.
    for j in range(iter):
        z = z ** 2 + c
        index = np.abs(z) < np.inf
        F[index] = F[index] + 1
    return np.linspace(x0, x1, N), np.linspace(y0, y1, N), F

x, y, F = julia_set(c=0.285 + 0.01 * 1j, iter=200, N=1000, X0=np.array([-
1.5, 1.5, -1.5, 1.5]))
plt.figure(figsize=(10, 10))
plt.pcolormesh(x, y, F, cmap="twilight_shifted")
plt.axis('equal')
plt.axis('off')
plt.show()
```

Program 14: MANDELBROT

\# MANDELBROT draws the Mandelbrot set. We generate a grid in the complex plane,
\# similarly to the JULIA program. Using $z = 0$ as an initial condition, we apply the map
\# $z = z^2 + c$ for all values of c in the grid. Points for which the iteration does not
\# diverge are then plotted.

```
import matplotlib.pyplot as plt
import numpy as np

def is_stable(c, num_iterations):
    z = 0
    for _ in range(num_iterations):
        z = z ** 2 + c
    return abs(z) < np.inf

x, y = np.meshgrid(np.linspace(-2, 0.5, 1000), np.linspace(-1.5, 1.5, 1000) * 1j)
c = x + y
plt.imshow(is_stable(c, num_iterations=200), cmap="binary")
plt.gca().set_aspect("equal")
plt.axis("off")
plt.tight_layout()
plt.show()
```

B

Answers to Selected Exercises

CHAPTER 1

Section 1.1

1. 1. a. 0.7390851332   b. 0.7390851332   c. 0.7390851332

3. They do; 0.7390851332   5. They oscillate between $1/3$ and $2/3$.

7. The limit exists precisely when $|a| < 1$; then its value is $b/(1-a)$.

9. 2.0946

11. (a) Approximate zero of f : -0.2554228711
     (b) Error message because $f'(1) = 0$
     (c) Overflow message
     (d) Approximate zero of f : -0.2554228711

Section 1.2

1. 0, 3; both are repelling   3. 0 is repelling; 1 is attracting

5. 0 is repelling   9. (b) No; $f'(x) \approx 1$ if $x \approx 0$

11. No, because $f'(0)$ does not exist

13. (a) $p = b/(1-a)$; it is attracting if $|a| < 1$ and is repelling if $|a| > 1$

23. The interval $(-1, 1)$ 25. The interval $(-1, 2)$

29. (a) $\frac{1}{2} - \frac{1}{4}\sqrt{2}$ is attracting, and $\frac{1}{2} + \frac{1}{4}\sqrt{2}$ is repelling

(b) $(-\frac{1}{2} - \frac{1}{4}\sqrt{2}, \frac{1}{2} + \frac{1}{4}\sqrt{2})$

31. (a) Neither   (b) $f(x) = \sin^2 x + x$

DOI: 10.1201/9781032678757-B

Section 1.3

5. $\{0.5130445095, 0.7994554905\}$

7. (a) $1/7$ (b) $1/17$ (c) $1/11$

11. Fixed points: 0 and 1;   period-2 points: $1/3$ and $2/3$

13. 2^n

19. $c = \pm\sqrt{2}$. The cycle $\{0, c\}$ is repelling.   23. (c) $f(x) = -x$

Section 1.4

1. (a) Eventually periodic   (b) Periodic

   (c) Eventually periodic   (b) Periodic

   (e) Eventually fixed

3. (a) 30   (b) 2182

7. (a) $2^{n-1}\mu^n$ (b) μ

11. (a) 1

   (b) If $x \leq 1$, then $\lim_{n\to\infty} E_\mu^{[n]}(x) = 1$; if $x > 1$, then $\lim_{n\to\infty} E_\mu^{[n]}(x) = \infty$.

13. q_μ is attracting; p_μ is repelling

Section 1.5

5. $\mu^2(4-\mu)/16$ 11. (a) $\dfrac{3}{2} + \dfrac{1}{2}\sqrt{17}$

Section 1.6

1. Tangent bifurcation   3. 1; neither

5. The interval is approximately $(3.737, 3.745)$.

Section 1.7

1. (a) Approximately 3.839995

5. (a) Let f have domain $\{0, 1, 2, 3, 4\}$, and let $f(0) = 0, f(1) = 2, f(2) = 1, f(3) = 4, f(4) = 5$, and $f(5) = 3$.

   (b) No, by the Li-Yorke Theorem

Section 1.8

1. $-1/2$ 7. $a > 1$

9. $(-1/\sqrt{2}, 1/\sqrt{2})$

11.   (c) 1   (d) No: $Sf(x) > 0$ for an x in $(0, 1)$ (f) 2-cycle

13. 1   15. 1

CHAPTER 2

Section 2.1

3.   (a) $\lambda(x) = \ln 2$ if x is not a dyadic rational   (b) $2 \ln 2$

7.   (b) $\ln |\mu - 2|$

Section 2.2

1. $B_n = [n, \infty)$ for $n = 1, 2, \ldots$

7.   (a) x such that $h(x) = 0\,100\,1000\,10000 \cdots$

13.   (a) $2/n, 4/n, 6/n, \ldots, (n-1)/n$

Section 2.3

1. $[1/\mu - 1, 1/\mu]$

5. (a) 0.165448   (b) 0.292293

9. No. If $h \circ Q_\mu = T_\lambda \circ h$ with h linear, then $h(x) = x$ or $h(x) = 1 - x$, neither of which is possible.

13. $g_\mu(x) = f_\mu(x + x_\mu) - x_\mu$

Section 2.4

5. (a) Neither   (b) Totally disconnected

7. $D_n(C) = \left(\dfrac{2}{3}\right)^n$; $\displaystyle\lim_{n \to \infty} D_n(C) = 0$

CHAPTER 3

Section 3.1

1. Normal form: $\begin{pmatrix} 1 & 0 \\ 0 & -1 \end{pmatrix}$; eigenvalue 1 has eigenvector $\begin{pmatrix} 1 \\ 1 \end{pmatrix}$; eigenvalue -1 has eigenvector $\begin{pmatrix} 1 \\ -1 \end{pmatrix}$

3. Normal form: $\begin{pmatrix} 1 & 1 \\ -1 & 1 \end{pmatrix}$; eigenvalues are $1 + i$ and $1 - i$

5. $a^2 + b \geq 0$ 7. $a = d$ and $b = 0 = c$

Section 3.2

3. Eigenvalue $1/3$ has eigenvector $\begin{pmatrix} 1 \\ 0 \end{pmatrix}$; eigenvalue 3 has eigenvector $\begin{pmatrix} 0 \\ 1 \end{pmatrix}$

5. Eigenvalue 2 has eigenvector $\begin{pmatrix} 1 \\ 1 \end{pmatrix}$; eigenvalue 3 has eigenvector $\begin{pmatrix} 1 \\ 2 \end{pmatrix}$

15. (b) It must be.

Section 3.3

1. Fixed points: $\begin{pmatrix} 1/2 \\ 1 \end{pmatrix}$ and $\begin{pmatrix} 1/2 \\ -2 \end{pmatrix}$; area-expanding at both

3. Fixed points: $\begin{pmatrix} 0 \\ 0 \end{pmatrix}$ and $\begin{pmatrix} 6 \\ 3 \end{pmatrix}$; area-contracting at $\begin{pmatrix} 0 \\ 0 \end{pmatrix}$, area-expanding at $\begin{pmatrix} 6 \\ 3 \end{pmatrix}$

5. Both are repelling.   7. $\begin{pmatrix} 0 \\ 0 \end{pmatrix}$ is attracting; $\begin{pmatrix} 6 \\ 3 \end{pmatrix}$ is repelling

9. No, since $L^{[2]}(\mathbf{v}) = \mathbf{0}$ for all $\mathbf{v}$

13. $\begin{pmatrix} 3/4 \\ 1 \end{pmatrix}$ 17. $\begin{pmatrix} 0 \\ 0 \end{pmatrix}$ and $\begin{pmatrix} 1/[2(1-c)] \\ 1 \end{pmatrix}$

Section 3.4

3. (a) $\begin{pmatrix} 0.6314 \\ 0.1894 \end{pmatrix}$ (b) $\lambda \approx -1.9239$ and $\mu \approx 0.1559$

(c) Eigenvector for λ is $\begin{pmatrix} 1 \\ -\mu \end{pmatrix}$; eigenvector for μ is $\begin{pmatrix} 1 \\ -\lambda \end{pmatrix}$

7. (a) $\begin{pmatrix} 1/(1-b) \\ b/(1-b) \end{pmatrix}$ (b) Attracting

(c) For eigenvalue $\sqrt{b}$, an eigenvector is $\begin{pmatrix} 1 \\ \sqrt{b} \end{pmatrix}$; for eigenvalue $-\sqrt{b}$, an

eigenvector is $\begin{pmatrix} -1 \\ \sqrt{b} \end{pmatrix}$

Section 3.5

3. (b) Let Let $x_k = z_k$ for $|k| \le n$; let $x_k = 0$ and $z_k = 1$ for $|k| > n$.

(c) $x = \cdots \overline{0.0} \cdots$ and $z = \cdots \overline{1.1} \cdots$

CHAPTER 4

Section 4.1

1. $-5, 1$; unstable saddle point

3. -2; asymptotically stable degenerate node

5. $-5, -2$; asymptotically stable node

7. $-2 + 2i, -2 - 2i$; asymptotically stable spiral point

9. 1; unstable degenerate node

11. $1 + \sqrt{6}, 1 - \sqrt{6}$; unstable saddle point

13. (a) $\dfrac{dQ}{dt} = 1$

$$\frac{dI}{dt} = -\frac{1}{LC}Q - \frac{R}{L}I$$

(b) $\pi LC/2$

(c) Asymptotically stable for all constants; spiral point if $R^2 < 4L/C$, and node if $R^2 > 4L/C$

Section 4.2

1. Almost linear; asymptotically stable node

3. Almost linear at **0**, which is an unstable critical point

   Almost linear at $\begin{pmatrix} 1 \\ 0 \end{pmatrix}$; cannot tell about stability

   Almost linear at $\begin{pmatrix} -1 \\ 0 \end{pmatrix}$; cannot tell about stability

5. Almost linear at **0**, which is an unstable saddle point

   Almost linear at $\begin{pmatrix} 1 \\ 1 \end{pmatrix}$, which is an asymptotically stable spiral point

   Almost linear at $\begin{pmatrix} 2 \\ 0 \end{pmatrix}$, which is an unstable saddle point

9. (b) $0 < \epsilon < 2$

11. Critical points: $\begin{pmatrix} n\pi \\ 0 \end{pmatrix}$; if n is even, then the critical point is a saddle point; if n is odd, then the critical point is asymptotically stable. It is a node if $a^2 > 4b$, a spiral point if $a^2 < 4b$ and is a degenerate node if $a^2 = 4b$.

Section 4.3

7. $\theta_0 = \arcsin\left(\frac{\omega}{2}\sqrt{L/g} \right)$

Section 4.4

9. (b) $r_0 = \dfrac{961}{72}$

11. $r_0 = \dfrac{\sigma(\sigma + b + 3)}{\sigma - b - 1}$

CHAPTER 5

Section 5.1

1. Neither given shape is the union of a finite number of exact miniatures of the shape.

3. (a) $.\overline{2}\cdots = 1$ (b) $.0\overline{12}\cdots = \dfrac{5}{26}$

5. $x_1 x_2 x_3 \cdots x_n \overline{0} \cdots$, where $x_k = 0, 1,$ or 2, for $k = 1, 2, \ldots, n$.

9. $2/3$

Section 5.2

1. (a) $0.0563135147\cdots$ (b) $0.0100225958\cdots$

5. 0

7. $B(p) = \infty$ since the set of rationals is dense in $(0, 1)$.

9. $A = \dfrac{2}{5}\sqrt{3}$

11. The Sierpiński gasket seems to appear.

Section 5.3

1. (a) $f_2(x) = (3x - \dfrac{2}{3}, 2 - 3x)$, for $1/3 < x \le 4/9$

  (b) $f_2(x) = (\dfrac{10}{3} - 3x, -2 + 3x)$, for $7/9 < x \le 8/9$

Section 5.4

1. $m = 8, s = 1/3$; $\dim_s S = \dfrac{\ln 8}{\ln 3} \approx 1.8928$

3. $m = 20, s = 1/3$; $\dim_s M = \dfrac{\ln 20}{\ln 3} \approx 2.7268$

5. Approximately 1.8713 9. 0

11. $S = $ the rational numbers in $(0, 1)$.

CHAPTER 6

Section 6.1

3. $\dfrac{1}{9}(16 + 5\sqrt{7})$

5. (c) They are equivalent.

7. Let $x_n = n + 1$ and $y_n = n$. Then $|x_n - y_n| = 1$ for all n, but $\lim_{n \to \infty} d(x_n, y_n) = \lim_{n \to \infty} |\sqrt{x_n} - \sqrt{y_n}| = 0$. Thus the two metrics are not equivalent.

13. (b) Let $A_n = [0, (n-1)/n]$, so A_n is compact, for $n = 1, 2, \ldots$. But $A = \bigcup_{n=1}^{\infty} A_n = [0, 1)$ is not compact.

Section 6.2

7. Let $A = \{\frac{1}{n}\}_{n=1}^{\infty}$, on the real line. Then $\lim_{n \to \infty} \left| \frac{1}{n} - \frac{1}{n+1} \right| = 0$. The minimum value would be, but cannot be, 0.

Section 6.3

1. Let $x = a/2$, so $0 < x < 1/3$.

3. $\lim_{x \to 0} (\sin x)/x = 1$, so $s = 1$.

9. $T \begin{pmatrix} x \\ y \end{pmatrix} = \begin{pmatrix} 1/2 \\ 1/3 \end{pmatrix}$, for real x and y

11. $\begin{pmatrix} 2e \\ 2f \end{pmatrix}$

Section 6.4

3. $T_1(x) = \frac{1}{3}x$ and $T_2(x) = \frac{2}{3} + \frac{1}{3}x$, for $0 \le x \le 1$.

5. (a) $T_1 \sim (\frac{1}{2} \ 0 \ 0 \ \frac{1}{2} \ 0 \ 0)$, $T_2 \sim (\frac{1}{2} \ 0 \ 0 \ \frac{1}{2} \ \frac{1}{2} \ 0)$,

   $T_3 \sim (\frac{1}{2} \ 0 \ 0 \ \frac{1}{2} \ \frac{1}{2} \ \frac{1}{2})$

   (b) $T_1 \sim (\frac{1}{2} \ 0 \ 0 \ \frac{1}{2} \ 0 \ 0)$, $T_2 \sim (\frac{1}{2} \ 0 \ 0 \ \frac{1}{2} \ \frac{1}{2} \ 0)$,

   $T_3 \sim (0 \ -\frac{1}{2} \ \frac{1}{2} \ 0 \ 1 \ \frac{1}{2})$

   (c) $T_1 \sim (\frac{1}{4} \ 0 \ 0 \ \frac{1}{4} \ 0 \ 0)$, $T_2 \sim (\frac{1}{2} \ 0 \ 0 \ \frac{1}{2} \ \frac{1}{2} \ 0)$,

   $T_3 \sim (\frac{1}{2} \ 0 \ 0 \ \frac{1}{2} \ \frac{1}{4} \ \frac{\sqrt{3}}{4})$

7. It looks like the Barnsley castle (see Figure 5.28).

Section 6.5

1. About 3.1688×10^{11} years.

CHAPTER 7

Section 7.1

1. $\frac{1}{2}\sqrt{2}\,(1\pm i),\ \frac{1}{2}\sqrt{2}\,(-1\pm i)$

3. $e^{2\pi i(1/16)}, e^{2\pi i(5/16)}, e^{2\pi i(9/16)}, e^{2\pi i(13/16)}$

7. $1/2$; it is neither attracting nor repelling.

Section 7.2

3. There are 2 3-cycles.

5. c is in the intervals $(-\infty, -3/4)$ or $(1/4, \infty)$.

Section 7.3

5. Yes: 0 is an eventually period-2 point.

Bibliography

[1] Abbott, Edwin, Flatland, 1884.

[2] Alligood, Kathleen, Tim Sauer, and James A. Yorke, *Chaos: An Introduction to Dynamical Systems*, Springer-Verlag, New York, 1997.

[3] Akin, Ethan, "The General Topology of Dynamical Systems", Springer 1993.

[4] Andres, J., "Randomization of Sharkovskii-type theorems", Proc. Amer. Math. Soc. 136, 1385–1395, 2008.

[5] Andres, J., Fiser, J., and Juttner, L., "On a multivalued version of the Sharkovskii theorem and its application to differential inclusions", Set-Valued Anal. 10, 1–14, 2002.

[6] Banach, Stefan. "Sur les opérations dans les ensembles abstraits et leur application aux équations intégrales" (PDF). Fundamenta Mathematicae. 3, 133–181, 1922.

[7] Barnsley, Michael, *Fractals Everywhere*, Second Edition, Elsevier Academic Press, 1993.

[8] Bendixson, Ivar, "Sur les courbes définies par des équations différentielles" (PDF), Acta Mathematica, 24 (1): 1–88, 1901.

[9] Betti, L. et al., "Fractal dimension, approximation and data sets", https://doi.org/10.48550/arXiv.2209.12079

[10] Bhatia, Nam Parshad and Giorgio P. Szegö, *Stability Theory of Dynamical Systems*, Springer, 1970

[11] Bonilla, Luis, Albar, M. and Carretero, M., "Chaos-based true random number generators", Journal of Mathematics in Industry, 7, 2016.

[12] Boyce, William, and Richard Di Prima, *Elementary Differential Equations and Boundary Value Problems*, Ninth Edition, Wiley, New York, 2008.

[13] Burckel, Robert B., *An Introduction to Classical Complex Analysis, Volume 1*, Birkhäuser, Basel, Switzerland, 1979.

[14] Cantini, Andrea and Riccardo Bruni, "Paradoxes and Contemporary Logic", The Stanford Encyclopedia of Philosophy (Fall 2021), Edward N. Zalta (ed.).

[15] Georg Canto. "Ueber eine elementare Frage der Mannigfaltigkeitslehre". Jahresbericht der Deutschen Mathematiker-Vereinigung, 1891.

[16] Chen, Yanguang "Modeling Fractal Structure of City-Size Distributions Using Correlation Functions". PLOS ONE. 6 (9), 2011.

[17] Collet, Pierre, and Jean-Pierre Eckmann, *Iterated Maps on the Interval as Dynamical Systems*, Birkhäuser, Boston, 1980.

[18] Coddington, Earl & Norman Levinson, *Theory of Ordinary Differential Equations* (1955).

[19] Derrida, B., A. Gervois, and Y. Pomeau, "Universal Metric Properties of Bifurcations of Endomorphisms," J. Physics A: Math. Gen., 12 No. 3 (1979), pp. 269–296.

[20] Devaney, Robert L., *An Introduction to Chaotic Dynamical Systems*, Second Edition, Addison-Wesley, 1989; now published by Westview Press, 2003.

[21] Devaney, Robert L., *Chaos, Fractals, and Dynamics*, Addison-Wesley, Menlo Park, California, 1990.

[22] Devaney, Robert, Morris Hirsch, and Stephen Smale, *Differential Equations, Dynamical Systems, and an Introduction to Chaos*, Elsevier Academic Press, 2004.

[23] Devaney, Robert and Paul Blanchard, *Differential Equations*, Fourth Edition, Brooks/Cole, Cengage Learning, 2011.

[24] Douady, A., and J. H. Hubbard, "Itération des polynômes quadratiques complexes," C. R. Acad. Sci. Paris I, 294 (1982), pp. 123–126.

[25] Du, Bau-Sen, "The Minimal Number of Periodic Orbits of Periods Guaranteed in Sharkovskii's Theorem," Bull. Austral. Math. Soc., 31 (1985), pp. 89–103.

[26] Edgar, Gerald A. *Classics on Fractals*, Addison-Wesley, 1993 (paperback: Westview Press, 2004).

[27] Farmer, J. Doyne, Edward Ott, and James A. Yorke, "The Dimension of Chaotic Attractors," Physica 7D (1983), pp. 153–180.

[28] Fatou, Pierre, "Sur les equations fonctionnelles," Bull. Soc. Math. France 47 (1919), pp. 161–271.

[29] Feigenbaum, Mitchell, "Quantitative Universality for a Class of Nonlinear Transformations," J. of Stat. Physics, 19 No. 1 (1978), pp. 25–52.

[30] Feigenbaum, Mitchell, "Universal Behavior in Nonlinear Systems," Physica 5D, (1983), pp. 16–39.

[31] Fitzpatrick, Patrick, *Advanced Calculus,* Second Edition, Amer. Math. Soc. 2009.

[32] Frame, Michael and Benoit Mandelbrot, *Fractals, Graphics, and Mathematics Education*, MAA, Washington, D.C., 2002.

[33] Franceschini, V., "A Feigenbaum Sequence of Bifurcations in the Lorenz Model," J. Stat Physics, 22 No. 3 (1980), pp. 397–406.

[34] Garfinkel, A. J., D. O. Walter, R. B. Trelease, R. K. Harper, and R. M. Harper, "Non-linear Dynamics of Electrocardiographic Waveforms Following Cocaine Administrations," Life Sciences, 48 No. 22 (1991), pp. 2189–2193.

[35] Gelbaum, Bernard, John M.H. Olmsted, *Counterexamples in Analysis*, Dover, 1964.

[36] Glass, Leon, Michael R. Guevara and Alvin Shrier, "Bifurcation and Chaos in a Periodically Stimulated Cardiac Oscillator," Physica 7D (1983), pp. 89–101.

[37] Gleick, James, *Chaos: Making a New Science*, Viking, New York, 1987.

[38] Goodson, Geoffrey, "Chaotic Dynamics", Cambridge Press, 2019.

[39] Grebogi, Celso, Edward Ott, Steven Pelikan, and James A. Yorke, "Strange Attractors that are not Chaotic," Physica 13D (1984), pp. 261–268.

[40] Grebogi, Celso, Edward Ott, and James A. Yorke, "Fractal Basin Boundaries," Lecture Notes in Physics, Vol. 278, Springer-Verlag (1986), pp. 28–32.

[41] Guckenheimer, John, and Philip Holmes, *Nonlinear Oscillations, Dynamical Systems, and Bifurcations of Vector Fields*, Springer-Verlag, New York, 1983.

[42] Hénon, Michel, "A Two-Dimensional Mapping with a Strange Attractor," Commun. Math. Phys. 5D (1976), pp. 69–77.

[43] Herschel, John Frederick William, emphOn a Remarkable Application of Cotes's Theorem. Philosophical Transactions of the Royal Society of London. London, 1813.

[44] Herschel, John Frederick William *Part III. Section I. Examples of the Direct Method of Differences.* A Collection of Examples of the Applications of the Calculus of Finite Differences, 1820.

[45] Ho, Chung-Wu, and Charles Morris, "A Graph Theoretic Proof of Sharkovsky's Theorem on the Periodic Points of Continuous Functions," Pacific J. of Math. 96 No. 2 (1981), pp. 361–370.

[46] Hutchinson, John, "Fractals and Self-Similarity," Indiana Univ. J. Math. 30 (1981), pp. 713–747.

[47] Isaeva, Olga, Kuznetsov, Sergey, Sataev, Igor, Savin, Dmitry, Seleznev, E.P., Nikov, Kotel, "Hyperbolic Chaos and Other Phenomena of Complex Dynamics Depending on Parameters in a Nonautonomous System of Two Alternately Activated Oscillators", International Journal of Bifurcation and Chaos, 25, 2015.

[48] Jacobi, C.G.J, (translated by G.W. Stewart) "On a New Way of Solving the Linear Equations that Arise in the Method of Least Squares" Astronomische Nachrichten, 22 (1845). Translation available at *http://hdl.handle.net/1903/568*

[49] Julia, Gaston, "Memoire sur l'itération des fonctions rationnelles," J. Math. 8 (1918), pp. 47–245.

[50] Kaplan, James L., and James A. Yorke, "Chaotic Behavior of Multidimensional Difference Equations," Springer Lecture Notes, Vol. 730, H. O. Peitgen and H. O. Walther, Editors, Springer-Verlag, New York, 1978.

[51] Kaye, Brian, *A Random Walk Through Fractal Dimensions*, Wiley, New York, 1994.

[52] Lewis, Wayne and Piotr Minc, "Drawing the pseudo-arc", Houston Journal of Mathematics, 36, 905–934, 2010.

[53] Li, Tien-Yien, and James A. Yorke, "Period Three Implies Chaos," Amer. Math. Monthly 82 (1975), pp. 985–992.

[54] Lind, Douglas and Brian Marcus, "Symbolic Dynamics and Coding". Cambridge 1999.

[55] Lofaro, Thomas and Jeff Ford, *Discovering Dynamical Systems Through Experiment and Inquiry*, CRC Press, 2021.

[56] Lorenz, Edward N., "Deterministic Nonperiodic Flow," J. Atmospheric Sciences 20 (1963), pp. 130–141.

[57] Lorenz, Edward N., "Predictability: Does the flap of a butterfly's wing in Brazil set of a tornado in Texas," American Association for the Advancement of Science (1972).

[58] Liu, Jing Z.; Zhang, Lu D.; Yue, Guang H. "Fractal Dimension in Human Cerebellum Measured by Magnetic Resonance Imaging". Biophysical Journal. 85 (6): 4041–4046, (2003).

[59] Luo, Dingjun, "Bifurcation Theory and Methods of Dynamical Systems." World Scientific. (1997). p. 26.

[60] Mandelbrot, Benoit, *The Fractal Geometry of Nature*, W. H. Freeman, 1982.

[61] Mandelbrot, Benoit, *Fractals and Chaos: The Mandelbrot Set and Beyond*, Springer, Berlin, 2004.

[62] Manning, Henry, "Geometry of Four Dimensions", Macmillian, 1914.

[63] Mauldin, R. Daniel, "The Scottish Book: Mathematics from the Scottish Cafe", Birkhauser, 2015.

[64] May, Robert M., "Simple Mathematical Models with Very Complicated Dynamics," Nature 261 (1976), pp. 459–467.

[65] Moon, Francis C., *Chaotic Vibrations*, Wiley, New York, 1987.

[66] Murray, Carl, "Is the Solar System Stable?" New Scientist, 25 November 1989, pp. 60–63.

[67] Nadler, Sam, "Continuum Theory", CRC Press, 1992.

[68] Palmer, Tim, "A Weather Eye on Unpredictability," New Scientist, 11 November 1989, pp. 56–59.

[69] Peitgen, H.-O., H. Jürgens, and D. Saupe, *Chaos and Fractals: New Frontiers of Science*, Springer-Verlag, New York, 1992.

[70] Peitgen, H.-O., and P. H. Richter, *The Beauty of Fractals*, Springer-Verlag, Berlin, 1986.

[71] Peters, Edgar, Chaos and order in the capital markets : a new view of cycles, prices, and market volatility. Wiley, 1996.

[72] Pikovsky A., M Rosenblum and J Kurths, 'Synchronization. A Universal Concept in Nonlinear Science'. 2001.

[73] Poincaré, Henri, "L'Équilibre d'une masse fluide animée d'un mouvement de rotation". Acta Mathematica, vol.7, Sept 1885, pp. 259–380.

[74] Poincaré, Henri, "Sur les courbes définies par une équation différentielle", Oeuvres, vol. 1, 1892.

[75] Preston, Chris, *Iterates of Maps on an Interval*, Lecture Notes in Mathematics, Vol. 999, Springer-Verlag, New York, 1983.

[76] Putnam, Ian, "A homology theory for Smale Spaces", Memoirs A.M.S., 232 (2014).

[77] Rasband, S. Neil, *Chaotic Dynamics of Nonlinear Systems*, Wiley, New York, 1990.

[78] Rayleigh, Lord, "On Convective Currents in a Horizontal Layer of Fluid When the Higher Temperature is on the Under Side," Phil. Mag. 32 (1916), pp. 529–546.

[79] Ruelle, David, *Thermodynamic Formalism*, Cambridge University Press, 2004.

[80] Saltzman, Barry, "Finite Amplitude Free Convection as an Initial Value Problem," J. Atmospheric Sci. 19 (1962), pp. 329–341.

[81] Sanchez, David A., Richard C. Allen, and Walter T. Kyner, *Differential Equations*, Addison-Wesley, Reading, Massachusetts, 1983.

[82] Schroeder, Manfred, *Fractals, Chaos, Power Laws*, Dover Publications, 2009.

[83] Scott, Stephen, "Clocks and Chaos in Chemistry," New Scientist, 2 Dec. 1989, pp. 50–59.

[84] Shanks, Daniel, *Solved and Unsolved Problems in Number Theory*, Fourth Edition, AMS Chelsea, New York, 2002.

[85] Sharkovsky, A. N., "Co-existence of Cycles of a Continuous Mapping of a Line into Itself," Ukranian Math. Z., 16 (1964), pp. 61–71.

[86] Singer, David, "Stable Orbits and Bifurcation of Maps of the Interval," SIAM J. Appl. Math., 35 No. 2 (1978), pp. 260–267.

[87] Smale, Stephen, "Differentiable Dynamical Systems," Bull. Amer. Math. Soc., 73 (1967), pp. 747–817.

[88] Smith, Henry J.S., "On the integration of discontinuous functions". Proceedings of the London Mathematical Society, 6: 140–153, 1874.

[89] Sokol, James, "The Hidden Heroines of Chaos", Quanta Magazine, May 2019.

[90] Song C, Phenix H, Abedi V, Scott M, Ingalls BP, et al. 'Estimating the Stochastic Bifurcation Structure of Cellular Networks'. PLoS Comput Biol 6(3) 2010.

[91] Sparrow, Colin, *The Lorenz Equations: Bifurcations, Chaos, and Strange Attractors*, Springer-Verlag, New York, 1982.

[92] Straffin, P. D., Jr., "Periodic Points of Continuous Functions," Math. Mag. 51 (1978), pp. 99–105.

[93] Steen, Lynn Arthur, J. Arthus Seebach Jr., *Counterexamples in Topology*, Dover, 1995.

[94] Szuca, P., 'Sharkovskii's theorem holds for some discontinuous functions', Fund. Math. 179, no. 1, 27–41, 2003.

[95] Teschl, Gerald, "Ordinary Differential Equations and Dynamical Systems" Graduate Studies in Mathematics, Volume 140, Amer. Math. Soc., Providence, 2012.

[96] University of Maryland Chaos Group, "Chaos and Fractals in Simple Physical Systems," videotape, University of Maryland, College Park, 1991.

[97] Van Buskirk, Robert, and Carson Jeffries, "Observation of Chaotic Dynamics of Coupled Nonlinear Oscillators," Physical Review A 31 No. 5 (1985), pp. 3332–3357.

[98] Wiggins, Stephen, *Global Bifurcations and Chaos*, Springer-Verlag, New York, 1988.

[99] Williams, Robert F., "Expanding attractors", Publ. Math. IHES, t. 43 (1974), pp. 169–203.

[100] Wisdom, Jack, "Urey Prize Lecture: Chaotic Dynamics in the Solar System," Icarus, 72 (1987), pp. 241–275.

[101] Yorke, James A., *Dynamics: An Interactive Program for IBM PC's and Compatibles*, University of Maryland, College Park, 1991.

Index

Printed in the United States
by Baker & Taylor Publisher Services